Sustainable Soil and Land Management and Climate Change

Footprints of Climate Variability on Plant Diversity
Series Editor: Shah Fahad

Climate Change and Plants: Biodiversity, Growth and Interactions
Shah Fahad, Osman Sönmez, Shah Saud, Depeng Wang, Chao Wu, Muhammad Adnan and Veysel Turan

Developing Climate Resilient Crops: Improving Global Food Security and Safety
Shah Fahad, Osman Sönmez, Shah Saud, Depeng Wang, Chao Wu, Muhammad Adnan and Veysel Turan

Sustainable Soil and Land Management and Climate Change
Shah Fahad, Osman Sönmez, Shah Saud, Depeng Wang, Chao Wu, Muhammad Adnan, and Veysel Turan

Sustainable Soil and Land Management and Climate Change

Edited by

Dr. Shah Fahad
Prof. Dr. Osman Sönmez
Dr. Shah Saud
Dr. Depeng Wang
Dr. Chao Wu
Dr. Muhammad Adnan
Dr. Veysel Turan

CRC Press
Taylor & Francis Group
Boca Raton London New York

CRC Press is an imprint of the
Taylor & Francis Group, an **informa** business

First edition published 2021
by CRC Press
6000 Broken Sound Parkway NW, Suite 300, Boca Raton, FL 33487-2742

and by CRC Press
2 Park Square, Milton Park, Abingdon, Oxon, OX14 4RN

ISBN: 978-0-367-62318-0 (hbk)
ISBN: 978-0-367-62322-7 (pbk)
ISBN: 978-1-003-10889-4 (ebk)

Typeset in Times
by SPi Global, India

Contents

Acknowledgements

Words are bound and knowledge is limited to praise ALLAH, the Instant and Sustaining Source of all Mercy and Kindness, and the Sustainer of the Worlds. My greatest and ultimate gratitude is due to ALLAH (Subhanahu wa Taqadus). I thank ALLAH with all my humility, for everything that I can think of. His generous blessing and exaltation succeeded my thoughts and drove my ambition to have the cherished fruit of my modest efforts in the form of this piece of literature from the blooming spring of blossoming knowledge. May ALLAH forgive my failings and weaknesses, strengthen and enliven my faith in HIM and endow me with knowledge and wisdom. All praises and respects are for **Holy Prophet Muhammad** *Salle Allah Alleh Wassalam*, the greatest educator, the everlasting source of guidance and knowledge for humanity. He taught the principles of morality and eternal values and enabled us to recognize our Creator. I have a deep sense of obligation to my parents, my brothers, my sisters and my son. Their unconditional love, care, and confidence in my abilities helped me achieve this milestone in my life. For this and much more, I am forever in their debt. It is to them that I dedicate this book. In this arduous time, I also appreciate the patience and serenity of my wife, who brought joy to my life in so many different ways. It is indeed on account of her affections and prayers that I was able to achieve something in my life.

Shah Fahad

Notes on Editors

Dr. Shah Fahad is an Assistant Professor in the Department of Agronomy, University of Haripur, Khyber Pakhtunkhwa, Pakistan. He obtained his PhD in Agronomy from Huazhong Agriculture University, China, in 2015. After doing his postdoctoral research in Agronomy at the Huazhong Agriculture University (2015–17), he accepted the position of Assistant Professor at the University of Haripur. He has published over 260 peer-reviewed papers (Impact factor 723.45) with more than 230 research and 30 review articles, on important aspects of climate change, plant physiology and breeding, plant nutrition, plant stress responses and tolerance mechanisms, and exogenous chemical priming-induced abiotic stress tolerance. He has also contributed 50 book chapters to various book editions published by Springer, Wiley-Blackwell, and Elsevier. He has edited 15 book volumes, including this one, published by CRC Press, Springer, and Intech Open. He won the Young Rice International Scientist Award and Distinguish Scholar Award in 2014 and 2015, respectively. He has also won 13 projects from international and national donor agencies. Dr Shah Fahad's name figured among the top 2% of scientists in a global list compiled by Stanford University, USA. He has worked on, and is presently continuing to research, a wide range of topics, including climate change, greenhouse emission gases, abiotic stresses tolerance, the roles of phytohormones and their interactions in abiotic stress responses, heavy metals, and the regulation of nutrient transport processes.

Prof. Dr. Osman Sönmez is a Professor in the Department of Soil Science, Faculty of Agriculture, Erciyes University, Kayseri, Turkey. He obtained his MS and PhD in Agronomy from Kansas State University, Manhattan-KS, USA in the period 1996–2004. In 2014 he accepted the position of Associate Professor at the University of Erciyes. Since 2014, he has worked in the Department of Soil Science, Faculty of Agriculture at Erciyes University. He has published over 90 peer-reviewed papers, research and review articles on soil pollution, plant physiology and plant nutrition.

Dr. Shah Saud received his PhD in Turf Grasses (Horticulture) from Northeast Agricultural University, Harbin, China. He is currently working as a Post-Doctoral researcher in the Department of Horticulture, Northeast Agricultural University, and Harbin, China. Dr. Shah Saud has published over 125 research publications in peer-reviewed journals. He has also edited 3 books and written 25 book chapters on important aspects of plant physiology, plant stress responses, and environmental problems in relation to agricultural plants. According to Scopus®, Dr. Shah Saud's publications have received roughly 2500 citations with an h-index of 24.

Dr. Depeng Wang completed his PhD in 2016 in the field of Agronomy and Crop Physiology from Huazhong Agriculture University, Wuhan, China. Presently, he is serving as a professor in the College of Life Science, Linyi University, Linyi, China. He is also the principal investigator of Crop Genetic Improvement, Physiology & Ecology Center in Linyi University. His current research focuses on crop ecology and physiology and agronomy. These have included the key characteristics associated with high-yield crops, the effects of temperature on crop grain yield and solar radiation utilization, morphological plasticity to agronomic manipulation in leaf dispersion and orientation, and optimal integrated crop management practices for maximizing crop grain yield. Dr. Depeng Wang has published more than 36 papers in reputed journals.

Dr. Chao Wu engages in research in the areas of field crop cultivation and physiology, and plant phenomics. He has completed his PhD during the period 2013–2016 from Huazhong Agricultural University, Wuhan, China, and completed his post PhD studies during 2017–2019 from Nanjing Agricultural University, Nanjing, China. He is now an associate research fellow in Guangxi Institute of Botany,

Guangxi Zhuang Autonomous Region and the Chinese Academy of Sciences, Guilin, China. He chairs the Natural Science Foundation of Jiangsu Province, and two Postdoctoral Science Foundation research groups, and focuses mainly on physiological mechanisms of abiotic stress tolerance (heat, drought) in crops and medicinal plants.

Dr. Muhammad Adnan is a lecturer in the Department of Agriculture at the University of Swabi (UOS), Pakistan. He completed his PhD (soil fertility and microbiology) from the Department of Soil and Environmental Sciences (SES), the University of Agriculture, Peshawar, Pakistan and also the Department of Plant, Soil and Microbial Sciences, Michigan State University, USA. He has received his MSc and BSc (Hons) in Soil and Environmental Sciences from Department of SES, the University of Agriculture, Peshawar, Pakistan.

Dr. Veysel Turan is an Associate Professor in the Department of Soil Science and Plant Nutrition, Bingöl University, Turkey. He obtained his PhD in Soil Science and Plant Nutrition from Atatürk University, Turkey, in 2016. Since doing his postdoctoral research in the Department of Microbiology, University of Innsbruck, Austria (2017–18), he has been working at Bingöl University. He has worked on, and continues to research, a wide range of topics, soil–plant interaction, and heavy metals accumulation, as well as the bioremediation of soil by some plant and soil amendments.

Contributors

Muhammad Aamer
Research Centre on Ecological Sciences
Jiangxi Agricultural University
Nanchang, China

Mohd Abas Shah
Potato Research Station
Jalandhar, Punjab, India

Muhammad Abbas Qazi
Kunming Institute of Botany
Chinese Academy of Sciences
Kunming, China

Ahmed Abu Zaiton
Department of Biology
Faculty of Science
Al A lbayt University
Jordan

Yousef M Abu-Zaitoon
Department of Biological Sciences
Al Hussein Bin Talal University
Ma'an, Jordan

Muhammad Adnan
Department of Agriculture
University of Swabi
Khyber Pakhtunkhwa
Pakistan

Saeed Ahmad
Kunming Institute of Botany
Chinese Academy of Sciences
Kunming, China

Muhammad Akmal
Institute of Soil Science
PMAS- Arid Agriculture University Rawalpindi
Pakistan

Muhammad Akram
Kunming Institute of Botany
Chinese Academy of Sciences
Kunming, China

Duraid K. A. Al-Taey
Department of Horticulture
Faculty of Agriculture
Al-Qasim Green University
Babylon Province, Baghdad, Iraq

Wafa'a A. Al-Taisan
Department of Biology
College of Science
Imam Abdulrahman Bin Faisal University
Dammam, Saudi Arabia

Abdel Razzaq Al-Tawaha
Department of Crop Science
Faculty of Agriculture
University Putra Malaysia
Selangor, Malaysia

Abdel Rahman Mohammad Al-Tawaha
Department of Biological Sciences
Al Hussein Bin Talal University
Ma'an, Jordan

Ezz Al-Dein Al-Ramamneh
Department of Agricultural Sciences
AL-Shouback University College
Al-Balqa Applied University
AL-Shouback, Maan, Jordan

Khurshid Alam
Kunming Institute of Botany
Chinese Academy of Sciences
Kunming, China

Alla Aleksanyan
Department of Geobotany and Plant Eco-Physiology
Institute of Botany aft. A.L. Takhtajyan NAS RA
Yerevan, Armenia

Mohammad Ali Shariati
Laboratory of Biocontrol and Antimicrobial Resistance
Orel State University Named After I.S. Turgenev
Orel, Russia

Mazhar Ali
Department of Environmental Sciences
COMSATS University Islamabad
Pakistan

Amanullah
Department of Agronomy
the University of Agriculture
Peshawar, Pakistan

Shazia Anjum
Kunming Institute of Botany
Chinese Academy of Sciences
Kunming, China

Muhammad Arif
Department of Agronomy
the University of Agriculture
Peshawar, Pakistan

Muhammad Asad Salim
Kunming Institute of Botany
Chinese Academy of Sciences
Kunming, China

Muhammad Asim
Plants Sciences Division PARC Islamabad
Pakistan

Muhammad Ayub Khan
Plants Sciences Division PARC Islamabad
Pakistan

Maria Batool
College of Plant Science and Technology
Huazhong Agricultural University
Wuhan, China

Rajan Bhatt
Scientist (Soil Science)
Regional Research Station
Kapurthala, Punjab Agricultural University
Ludhiana, Punjab, India

Ismail Celik
Faculty of Agriculture
Department of Soil Science and Plant
Cukurova University, Turkey

Cristina Halbac Cotoara Zamfir
Faculty of Civil Engineering
Politehnica University Timisoara
Timisoara, Romania

Mayur S. Darvhankar
School of Agriculture
Lovely Professional University
Phagwara, Punjab, India

Aman Deep Ranga
School of Agriculture
Lovely Professional University
Phagwara, Punjab, India

Ayman EL Sabagh
Department of Agronomy
Faculty of Agriculture
University of Kafrelsheikh
Egypt

Shah Fahad
Department of Agriculture
University of Haripur
Khyber Pakhtunkhwa, Pakistan

Khalid Fandi
Bangladesh Wheat and Maize Research Institute
Dinajpur, Bangladesh

Muhammad Farooq
Department of Crop Sciences
College of Agricultural and Marine Sciences
Sultan Qaboos University
Al-Khoud, Oman

Hikmet Günal
Department of Soil Science and Plant Nutrition
Gaziosmanpaşa University
Tokat, Turkey

Rares Halbac Cotoara Zamfir
Politehnica University Timisoara
Faculty of Civil Engineering
Timisoara, Romania

Mirza Hasanuzzaman
Department of Agronomy
Faculty of Agriculture
Sher-e-Bangla Agricultural University
Dhaka, Bangladesh

Akbar Hossain
Bangladesh Wheat and Maize Research Institute
Dinajpur, Bangladesh

Nazish Huma Khan
Department of Environmental Sciences
University of Swabi
Pakistan

Qaiser Hussain
Institute of Soil Science
PMAS- Arid Agriculture University Rawalpindi
Pakistan

Shahid Ibni Zamir
Department of Agronomy
University of Agriculture Faisalabad
Pakistan

Muhammad Ilyas
Department of Plant Breeding and Molecular Genetics
Faculty of Agriculture
University of Poonch, Rawalakot
Azad Jammu and Kashmir
Pakistan

Imran
Department of Agronomy
the University of Agriculture
Peshawar, Pakistan

Shah Khalid
Department of Agronomy
the University of Agriculture
Peshawar, Pakistan

Josef Křeček
Department of Hydraulics and Hydrology (FSV)
Czech Technical University
Prague, Czech

Huijie Li
College of Humanity and Public Administration
Jiangxi Agricultural University
Nanchang, China

David L. McNeil
Department of Primary Industries and Regional Development
Kununurra, Western Australia

Huma Naz
Mohammad Ali Nazeer Fatima Degree College
Hardoi, Uttar Pradesh, India

Patel H. K.
Department of Agricultural Microbiology B. A. College of Agriculture
Anand Agricultural University
Anand, Gujarat, India

Alireza Pour-Aboughadareh
Seed and Plant Improvement Institute
Agricultural Research, Education and Extension Organization (AREEO)
Karaj, Iran

Ali M. Qaisi
Department of Pharmaceutical Sciences
School of Pharmacy
University of Jordan
Amman, Jordan

Hafiz Muhammad Rashad Javeed
Department of Environmental Sciences
COMSATS University Islamabad
Pakistan

Adnan Rasheed
Key Laboratory of Crop Physiology, Ecology and Genetic Breeding
Ministry of Education/Collage of Agronomy
Jiangxi Agriculture University
Nanchang, PR China

Abdur Rauf
Department of Chemistry
University of Swabi
Anbar, Khyber Pakhtunkhwa
Pakistan

Tooba Saeed
National Centre of Excellence in Physical Chemistry
University of Peshawar
Pakistan

Khalid Saifullah Khan
Institute of Soil Science
PMAS-Arid Agriculture University Rawalpindi
Pakistan

Nezar Samarah
Department of Plant Production
Jordan University of Science and Technology
Irbid, Jordan

Jeyabalan Sangeetha
Department of Environmental Science
Central University of Kerala
Kasaragod, Kerala, India

P. Saranraj
Department of Microbiology
Sacred Heart College (Autonomous)
Tirupattur, Tamil Nadu, India

Fang Shang
Key Laboratory of Crop Physiology, Ecology and Genetic Breeding
Ministry of Education/Collage of Agronomy
Jiangxi Agriculture University
Nanchang, PR China

Pradeep Sharma
ICAR-Indian Institute of Wheat and Barley Research
Karnal, India

Sonia Sheoran
ICAR-Indian Institute of Wheat and Barley Research
Karnal, India

Kadambot HM Siddique
The UWA Institute of Agriculture
The University of Western Australia
Perth, WA, Australia

Garima Singroh
ICAR-Indian Institute of Wheat and Barley Research
Karnal, India

Muhammad Subtain Abbas
Institute of Soil Science
PMAS-Arid Agriculture University Rawalpindi
Pakistan

Sikander Khan Tanveer
Wheat Program
PARC-National Agricultural Research Center (NARC)
Park Road Islamabad Pakistan

Devarajan Thangadurai
Department of Botany
Karnatak University
Dharwad, Karnataka, India

Laila Trioui
Faculty of Sciences Ain Chock
Laboratory of Microbiology, Pharmacology, Biotechnology, and Environment
University Hassan II of Casablanca
Morocco

Hafeez Ullah Rafa
Institute of Soil Science
PMAS-Arid Agriculture University Rawalpindi
Pakistan

Muhammad Umair Hassan
Research Centre on Ecological Sciences
Jiangxi Agricultural University
Nanchang, China

Vyas R. V.
Department of Agricultural Microbiology B. A. College of Agriculture
Anand Agricultural University
Anand, Gujarat, India

Hafiz Muhammad Wariss
Kunming Institute of Botany
Chinese Academy of Sciences
Kunming, China

Ziming Wu
Key Laboratory of Crop Physiology, Ecology and Genetic Breeding
Ministry of Education/Collage of Agronomy
Jiangxi Agriculture University
Nanchang, PR China

Fazli Zuljalal
Department of Environmental Sciences
University of Swabi
Pakistan

1

Consequences of Salt and Drought Stresses in Rice and Their Mitigation Strategies through Intrinsic Biochemical Adaptation and Applying Stress Regulators

Akbar Hossain
Bangladesh Wheat and Maize Research Institute, Bangladesh

Ayman EL Sabagh
University of Kafrelsheikh, Egypt

Rajan Bhatt
Punjab Agricultural University, India

Muhammad Farooq
College of Agricultural and Marine Sciences, Oman; University of Agriculture, Pakistan; The University of Western Australia, Australia

Mirza Hasanuzzaman
Sher-e-Bangla Agricultural University, Bangladesh

CONTENTS

1.1 Introduction

In terms of global cereal production, rice is second only to wheat and feeds over half of the world's population. Over 400 million people in the rice-producing areas of Asia, Africa, and South America rely on rice and its derived products to meet their dietary requirements (Joseph et al. 2010). An area of approximately 162.3 million hectares is under rice cultivation globally, with a total production of 738.1 million tons (FAOSTAT 2014). Due to increasing population, the demand for rice is increasing rapidly. In terms of rice-consuming countries, China is at the top, with a consumption of 141.45 million metric tons per year (Statista 2018). The world's population is increasing, and is projected to reach 9 billion by 2050.

Most of this population increase will be increased in the developing world, where rice is a major foodstuff. This situation demands an increase in investments in agricultural research, even without the challenges of climate change, to meet the food demands of an increasing population (Nelson et al., 2009; Adnan et al. 2018a, 2018b, 2019; Akram et al. 2018a, 2018b; Aziz et al. 2017a, 2017b; Habib et al. 2017; Hafiz et al. 2016, 2019; Kamarn et al. 2017; Muhmmad et al. 2019; Sajjad et al. 2019; Saud et al. 2013, 2014, 2016, 2017, 2020; Shah et al. 2013; Qamar-uz et al. 2017; Wajid et al. 2017; Yang et al. 2017; Zahida et al. 2017; Depeng et al. 2018; Hussain et al. 2020; Hafiz et al. 2020a, 2020b; Shafi et al. 2020; Fahad and Bano 2012; Fahad et al. 2013, 2014a, 2014b, 2015a, 2015b, 2016a, 2016b, 2016c, 2016d, 2018, 2017, 2019a, 2019b; Hesham and Fahad 2020).

At the same time, the IPCC (Intergovernmental Panel on Climate Change 2007) predicted that extreme weather events would be more frequent and severe in the future. This is expected to reduce the rice yield by 14% in South Asia, 10% in East Asia and the Pacific, and 15% in Sub-Saharan Africa (RicePedia 2018; Khan et al. 2014). Using data from the International Food Policy Research Institute, Nelson et al. (2009) predicted a 32 to 37% increase in rice price by the year 2050. The International Rice Research Institute also predicted an increase in temperature in the near future, causing incidences of heat stress, drought and salinity, and flooding owing to sea level rise (IRRI 2018). Increase in incidences of the above stresses will have a drastic impact on the productivity of rice. This demands developing strategies to mitigate the adversities of environmental stresses and improving rice yield to feed the masses.

Heat stress causes diverse and adverse changes in growth, development, physiological processes, and plant yield (Fahad et al. 2016a, 2016b, 2016c, 2016d, 2017, 2018). One of the major influence of abiotic stresses is the overproduction of reactive oxygen species (ROS), which causes oxidative damage to the structural components of plants (Nievola et al. 2017). However, these ROS, in addition to oxidative damages, also act as signaling agents in plants to persist under conditions of stress (Krishnamurthy and Rathinasabapathi 2013; Ali et al. 2013).

The expression of osmoprotectants, such as Pro and polyamines, can be a useful screening criterion for tolerance against abiotic stresses (Bohnert et al. 1995). In this respect, the exogenous application of stress regulators may also help to reduce oxidative damage maintaining the cellular pH, scavenging of ROS, the inherent protein structure of plants and also protecting macromolecules under stress conditions (Sengupta et al. 2016).

In this chapter, the biochemical adaptation and foliar application of stress regulators for improving tolerance against abiotic stresses in rice are discussed.

1.2 General Aspects of Abiotic Stresses on Growth and Development of Rice

Among the abiotic stresses, heat stress, drought, and salinity are the most important, having an adverse effect on sustainable rice production in rice-growing countries in the world (Gao et al. 2007; Liu et al. 2010; Munns 2011; Wahid et al. 2020; Subhan et al. 2020; Zafar-ul-Hye et al. 2020a, 2020b; Adnan et al. 2020; Ilyas et al. 2020; Saleem et al. 2020a, 2020b, 2020c; Rehman et al. 2020; Farhat et al. 2020; Wu et al. 2020; Mubeen et al. 2020; Farhana et al. 2020; Jan et al. 2019; Wu et al. 2019; Ahmad et al. 2019; Baseer et al. 2019; Hafiz et al. 2018; Tariq et al. 2018). Due to climate change, IPCC projected that global air temperature has increased by ~0.74°C in the last decade and is projected to increase by ~1.1–6.4°C by the end of 2100 (IPCC 2007). The increasing trend in temperature will have a strong impact on rice production. For instance, Peng et al. (2004) has suggested that there would be a 10% reduction in rice yield from a 1°C increase in night air temperature. Heat stress alters the physiological processes and morphological responses of plants, leading to a reduction in yields. One of the consequences of abiotic stresses is the overproduction of ROS, which causes oxidative damages in the plants (Tariq et al. 2018). However, plants expressing heat-shock proteins (HSPs) may cope with the heat-induced damages to some extent (Gurley 2000). Gong et al. (1997a, 1997b) showed an indication that heat stress induces the oxidative stress, which leads to an induction of the pathways for an accumulation of some HSPs (Tariq et al. 2013; Storozhenko et al. 1998; Schett et al. 1999; Ahmad et al. 2015). In addition, oxidative stress also induces heat shock genes (Sun et al. 2002; Mittler et al. 2004).

Drought stress is also a major constraint for limiting global agricultural productivity (Zheng et al. 2001; Vinocur and Altman 2005; Mittler 2006). The severity of drought stress varies both within species and between species (Nakayama et al. 2007). During drought stress, plants close the stomata, and also limit the gas exchange, leading to a decrease in photosynthesis (Jaleel et al. 2007). Rice is very sensitive to drought (Lefèvre et al. 2001), particularly during the initial stage (seedling) and reproductive stages (Khan and Abdullah 2003). It has been reported that 50% of global rice production is highly influenced by drought stress worldwide (Bouman et al. 2005). Thus, drought stress currently causes extensive loss to global rice production and is a severe threat to sustainable crop production. Therefore, it is imperative to develop drought-tolerant rice varieties which are highly adapted to dry environments in order to meet the food security of increasing populations across the globe (Foley et al. 2011; IRRI 2008; Mohd Zain and Ismail 2016).

Salinity is also a major abiotic stress overwhelming crop production worldwide after drought, particularly in the arid and semi-arid regions (Gregorio 1997). It has been estimated that almost 20% of arable and 33% of arable irrigated lands across the globe are suffering from a high level of salinity stress. About 1 billion ha lands globally, signifying nearly 7% of the Earth's landmass, is under salinity and this area is increasing at a global rate of around 10% (Yensen 2008; Shrivastava and Kumar 2015). This indicates that more than 50% of the cultivated land would be affected by a high level of salinity by the year 2050 (Jamil et al. 2011; Shrivastava and Kumar 2015). Saline soils have electrical conductivity (EC_e) in the root zone of soils >4 dS m^{-1} with exchangeable sodium ≤15% at 25°C. An increase in salt concentration in the root zones poses osmotic effect to the plants, which is followed by ionic stress at the cellular, organ, and whole plant levels (Munns 2005; Munns and Tester 2008). Plants respond to salinity in two phases *viz*., a rapid, osmotic phase that inhibits the growth of young leaves and a slower, ionic phase that accelerates the senescence of mature leaves (Akbar et al. 1986; Khatun and Flowers 1995). Similar to drought stress, rice is also highly sensitive to salinity in both seedling and early reproductive stages (Shannon 1998; Lefèvre et al. 2001; and Abdullah 2003) and more resistant at late reproductive to grain filling (Heenan et al. 1988; Djanaguiraman et al. 2006).

1.3 Biochemical Mechanisms of Rice to Survive under Abiotic Stress Conditions

1.3.1 Survival Mechanism under Heat Stress

Heat stress (HS) causes several modifications in plants at biochemical levels, resulting in a drastic loss of economic yield. The effect of HS starts from the seed germination to flowering and maturity, at various stages of plants, i.e. the growth stages are temperature-dependent. Both seed germination and seed vigor are adversely affected by HS, which causes heat-related damage or even the death of the seed (Siddiqui et al. 2015). HS impairs the photosynthetic activity, and the reduced water content caused by heat produces negative effects on cell division and growth. Thus, it is imperative to determine the biochemical mechanisms involved in the HS effects in plants to understand how plants respond and adapt to heat stress leading to in-breed crop varieties having desired heat tolerance. The modification of biochemical processes by gene expression changes gradually and leads to the development of heat tolerance in the form of acclimation, or, in the ideal case, for adaptation (Nievola et al. 2017).

During conditions of HS, plants produce reactive oxygen species (ROS), which leads to the initiation of oxidative stress damage in plants (Suzuki and Mittler 2006). HS-tolerant plants create a signaling role in the plant for adaptation against ROS damage. Among the growth regulators, auxin has been found to link many developmental activities in plants, such as cell division, cell elongation, meristem development and the maintenance of cell polarity under stress conditions, including extremes of heat (Woodward and Bartel 2005; Mockaitis and Estelle 2008). Therefore, it is confirmed that the function of auxin has been associated with the survival mechanism of plants under stress. The greatest consequence of ROS during drought (Moran et al. 1994), high-temperature stress (Larkindale and Knight 2002), and salinity (Hernandez et al. 1993) is oxidative stress, which lead to the oxidation of DNA, proteins, and lipids. Scientists found that during oxidative stress damage, auxin is linked to regulating the level of

ROS that leads either to oxidative damage (susceptible plants) or to signaling in plants to survive under stress (tolerant plants) (Krishnamurthy and Rathinasabapathi 2013). They also found that many processes are sensitive to high-temperature stress, such as photosynthesis, respiration, and physiological mechanisms (Krishnamurthy and Rathinasabapathi 2013). Within the plant cells, when different pathways are uncoupled, electrons that have a high-energy state are transferred to molecular oxygen (O_2) to form ROS (Mittler 2002). ROS, such as O_2, H_2O_2, O_2^-, and HO,· are toxic molecules capable of causing oxidative damage to proteins, DNA, and lipids (Apel and Hirt 2004). Under optimal growth conditions, they are mainly produced at a low level in organelles such as chloroplasts, mitochondria, and peroxisomes. Therefore, during stress, their rate of production is elevated dramatically (Essemine et al. 2010).

Other studies have demonstrated that temperature stress such as heat, cold or freezing is the main cause for yield reduction in crops and ROS generated by these stresses have been shown to injure cell membranes and proteins (Larkindale and Knight 2002). Earlier studies demonstrated that ROS-scavenging mechanisms have an important role in protecting plants against temperature stresses and a combination of high light and temperature stress (Yoshimura et al. 2004). Heat stress impairs the stability of proteins, membrane integrity, RNA and activity of enzymes in chloroplast and mitochondria, resulting in an imbalance in the metabolic homeostasis (Hemantaranjan et al. 2014; Siddiqui et al. 2015). Recently, it has been reported that the heat shock proteins (HSPs) and ROS-scavenging enzymes are major functional proteins in plants that are induced by heat stress (Jacob et al. 2017). Proteomic studies have identified HSPs that may be involved in improving tolerance to heat in the most heat-tolerant rice cultivars (N22) (Jagadish et al. 2010). In addition, González-Schain et al. (2015) reported that the expression of genes encoding HSPs and heat shock factors was highly activated at late stages of anther development in the heat-tolerant N22 rice plants. Therefore, the increase of HSPs could be correlated with heat tolerance responses in order to maintain productivity in plants.

1.3.2 Survival Mechanism under Drought Stress

Plant tolerance to unfavorable conditions, particularly water deficit, has been associated with Pro (formed in the leaf tissues of plants exposed to water stress) accumulation and increased antioxidative enzymes such as peroxidase (Ashraf and Foolad 2007; Kaur and Asthir, 2015; Per et al. 2017; Hesham and Fahad 2020). For example, Chandru et al. (2003) found that under conditions of drought stress, Pro concentrations, leaf lipid peroxidation, leaf hydrogen peroxide, and stomatal resistance, were increased in those rice cultivars which were resistant against drought stress. Although Pro can act not only as solute-compatible, but also as a protector to cells from oxidative damage under conditions of drought stress (Chaves et al. 2002). Sakamoto and Murata (2002); Hussain Wani et al. (2013); Sekmen et al. (2014) also observed that under conditions of drought soluble sugars such as sucrose can also perform as compatible solutes for protecting the integrity of the cell membrane, preventing the denaturation of proteins and also maintaining leaves' cell turgor as well as playing a significant role at both cellular and whole-organism levels. For example, sucrose is linked to the long-distance signals that coordinate with the modifications in physiological, developmental, and environmental responses (Lemoine et al. 2013; Mohammadkhani and Heidari 2008).

By contrast, Tripathi et al. (2009) found that proteins such as thioredoxin, glutaredoxin, and cyclophilin play a significant role in scavenging the ROS in plant cell during drought stress. It is well documented that initiation of antioxidant enzymes such as Pro, glycine betaine (GB), peroxidase, and CAT in plant cells leads to an overcoming of oxidative stresses (Foyer et al. 1994; Chaitanya et al. 2002). The enzymatic and non-enzymatic antioxidants are involved in many reactions inside plants, such as linking polysaccharides, the oxidation of indole-3-acetic acid, monomer connections, lignification, oxidation of phenols, pathogen defense, regulation of cell elongation and others (Kao 2003). A decrease in membrane stability reflects the extent of lipid peroxidation caused by ROS. Furthermore, lipid peroxidation is an indicator of the prevalence of free radical reaction in tissues. The increased application of potassium fertilizer would increase these proteins (Chaitanya et al. 2002). Moreover, oxygen uptake loading on the tissues as both processes generate ROS, particularly H_2O_2, which is produced at very high rates by the glycollate oxidase reaction in the peroxisomes in photorespiration (Anjum et al. 2011).

Some osmoprotectants, such as sucrose and polyols, protect cellular components of plants from dehydration damage (Per et al. 2017; Kahlaoui et al. 2018). While plants which are tolerant against drought, several compatible solutes, such as quercitol, pinitol or quebrachitol, were slightly different from the primary metabolites (Chaves et al. 2003; Marchant and Tausz 2006). However, under drought stress, hydrogen peroxide (H_2O_2), membrane damage, ascorbate peroxidase (AP), catalase (CAT) are generally increased in affected plants (Yildizli et al. 2018). By contrast, plants which are able to survive under drought stress could accumulate compatible solutes (cyclitols), including myo-inositol, pinitol, and quercitol in the cytosol of the plants, which lead to a reduction in H_2O_2, membrane damage, AP, CAT and increase water status, Pro, calcium level and glutathione reductase activities (Yildizli et al. 2018).

1.3.3 Survival Mechanism under Salt Stress

Salinity effects on many developmental activities in plants such as cell division, cell elongation, and meristem development, and also limit the cell polarity that leads to negatively influence the agricultural yields (Munns and Tester 2008). High salt levels in soils can induce a water deficit in plants and cause ion toxicity along with deficiency of some nutrients, leading to physiological and molecular damage and consequently the death of plants (Fahad and Bano 2012). Salinity affects seed germination, water uptake, ion balance of the cellular ions, resulting in osmotic stress and ion toxicity in plant cells (Khan and Panda 2008). Salt stress has been reported to cause an inhibition of growth and development, a reduction in photosynthesis, respiration and protein synthesis in sensitive species (Munns 2002; Pal et al. 2004; Gautam et al. 2013).

Under saline conditions, morphological parameters and growth of plants will be affected resulting in reduced vegetative growth (Rogers et al. 2009), leaf area (Saleh and Maftoun 2008), chlorophyll content (Netondo et al. 2004), plant height (Rahman et al. 2008), plant dry weight (Razzaque et al. 2009) and consequently crop yield (Zeng and Shannon 2000). Saline conditions provoke osmotic stress and oxidative stress which lead to similar drought adaptive responses, such as the accumulation of compatible solutes, the induction of stress proteins, and the activation of ROS-scavenging systems which comprise non-enzymatic and enzymatic antioxidants (Apel and Hirt 2004). Although several viable management options have been developed to improve the productivity of plants in saline areas, including irrigation and drainage management, these options have not proven successful in mitigating salinity (Lin and Kao 2001).

Therefore, improving the salt tolerance of crops by molecular and plant breeding approaches is the most attractive and sustainable option to support crop production in saline soils (Ondrasek et al. 2011). The generation of salt-tolerant crops could be carried out by a clear understanding of the plant mechanisms under saline conditions. Plants have developed a wide range of adaptive mechanisms, including morphological, anatomical, physical-biochemical, and molecular responses, allowing them to respond and properly adapt to salinity (Negrão et al. 2017).

1.4 Management of Abiotic Stresses by Using Osmoprotectants and Antioxidants

Under stress conditions, there is a serious imbalance between the production of ROS and antioxidant defense in plant cells, causing oxidative stress. The particularly high concentration of ROS in plant cells is the basis of oxidative damage. Therefore, to stabilize the destruction from the oxidative stage, plant cells should be fortified through an enormous enzymatic and non-enzymatic antioxidants defense mechanism. Among these enzymatic and non-enzymatic antioxidants, some are constitutive, and others are stimulated when they received a stress-specific signal. ROS enzymatic scavengers include catalase, glutathione reductase, guaiacol peroxidase, superoxide dismutase, and ascorbate peroxidase, and non-enzymatic scavengers are (antioxidants) are flavonoids, carotenoids, α-tocopherol, ascorbic acid and glutathione. Without these, Pro, GB, trehalose (Tre), polyphenols, and polysugars also accumulate by plant cells under stress conditions. Under stress conditions, plant cells also release multiple biogenic volatile organic compounds through the degradation of cellular structures of plants; of the compounds methanol, ethanol, formaldehyde, acetaldehyde, terpenes, and oxylipins are the most important (Anjum et al. 2012). Exogenous applications of osmoprotectants (Pro, GB, Tre etc.), phytohormones (abscisic acid, gibberellic acids, jasmonic acids, brassinosteroids, salicylic acid, etc.), signaling molecules (nitric oxide), polyamines (putrescine, spermidine and spermine), trace elements (selenium and silicon) and nutrients (nitrogen, phosphorus, potassium and calcium) have been found to be effective in mitigating heat stress-induced damage in plants.

While Yang et al. (2007) suggested that higher levels of free spermine and insoluble-conjugate putrescine in rice perform well under drought stress. Silicon also plays a significant role in growth, mineral nutrition, mechanical strength, and resistance to several stresses (Epstein 1994). Silicification of endodermal tissues of rice root was found to play an important role in water transport across the root, thereby enhancing drought tolerance (Lux et al. 1999).

Generally, drought-tolerant plants could accumulate cyclitols, including myo-inositol, pinitol, quercitol in the cytosol of the plants (Yildizli et al. 2018). At the same time the foliar application of synthesis myo-inositol leads to a fall in hydrogen peroxide, ascorbate peroxidase (AP), catalase (CAT), Pro and calcium in plants under drought stress (Yildizli et al. 2018).

The accumulation of osmolytes such as Pro, GB is a well-known adaptive mechanism in plants against abiotic stress conditions, including environment stress (Ashraf and Foolad 2007; Hayat et al. 2012). Since heat-sensitive plants lack the ability to accumulate these substances, environmental stress tolerance in such plants can be improved by the exogenous application of osmoprotectants (Lu et al. 2009). Similar to heat and drought stress, the exogenous application of compatible osmolytes such as Pro, GB, Tre, etc. had gained considerable attention in mitigating the effect of salt stress (Ashraf and Foolad 2007). The beneficial effects of Pro or GB application under salt or drought stress were also shown by Sobahan et al. (2012) and Kongngem et al. (2012). Proline and GB play a pivotal role in an osmotic adjustment in stressed plants (Hasegawa et al. 2000), and also contribute to the detoxification of ROS and the stabilization of enzymes/proteins (Bohnert and Jensen 1996; Verbruggen and Hermans 2008). Under conditions of salt stress, GB application maintains or enhances growth and yield in rice (Cha-Um et al. 2013).

Similarly, the exogenous application of nutrients has been found to improve crop performance under salt stress (Raza et al. 2006). Whereas, Zain et al. (2014) observed that the application of potassium fertilizer with periodical water stress could mitigate drought stress effects on rice, indicating that under drought K fertilizer helps plants to survive through maintaining the osmotic balance of the affected rice plants.

Ascorbic acid (AsA; vitamin C) is the main non-enzymatic antioxidant exploited by plants under stressful conditions to ameliorate the adverse effects through imposing by ROS (Mittler 2002; Ozgur et al. 2013). The foliar application of AsA helps plants to survive under salt stress (Zonouri et al. 2014; El-Bassiouny and Sadak 2015) and drought stress (Khalil et al. 2010) through maintaining photosynthesis pigments, cell-wall expansion and cell-cycle progression, gene expression. It also leads to the synthesis of many hormones, including anthocyanin and flavonoids (Gest et al. 2013). The AsA is an important primary metabolite in plants that act as antioxidants, enzyme cofactors and as a modulator of cell signaling in a variety of physiological and biochemical processes, including cell wall biosynthesis, secondary metabolites, and phytohormones, stress tolerance, photoprotection, and cell and cell wall expansion (Akram et al. 2017).

Similarly, salicylic acid (SA) also plays a significant role in the regulation of a number of vital physiological processes in plants under stress conditions (Singh and Usha 2003; Hayat et al. 2010), and also provides protection against heat, drought and salinity (Kaya et al. 2009; Hayat et al. 2013). It also has a role in germination under stressful conditions, although its definite role and the underlying physiological mechanisms have not been fully elucidated (Asadi et al. 2013; Bagheri 2014). Karlidag et al. (2009) and Elwana and El-Hamahmy (2009) found that seed treated with GB, kinetin and SA increased the yield of different crops, by reducing the stress-induced inhibition (Khan and Abdullah 2003).

1.5 Conclusions and Prospect

Rice plants are found to be highly sensitive to abiotic stress, particularly early growth and reproductive stages due to the fluctuations temperature, drought, and salinity. Therefore, physiological traits such as ion contents in leaves and stems, photosynthetic parameters and water relation parameters at different phenological stages under heat, drought and salt stresses should be evaluated in terms of grain yield and stress tolerance. On the other hand, in the case of the development of rice-tolerant cultivars against abiotic stress conditions, it is desirable for widespread information on molecular, biochemical and physiological responses of rice under abiotic stresses. Although the use of short duration with abiotic stress-tolerant rice varieties has a great potential for rice cultivation under adverse environmental conditions. However, the improvement of short duration crop cultivars tolerant to abiotic stress through breeding procure is time-consuming and has to follow a complex, lengthy protocol. Therefore, it is important to know the multiple reliable physiological and biochemical traits that can be useful as a quick, easy, and economic technique for the screening of crops genotypes to abiotic stress. The second direction is to apply exogenous osmotic and antioxidant protectors and/or plant growth regulators, which have been found to be effective in enhancing abiotic tolerance in rice. The integration of all these approaches can lead to the sustainable production of rice through the effective management of environmental stresses during this period of climate change.

REFERENCES

Adnan M, Fahad S, Muhammad Z, Shahen S, Ishaq AM, Subhan D, Zafar-ul-Hye M, Martin LB, Raja MMN, Beena S, Saud S, Imran A, Zhen Y, Martin B, Jiri H, Rahul D (2020) Coupling phosphate-solubilizing bacteria with phosphorus supplements improve maize phosphorus acquisition and growth under lime induced salinity stress. *Plants 9*(900). doi:10.3390/plants9070900

Adnan M, Fahad S, Khan IA, Saeed M, Ihsan MZ, Saud S, Riaz M, Wang D, Wu C (2019) Integration of poultry manure and phosphate solubilizing bacteria improved availability of Ca bound P in calcareous soils. *Biotech 9*(10):368.

Adnan M, Shah Z, Sharif M, Rahman H (2018b) Liming induces carbon dioxide (CO_2) emission in PSB inoculated alkaline soil supplemented with different phosphorus sources. *Environ Sci Poll Res 25*(10):9501–9509.

Adnan M, Zahir S, Fahad S, Arif M, Mukhtar A, Imtiaz AK, Ishaq AM, Abdul B, Hidayat U, Muhammad A, Inayat-Ur R, Saud S, Muhammad ZI, Yousaf J, Amanullah, Hafiz MH, Wajid N (2018a) Phosphate-solubilizing bacteria nullify the antagonistic effect of soil calcification on bioavailability of phosphorus in alkaline soils. *Sci Rep 8*: 4339. https://doi.org/10.1038/s41598-018-22653-7

Ahmad K, Adnan M, Khan MA, Hussain Z, Junaid K, Saleem N, Ali A (2015) Bioactive neem leaf powder enhances the shelf life of stored mungbean grains and extends protection from pulse beetle. *Pak J Weed Sci Res 21*(1): 71–81.

Ahmad S, Kamran M, Ding R, Meng X, Wang H, Ahmad I, Fahad S, Han Q (2019) Exogenous melatonin confers drought stress by promoting plant growth, photosynthetic capacity and antioxidant defense system of maize seedlings. *Peer J 7*:e7793. http://doi.org/10.7717/peerj.7793

Akbar M, Gunawardena IE, Ponnamperuma FN (1986) *Breeding for soil stresses In: Progress in rain fed lowland rice*. IRRI, Manila, Philippines, pp 263–272.

Akram NA, Shafiq F, Ashraf M (2017) Ascorbic acid-a potential oxidant scavenger and its role in plant development and abiotic stress tolerance. *Frontiers Plant Sci 8*: 613. https://doi.org/10.3389/fpls.2017.00613

Akram R, Turan V, Hammad HM, Ahmad S, Hussain S, Hasnain A, Maqbool MM, Rehmani MIA, Rasool A, Masood N, Mahmood F, Mubeen M, Sultana SR, Fahad S, Amanet K, Saleem M, Abbas Y, Akhtar HM, Waseem F, Murtaza R, Amin A, Zahoor SA, ul Din MS, Nasim W (2018a) Fate of organic and inorganic pollutants in paddy soils. In: Hashmi MZ, Varma A (ed.), *Environmental pollution of paddy soils, soil biology*. Cham: Springer International Publishing AG, pp 197–214.

Akram R, Turan V, Wahid A, Ijaz M, Shahid MA, Kaleem S, Hafeez A, Maqbool MM, Chaudhary HJ, Munis, MFH, Mubeen M, Sadiq N, Murtaza R, Kazmi DH, Ali S, Khan N, Sultana SR, Fahad S, Amin A, Nasim W (2018b) Paddy land pollutants and their role in climate change. In: Hashmi MZ, Varma A (ed.), *Environmental pollution of paddy soils, soil biology*. Cham: Springer International Publishing AG, pp 113–124.

Ali K, Arif M, Khan ZA, Tariq M, Waqas M, Gul B, Bibi M, Din Z, Ali M, Shafi B, Adnan, M (2013) Effect of cutting on productivity and associated weeds of canola. *Pak J Weed Sci Res 19*(4): 393–401.

Anjum NA, Umar S, Ahmad A (2012) *Oxidative stress in plants: causes, consequences and tolerance*. I.K. International Publishing House Pvt.

Anjum SA, Wang LC, Farooq M, Hussain M, Xue LL, Zou CM (2011) Brassinolide application improves the drought tolerance in maize through modulation of enzymatic antioxidants and leaf gas exchange. *J Agron Crop Sci 197*: 177–185. doi:10.1111/j.1439-037X.2010.00459.x.

Apel K, Hirt H (2004) Reactive oxygen species: metabolism, oxidative stress and signal transduction. *Annu Rev Plant Biol 55*: 373–399.

Asadi M, Heidari MA, Kazemi M., Filinejad AR (2013) Salicylic acid induced changes in some physiological parameters in chickpea (*Cicer arietinum* L.) under salt stress. *J Agric Technol 9*(2): 311–316.

Ashraf M, Foolad MR (2007) Roles of glycine betaine and proline in improving plant abiotic stress resistance. *Environ Exp Bot 59*(2): 206–216.

Aziz K, Daniel KYT, Fazal M, Muhammad ZA, Farooq S, Fan W, Fahad S, Ruiyang Z (2017a) Nitrogen nutrition in cotton and control strategies for greenhouse gas emissions: a review. *Environ Sci Pollut Res 24*: 23471–23487. https://doi.org/10.1007/s11356-017-0131-y

Aziz K, Daniel KYT, Muhammad ZA, Honghai L, Shahbaz AT, Mir A, Fahad S (2017b) Nitrogen fertility and abiotic stresses management in cotton crop: a review. *Environ Sci Pollut Res 24*: 14551–14566. https://doi.org/10.1007/s11356-017-8920-x

Bagheri MZ (2014) The effect of maize priming on germination characteristics, catalase and peroxidase enzyme activity, and total protein content under salt stress. *Int J Biosci 4*(2): 104–112.

Baseer M, Adnan M, Fazal M, Fahad S, Muhammad S, Fazli W, Muhammad A, Jr., Amanullah, Depeng W, Saud S, Muhammad N, Muhammad Z, Fazli S, Beena S, Mian AR, Ishaq AM (2019) Substituting urea by organic wastes for improving maize yield in alkaline soil. *J Plant Nutr*. doi.org/10.1080/01904167.2019.1659344

Bohnert HJ, Jensen RG (1996) Strategies for engineering water stress tolerance in plants. *Trends Biotechnol 14*: 89–97.

Bohnert HJ, Nelson DE, Jensen JE (1995) Adaptations to environmental stresses. *Plant Cell 7*: 1099–1111.

Bouman BAM, Peng S, Castaðeda AR, Visperas RM (2005) Yield and water use of irrigated tropical aerobic rice systems. *Agric Water Manag 74*(2): 87–105.

Chaitanya, K, Sundar D, Masilamani S, Ramachandra RA (2002) Variation in heat stress induced antioxidant enzyme activities among three mulberry cultivars. *Plant Growth Regul 36*: 175–180.

Chandru HK, Kim E, Kuk Y, Cho K, Han O (2003) Kinetics of wound induced activation of antioxidative enzymes in *Oryza sativa*: differential activation at different growth stages. *Plant Sci 164*(6): 935–941.

Cha-Um S, Samphumphuang T, Kirdmanee C (2013) Glycinebetaine alleviates water deficit stress in indica rice using proline accumulation, photosynthetic efficiencies, growth performances and yield attributes. *Aust J Crop Sci 7*: 213–218.

Chaves MM, Maroco JP, Pereira JS (2003) Understanding plant responses to drought-from genes to the whole plant. *Funct Plant Biol 30*: 239–264.

Chaves MM, Pereira JS, Maroco J, Rodrigues ML, Ricardo CPP, Osorio ML, Carvalho I, Faria T, Pinheiro C (2002) How plants cope with water stress in the field. Photosynthesis and growth? *Ann Bot 89*: 907–916.

Depeng W, Fahad S, Saud S, Muhammad K, Aziz K, Mohammad NK, Hafiz MH, Wajid N (2018) Morphological acclimation to agronomic manipulation in leaf dispersion and orientation to promote "Ideotype" breeding: evidence from 3D visual modeling of "super" rice (*Oryza sativa* L.). *Plant Physiol Biochem*. https://doi.org/10.1016/j.plaphy.2018.11.010

Djanaguiraman M, Sheeba JA, Shanker AK, Durga Devi D, Bangarusamy U (2006) Rice can acclimate to lethal level of salinity by pretreatment with sublethal level of salinity through osmotic adjustment. *Plant Soil 284*: 363–373.

El-Bassiouny H, Sadak MS (2015) Impact of foliar application of ascorbic acid and α-tocopherol on antioxidant activity and some biochemical aspects of flax cultivars under salinity stress. *Acta Biol Colomb 20*(2): 209–222.

Elwana MWM et al. (2009) Improved productivity and quality associated with salicylic acid application in greenhouse pepper. *Sci Horti 122*: 521.

Epstein E (1994) The anomaly of silicon in plant biology. *Proc Natl Acad Sci (USA) 91*: 11–17.

Essemine J, Ammar S, Bouzid S (2010) Impact of heat stress on germination and growth in higher plants: physiological, biochemical and molecular repercussions and mechanisms of defence. *J Biol Sci 10*: 565–572.

Fahad S, Bajwa AA, Nazir U, Anjum SA, Farooq A, Zohaib A, Sadia S, Nasim W, Adkins S, Saud S, Ihsan MZ, Alharby H, Wu C, Wang D, Huang J (2017) Crop production under drought and heat stress: plant responses and management options. *Front Plant Sci 8*: 1147. https://doi.org/10.3389/fpls.2017.01147

Fahad S, Bano A (2012) Effect of salicylic acid on physiological and biochemical characterization of maize grown in saline area. *Pak J Bot 44*: 1433–1438.

Fahad S, Chen Y, Saud S, Wang K, Xiong D, Chen C, Wu C, Shah F, Nie L, Huang J (2013) Ultraviolet radiation effect on photosynthetic pigments, biochemical attributes, antioxidant enzyme activity and hormonal contents of wheat. *J Food Agri Environ 11*(3&4): 1635–1641.

Fahad S, Hussain S, Bano A, Saud S, Hassan S, Shan D, Khan FA, Khan F, Chen Y, Wu C, Tabassum MA, Chun MX, Afzal M, Jan A, Jan MT, Huang J (2014a) Potential role of phytohormones and plant growth-promoting rhizobacteria in abiotic stresses: consequences for changing environment. *Environ Sci Pollut Res 22*(7): 4907–4921. https://doi.org/10.1007/s11356-014-3754-2

Fahad S, Hussain S, Matloob A, Khan FA, Khaliq A, Saud S, Hassan S, Shan D, Khan F, Ullah N, Faiq M, Khan MR, Tareen AK, Khan A, Ullah A, Ullah N, Huang J (2014b) Phytohormones and plant responses to salinity stress: a review. *Plant Growth Regul 75*(2): 391–404. https://doi.org/10.1007/s10725-014-0013-y

Fahad S, Hussain S, Saud S, Hassan S, Chauhan BS, Khan F et al (2016a) Responses of rapid viscoanalyzer profile and other rice grain qualities to exogenously applied plant growth regulators under high day and high night temperatures. *PLoS One 11*(7):e0159590. https://doi.org/10.1371/journal.pone.0159590

Fahad S, Hussain S, Saud S, Hassan S, Ihsan Z, Shah AN, Wu C, Yousaf M, Nasim W, Alharby H, Alghabari F, Huang J (2016c) Exogenously applied plant growth regulators enhance the morphophysiological growth and yield of rice under high temperature. *Front Plant Sci 7*: 1250. https://doi.org/10.3389/fpls.2016.01250

Fahad S, Hussain S, Saud S, Hassan S, Tanveer M, Ihsan MZ, Shah AN, Ullah A, Nasrullah KF, Ullah S, Alharby H, Nasim W, Wu C, Huang J (2016d) A combined application of biochar and phosphorus alleviates heat-induced adversities on physiological, agronomical and quality attributes of rice. *Plant Physiol Biochem 103*: 191–198.

Fahad S, Hussain S, Saud S, Khan F, Hassan S, Nasim W, Arif M, Wang F, Huang J (2016b) Exogenously applied plant growth regulators affect heat-stressed rice pollens. *J Agron Crop Sci 202*: 139–150.

Fahad S, Hussain S, Saud S, Tanveer M, Bajwa AA, Hassan S, Shah AN, Ullah A, Wu C, Khan FA, Shah F, Ullah S, Chen Y, Huang J (2015a) A biochar application protects rice pollen from high-temperature stress. *Plant Physiol Biochem 96*: 281–287.

Fahad S, Muhammad ZI, Abdul K, Ihsanullah D, Saud S, Saleh A, Wajid N, Muhammad A, Imtiaz AK, Chao W, Depeng W, Jianliang H (2018) Consequences of high temperature under changing climate optima for rice pollen characteristics-concepts and perspectives. *Archives Agron Soil Sci*. doi:10.1080/03650340.2018.1443213

Fahad S, Nie L, Chen Y, Wu C, Xiong D, Saud S, Hongyan L, Cui K, Huang J (2015b) Crop plant hormones and environmental stress. *Sustain Agric Rev 15*: 371–400.

Fahad S, Adnan M, Hassan S, Saud S, Hussain S, Wu C, Wang D, Hakeem KR, Alharby HF, Turan V, Khan MA, Huang J (2019b) Rice responses and tolerance to high temperature. In: Hasanuzzaman M, Fujita M, Nahar K, Biswas JK (ed.), *Advances in rice research for abiotic stress tolerance*. Abington Hall; Cambridge: Woodhead Publ Ltd, pp 201–224.

Fahad S, Rehman A, Shahzad B, Tanveer M, Saud S, Kamran M, Ihtisham M, Khan SU, Turan V, Rahman MHU (2019a) Rice responses and tolerance to metal/metalloid toxicity. In Hasanuzzaman M, Fujita M, Nahar K, Biswas JK (ed.), *Advances in rice research for abiotic stress tolerance*. Abington Hall; Cambridge: Woodhead Publ Ltd, pp 299–312.

FAOSTAT (2014) Food and Agriculture Organization of the United Nations "FAOSTAT: Production-Crops, 2012 data".

Farhana G, Ishfaq A, Muhammad A, Dawood J, Fahad S, Xiuling L, Depeng W, Muhammad F, Muhammad F, Syed AS (2020) Use of crop growth model to simulate the impact of climate change on yield of various wheat cultivars under different agro-environmental conditions in Khyber Pakhtunkhwa, Pakistan. *Arabian J Geosci 13*: 112. https://doi.org/10.1007/s12517-020-5118-1

Farhat A, Hafiz MH, Wajid I, Aitazaz AF, Hafiz FB, Zahida Z, Fahad S, Wajid F, Artemi C (2020) A review of soil carbon dynamics resulting from agricultural practices. *J Environ Manage 268*: 110319.

Foley JA, Ramankutty N, Braumann KA, Cassidy ES, Gerber JS, Johnston M, Mueller ND, O'Connell, C, Ray DK, West PC, Balzer, C, Bennett EM, Carpenter SR, Hill J, Monfreda C, Polasky S, Rockström J, Sheehan J, Siebert S, Tilman D, Zaks DP (2011) Solutions for a cultivated planet. *Nature 478*: 337–342.

Foyer CH, Decourviers P, Kenerk KJ (1994) Protection against oxygen radicals: an important defense mechanism studied in transgenic plants. *Plant Cell Environ 17*: 507–523.

Gao JP, Chao DY, Lin HX (2007) Understanding abiotic stress tolerance mechanisms: recent studies on stress response in rice. *J Integr Plant Biol 49*: 742–750.

Gautam AR, Khan AH, Chandra PK, Singh AK (2013) Effect of different levels of salinity on morphophysiological traits of rice (*Oryza sativa* L.) varieties relation with grain yield.
Gest N, Gautier H, Stevens R (2013) Ascorbate as seen through plant evolution: the rise of a successful molecule. *J Exp Bot 64*: 33–53. doi:10.1093/jxb/ers297
Gong M, Chen SN, Song YQ, Li ZG (1997a) Effect of calcium and calmodulin on intrinsic heat tolerance in relation to antioxidant systems in maize seedlings. *Aust J Plant Physiol 24*: 371–379.
Gong M, Li X-J, Dai X, Tian M, Li Z-G (1997b) Involvement of calcium and calmodulin in the acquisition of HS induced thermos-tolerance in maize seedlings. *J Plant Physiol 150*: 615–621.
González-Schain N, Dreni L, Lawas LM, Galbiati M, Colombo L, Heuer S, Jagadish KS and Kater MM (2015) Genome-wide transcriptome analysis during anthesis reveals new insights into the molecular basis of heat stress responses in tolerant and sensitive rice varieties. *Plant Cell Physiol 57*(1): 57–68.
Gregorio GB (1997) *Tagging salinity tolerance genes in rice using amplified fragment length polymorphism AFLP*. Los Baños, Philippines: University of Philippines.
Gurley WB (2000) HSP101: a key component for the acquisition of thermos-tolerance in plants. *Plant Cell 12*: 457–459.
Habib R, Ashfaq A, Aftab W, Manzoor H, Fahd R, Wajid I, Aminul I, Vakhtang S, Muhammad A, Asmat U, Abdul W, Syeda RS, Shah S, Shahbaz K, Fahad S, Manzoor H, Saddam H, Wajid N (2017) Application of CSM-CROPGRO-Cotton model for cultivars and optimum planting dates: evaluation in changing semi-arid climate. *Field Crops Res* http://dx.doi.org/10.1016/j.fcr.2017.07.007
Hafiz MH, Abdul K, Farhat A, Wajid F, Fahad S , Muhammad A, Ghulam MS, Wajid N, Muhammad M, Hafiz FB (2020b) Comparative effects of organic and inorganic fertilizers on soil organic carbon and wheat productivity under arid region. *Commn Soil Sci Plant Anal.* doi:10.1080/00103624.2020.1763385
Hafiz MH, Farhat A, Ashfaq A, Hafiz FB, Wajid F, Carol Jo W, Fahad S, Gerrit H (2020a) Predicting Kernel Growth of Maize under Controlled Water and Nitrogen Applications. *Int J Plant Prod* https://doi.org/10.1007/s42106-020-00110-8
Hafiz MH, Farhat A, Shafqat S, Fahad S, Artemi C, Wajid F, Chaves CB, Wajid N, Muhammad M, Hafiz FB (2018) Offsetting land degradation through nitrogen and water management during maize cultivation under arid conditions. *Land Degrad Dev* 1–10. doi:10.1002/ldr.2933
Hafiz MH, Muhammad A, Farhat A, Hafiz FB, Saeed AQ, Muhammad M, Fahad S, Muhammad A (2019) Environmental factors affecting the frequency of road traffic accidents: a case study of sub-urban area of Pakistan. *Environ Sci Pollut Res* https://doi.org/10.1007/s11356-019-04752-8
Hafiz MH, Wajid F, Farhat A, Fahad S, Shafqat S, Wajid N, Hafiz FB (2016) Maize plant nitrogen uptake dynamics at limited irrigation water and nitrogen. *Environ Sci Pollut Res 24*(3): 2549–2557. https://doi.org/10.1007/s11356-016-8031-0
Hasegawa PM, Bressan RA, Zhu JK, Bohnert HJ (2000) Plant cellular and molecular responses to high salinity. *Annu. Rev. Plant Physiol. Plant Mol Biol 51*: 463–499.
Hayat Q, Hayat S, Irfan M, Ahmad A (2010) Effect of exogenous salicylic acid under changing environment: a review. *Environ Exp Bot 68*(1): 14–25.
Hayat S, Ahmad A, Alyemeni MN (2013) *Salicylic acid: plant growth and development.* Dordrecht, The Netherlands: Springer.
Hayat S, Hayat Q, Alyemeni MN, Wani AS, Pichtel J, Ahmad A (2012) Role of proline under changing environments: a review. *Plant Signal Behav 7*(11): 1456–1466.
Heenan DP, Levin LG, McCaffery L (1988) Salinity tolerance in rice varieties at different growth stages. *Aust J Exper Agric 28*: 343–349.
Hemantaranjan A, Bhanu AN, Singh MN, Yadav DK, Patel PK, Singh R, Katiyar D (2014) Heat stress responses and thermos-tolerance. *Adv Plants Agric Res 3*: 1–10.
Hernandez JA, Corpas FJ, Gomez M, del Rio LA, Sevilla F (1993) Salt-induced oxidative stress mediated by activatedoxygenspeciesinpealeafmitochondria.*PhysiolPlant89*:103–110.doi:10.1111/j.1399-3054.1993.tb01792.x.
Hesham FA, Fahad S (2020) Melatonin application enhances biochar efficiency for drought tolerance in maize varieties: modifications in physio-biochemical machinery. *Agron J* 1–22.
Hussain MA, Fahad S, Rahat S, Muhammad FJ, Muhammad M, Qasid A, Ali A, Husain A, Nooral A, Babatope SA, Changbao S, Liya G, Ibrar A, Zhanmei J, Juncai H (2020) Multifunctional role of brassinosteroid and its analogues in plants. *Plant Growth Regul.* https://doi.org/10.1007/s10725-020-00647-8
Hussain Wani S, Brajendra Singh N, Haribhushan A, Iqbal Mir J (2013) Compatible solute engineering in plants for abiotic stress tolerance-role of glycine betaine. *Curr Genom 14*(3): 157–165.

Ilyas M, Mohammad N, Nadeem K, Ali H, Aamir HK, Kashif H, Fahad S, Aziz K, Abid U (2020) Drought tolerance strategies in plants: a mechanistic approach. *J Plant Growth Regulation*. https://doi.org/10.1007/s00344-020-10174-5

IPCC (Intergovernmental Panel on Climate Change, Climate change) (2007) The physical science basis. In *Contribution of Working Group I to the Fourth Assessment Report of the Intergovernmental Panel on Climate Change*; Cambridge, UK: Cambridge University Press.

IRRI (2018) *Rice and climate change*. 5th International Rice Congress in Singapore-2018. http://irri.org/news/hot-topics/rice-and-climate-change. Accessed 15 July 2019.

IRRI (International Rice Research Institute) (2008) *Annual Report for 2007*. International Rice Research Institute, Los Baños, Philippines.

Jacob P, Hirt H, Bendahmane A (2017) The heat-shock protein/chaperone network and multiple stress resistance. *Plant Biotechnol J 15*(4): 405–414.

Jagadish SVK, Muthurajan R, Oane R, Wheeler TR, Heuer S, Bennett J (2010) Physiological and proteomic approaches to address heat tolerance during anthesis in rice (*Oryza sativa* L.). *J Exp Bot 61*: 143–156. 10.1093/jxb/erp289

Jaleel CA, Manivannan P, Sankar B, Kishorekumar A, Gopi R, Somasundaram R, Panneerselvam R (2007) Induction of drought stress tolerance by ketoconazole in Catharanthus roseus is mediated by enhanced antioxidant potentials and secondary metabolite accumulation. *Colloids Surf B: Biointerfaces 60*: 201–206.

Jamil A, Riaz S, Ashraf M, Foolad MR (2011) Gene expression profiling of plants under salt stress. *Crit Rev Plant Sci 30*(5): 435–458.

Jan M, Anwar-ul-Haq M, Adnan NS, Muhammad Y, Javaid I, Xiuling L, Depeng W, Fahad S (2019) Modulation in growth, gas exchange, and antioxidant activities of salt-stressed rice (*Oryza sativa* L.) genotypes by zinc fertilization. *Arabian J Geosci 12*: 775 https://doi.org/10.1007/s12517-019-4939-2

Joseph B, Jini D, Sujatha S (2010) Biological and physiological perspectives of specificity in abiotic salt stress response from various rice plants. *Asian J Agric Sci 2*: 99–105.

Kahlaoui B, Hachicha M, Misle E, Fidalgo F, Teixeira J (2018) Physiological and biochemical responses to the exogenous application of proline of tomato plants irrigated with saline water. *J Saudi Soc Agric Sci 17*(1): 17–23.

Kamarn M, Wenwen C, Irshad A, Xiangping M, Xudong Z, Wennan S, Junzhi C, Shakeel A, Fahad S, Qingfang H, Tiening L (2017) Effect of paclobutrazol, a potential growth regulator on stalk mechanical strength, lignin accumulation and its relation with lodging resistance of maize. *Plant Growth Regul 84*: 317–332. https://doi.org/10.1007/ s10725-017-0342-8

Kao CH (2003) Differential effect of sorbitol and polyethylene glycol on antioxidant enzymes in rice leaves. *Plant Growth Reg 39*(1): 83–89.

Karlidag H, Yildirim E, Turan M (2009) Salicylic acid ameliorates the adverse effect of salt stress on strawberry. *J Agri Sci 66*: 271.

Kaur G, Asthir B (2015) Proline: a key player in plant abiotic stress tolerance. *Biol Plant 59*(4): 609–619.

Kaya, C, Tuna, A, Yokas I (2009) *Salinity and Water Stress*. Netherlands: Springer, pp 45–50.

Khalil SE, Nahed G, Aziz A, Bedour LAH (2010) Effect of water stress and ascorbic acid on some morphological and biochemical composition of *Ocimum basilicum* plant. *J Am Sci 6*: 33–46.

Khan A, Sharif M, Ali A, Shah SNM, Mian IA, Wahid F, Jan B, Adnan M, Nawaz S, Ali N (2014) Potential of AM fungi in phytoremediation of heavy metals and effect on yield of wheat crop. *Am J Plant Sci 5*: 1578–1586.

Khan MA, Abdullah Z (2003) Salinity-sodicity induced changes in reproductive physiology of rice (*Oryza sativa*) under dense soil conditions. *Environ Exp Bot 49*: 145–157.

Khan MH, Panda SK (2008) Alterations in root lipid peroxidation and antioxidative responses in two rice cultivars under NaCl salinity stress, *Acta Physiol Plant 30*: 81–89.

Khatun S, Flowers TJ (1995) Effects of salinity on seed set in rice. *Plant Cell Environ 18*: 61–67.

Kongngern K, Bunnag S, Theerakulpisut P (2012) Proline, hydrogen peroxide, membrane stability and antioxidant enzyme activity as potential indicators for salt tolerance in rice (*Oryza sativa* L.). *Int J Bot 8*: 54–65.

Krishnamurthy A, Rathinasabapathi B (2013) Oxidative stress tolerance in plants: novel interplay between auxin and reactive oxygen species signaling. *Plant Signal Behav 8*(10): e25761. doi:10.4161/psb.25761.

Larkindale J, Knight MR (2002) Protection against heat stress-induced oxidative damage in Arabidopsis involves calcium, abscisic acid, ethylene, and salicylic acid. *Plant Physiol. 128*: 682–695. doi:10.1104/pp.010320

Lefèvre L, Gratia E, Lutts S (2001) Discrimination between the ionic and osmotic components of salt stress in relation to free polyamine level in rice (*Oryza sativa*). *Plant Sci 161*: 943–952.
Lemoine R, Camera S, Atanassova R, Dédaldéchamp F, Allario T, Pourtau N, Bonnemain JL, Laloi M, Coutos-Thévenot P, Maurousset L, Faucher M (2013) Source-to-sink transport of sugar and regulation by environmental factors. *Front Plant Sci 4*: 272.
Lin CC, Kao CH (2001) Cell wall peroxidase against ferulic acid, lignin, and NaCl-reduced root growth of rice seedlings. *J Plant Physiol 158*(5): 667–671.
Liu SP, Zheng LQ, Xue YH, Zhang QA, Wang L, Shou HX (2010) Overexpression of OsVP1 and OsNHX1 increases tolerance to drought and salinity in rice. *Plant Biol 53*: 444–452.
Lu J, Zhang R, Zong XF, Wang SG, He GH (2009) Effect of salicylic acid on heat resistance of rice seedling under heat stress. *Chin J Eco-Agric 17*: 1168–1171.
Lux A, Luxová M, Morita S, Abe J, Inanaga S (1999) Endodermal silicification in developing seminal roots of lowland and upland cultivars of rice (*Oryza sativa* L.). *Can J Bot 77*: 955–960.
Marchant A, Tausz M (2006) Cyclitols and carbohydrates in leaves and roots of 13 eucalyptus species suggest contracting physiological responses to water deficit. *Plant Cell Environ 29*: 2017–2029.
Mittler R (2002) Oxidative stress, antioxidants and stress tolerance. *Trends Plant Sci 7*: 405–410.
Mittler R (2006) Abiotic stress, the field environment and stress combination. *Trend Plant Sci 11*: 15–19.
Mittler R, Vanderauwera S, Gollery M, Van Breusegem F (2004) Reactive oxygen gene network of plants. *Trends Plant Sci 9*(10): 490–498.
Mockaitis K, Estelle M (2008) Auxin receptors and plant development: a new signaling paradigm. *Annu Rev Cell Dev Biol 24*: 55–80.
Mohammadkhani N, Heidari R (2008) Drought-induced accumulation of soluble sugars and proline in two maize varieties. *World Appl Sci J 13*: 448–453.
Mohd Zain NA, Ismail MR (2016) Effects of potassium rates and types on growth, leaf gas exchange and biochemical changes in rice (*Oryza sativa*) planted under cyclic water stress. *Agric Water Manag 164*: 83–90.
Moran J, Becana M, Iturbe-Ormaetxe I, Frechilla S, Klucas R, Aparicio-Tejo P (1994) Drought induces oxidative stress in pea plants. *Planta 194*: 346–352.
Mubeen M, Ashfaq A, Hafiz MH, Muhammad A, Hafiz UF, Mazhar S, Muhammad Sami ul Din, Asad A, Amjed A, Fahad S, Wajid N (2020) Evaluating the climate change impact on water use efficiency of cotton-wheat in semi-arid conditions using DSSAT model. *J Water Clim Change.* doi:10.2166/wcc.2019.179/622035/jwc2019179.pdf
Muhammad Z, Abdul MK, Abdul MS, Kenneth BM, Muhammad S, Shahen S, Ibadullah J, Fahad S (2019) Performance of *Aeluropus lagopoides* (mangrove grass) ecotypes, a potential turfgrass, under high saline conditions. *Environ Sci Pollut Res.* https://doi.org/10.1007/s11356-019-04838-3
Munns R (2002) Comparative physiology of salt and water stress. *Plant Cell Environ 25*: 239–250.
Munns R (2005) Genes and salt tolerance: bringing them together. *New Phytol 167*: 645–663.
Munns R (2011) Plant adaptations to salt and water stress: differences and commonalities. In: Ismail T, Waltham MA (ed.), *Advances in Botanical Research*. Academic Press, pp 1–32.
Munns R, Tester M (2008) Mechanisms of salinity tolerance. *Annu Rev Plant Biol 59*: 651–681.
Nakayama N, Saneoka H, Moghaieb REA, Premachandra GS, Fujita K (2007) Response of growth, photosynthetic gas exchange, translocation of 13C-labelled photosynthate and N accumulation in two soybean (*Glycine max* L. Merrill) cultivars to drought stress. *Int J Agric Biol 9*: 669–674.
Negrão S, Schmöckel, SM, Tester M (2017) Evaluating physiological responses of plants to salinity stress. *Ann Bot 119*(1): 1–11. doi:10.1093/aob/mcw191
Nelson GC, Rosegrant MW, Koo J, Robertson R, Sulser T, Zhu T, Ringler C, Msangi S, Palazzo A, Batka M, Magalhaes M, Valmonte-Santos R, Ewing M, Lee D (2009) *Climate change: impact on agriculture and costs of adaptation* (Vol. *21*). International Food Policy Research Institute; pp 1–19. doi:10.2499/0896295354
Netondo GW, Onyango JC, Beck E (2004) Sorghum and salinity: II. Gas exchange and chlorophyll fluorescence of sorghum under salt stress. *Crop Sci 44*: 806–811.
Nievola CC, Carvalho CP, Carvalho V, Rodrigues E (2017) Rapid responses of plants to temperature changes. *Temperature 4*(4): 371–405.

Ondrasek G, Rengel Z, Veresz S (2011) Soil salinisation and salt stress in crop production. In: Shanker AK, Venkateswarlu B (eds.), *Abiotic stress in plants: mechanisms and adaptations*. Intech Open, pp 171–190.

Ozgur R, Uzilday B, Sekmen AH, Turkan I (2013) Reactive oxygen species regulation and antioxidant defence in halophytes. *Funct Plant Biol 40*(9): 832–847.

Pal AJ, Mansour MM, Salama FZ (2004) Water relations and xylem transport of nutrients. *Climate Change 66*: 89–107.

Peng S, Huang J, Sheehy JE, Laza RC, Visperas RM, Zhong X (2004) Rice yield declines with higher night temperature from global warming. *Proc Natl Acad Sci USA 101*: 9971–9975. doi:10.1073/pnas.0403720101.

Per TS, Khan NA, Reddy PS, Masood A, Hasanuzzaman M, Khan MIR, Anjum NA (2017) Approaches in modulating proline metabolism in plants for salt and drought stress tolerance: phytohormones, mineral nutrients and transgenics. *Plant Physiol Biochem 115*: 126–140.

Qamar-uz Z, Zubair A, Muhammad Y, Muhammad ZI, Abdul K, Fahad S, Safder B, Ramzani PMA, Muhammad N (2017) Zinc biofortification in rice: leveraging agriculture to moderate hidden hunger in developing countries. *Arch Agron Soil Sci 64*: 147–161. https://doi.org/10.1080/03650340.2017.1338343

Rahman M, Soomro UA, Haq MZ, Gul S (2008) Effects of NaCl salinity on wheat (*Triticum aestivum* L.) cultivars. *World J Agric Sci 4*(3): 398–400.

Raza SH, Athar HR, Ashraf M (2006) Influence of exogenously applied glycinebetaine on the photosynthetic capacity of two differently adapted wheat genotypes under salt stress. *Pak J Bot 38*: 341–351.

Razzaque MA, Talukder NM, Islam MS, Bhadra AK, Dutta RK (2009) The effect of salinity on morphological characteristics of seven rice (*Oryza sativa*) genotypes differing in salt tolerance. *Pak J Bio Sci 12*(5): 406–412.

Rehman M, Fahad S, Saleem MH, Hafeez M, Habib R, Liu F, Deng G (2020) Red light optimized physiological traits and enhanced the growth of ramie (*Boehmeria nivea* L.). *Photosynthetica 58*(4): 922–931.

RicePedia (2018) *Rice Almanac (fourth edition): impact of climate change on rice*. http://ricepedia.org/challenges/climate-change. Accessed on 15 July 2019.

Rogers ME, Colmer TD, Frost K, Henry D, Cornwall D, Hulm E, Hughes S, Nichols, PGH, Craig AD (2009) The influence of NaCl salinity and hypoxia on aspects of growth in Trifolium species. *Crop Pasture Sci 60*: 71–82.

Sajjad H, Muhammad M, Ashfaq A, Waseem A, Hafiz MH, Mazhar A, Nasir M, Asad A, Hafiz UF, Syeda RS, Fahad S, Depeng W, Wajid N (2019) Using GIS tools to detect the land use/land cover changes during forty years in Lodhran district of Pakistan. *Environ Sci Pollut Res*. https://doi.org/10.1007/s11356-019-06072-3

Sakamoto A, Murata N (2002) The role of glycine betaine in the protection of plants from stress: clues from transgenic plants. *Plant Cell Environ 25*(2): 163–171.

Saleem MH, Fahad S, Adnan M, Mohsin A, Muhammad SR, Muhammad K, Qurban A, Inas AH, Parashuram B, Mubassir A, Reem MH (2020a) Foliar application of gibberellic acid endorsed phytoextraction of copper and alleviates oxidative stress in jute (*Corchorus capsularis* L.) plant grown in highly copper-contaminated soil of China. *Environ Sci Pollution Res*. https://doi.org/10.1007/s11356-020-09764-3

Saleem MH, Fahad S, Shahid UK, Mairaj D, Abid U, Ayman ELS, Akbar H, Analía L, Lijun L (2020c) Copper-induced oxidative stress, initiation of antioxidants and phytoremediation potential of flax (*Linum usitatissimum* L.) seedlings grown under the mixing of two different soils of China. *Environ Sci Poll Res*. https://doi.org/10.1007/s11356-019-07264-7

Saleem MH, Rehman M, Fahad S, Tung SA, Iqbal N, Hassan A, Ayub A, Wahid MA, Shaukat S, Liu L, Deng G (2020b) Leaf gas exchange, oxidative stress, and physiological attributes of rapeseed (*Brassica napus* L.) grown under different light-emitting diodes. *Photosynthetica 58*(3): 836–845.

Saleh J, Maftoun M (2008) Interactive effect of NaCl levels and Zinc sources and levels on the growth and mineral composition of rice. *J Agric Sci Tech 10*: 325–336.

Saud S, Chen Y, Fahad S, Hussain S, Na L, Xin L, Alhussien SA (2016) Silicate application increases the photosynthesis and its associated metabolic activities in Kentucky bluegrass under drought stress and post-drought recovery. *Environ Sci Pollut Res 23*(17): 17647–17655. https://doi.org/10.1007/s11356-016-6957-x

Saud S, Chen Y, Long B, Fahad S, Sadiq A (2013) The different impact on the growth of cool season turf grass under the various conditions on salinity and drought stress. *Int J Agric Sci Res 3*: 77–84.

Saud S, Fahad S, Cui G, Chen Y, Anwar S (2020) Determining nitrogen isotopes discrimination under drought stress on enzymatic activities, nitrogen isotope abundance and water contents of Kentucky bluegrass. *Sci Rep 10*: 6415. https://doi.org/10.1038/s41598-020-63548-w

Saud S, Fahad S, Yajun C, Ihsan MZ, Hammad HM, Nasim W, Amanullah Jr, Arif M, Alharby H (2017) Effects of nitrogen supply on water stress and recovery mechanisms in Kentucky Bluegrass Plants. *Front Plant Sci 8*: 983. doi:10.3389/fpls.2017.00983

Saud S, Li X, Chen Y, Zhang L, Fahade S, Hussain S, Sadiq A, Chen Y (2014) Silicon application increases drought tolerance of Kentucky bluegrass by improving plant water relations and morph physiological functions. *Sci World J 2014*: 1–10. https://doi.org/10.1155/2014/368694

Schett G, Steiner CW, Groger M, Winkler S, Graninger W, Smolen J, Xu Q, Steiner G (1999) Activation of Fas inhibits heat induced activation of HSF1 and upregulation of HSP70. *FASEB J 13*: 833–842.

Sekmen AH, Ozgur R, Uzilday B, Turkan I (2014) Reactive oxygen species scavenging capacities of cotton (*Gossypium hirsutum*) cultivars under combined drought and heat induced oxidative stress. *Environ Exp Bot 99*: 141–149.

Sengupta A, Chakraborty M, Saha J, Gupta B, Gupta K (2016) Polyamines: osmoprotectants in plant abiotic stress adaptation. In: Iqbal N, Nazar RA, Khan N (eds.), *Osmolytes and plants acclimation to changing environment: emerging omics technologies*. New Delhi: Springer. doi:10.1007/978-81-322-2616-1_7

Shafi MI, Adnan M, Fahad S, Fazli W, Ahsan K, Zhen Y, Subhan D, Zafar-ul-Hye M, Martin B, Rahul D (2020) Application of single superphosphate with humic acid improves the growth, yield and phosphorus uptake of wheat (*Triticum aestivum* L.) in calcareous soil. *Agron* (10): 1224; doi:10.3390/agronomy10091224

Shah F, Lixiao N, Kehui C, Tariq S, Wei W, Chang C, Liyang Z, Farhan A, Fahad S, Huang J (2013) Rice grain yield and component responses to near 2°C of warming. *Field Crop Res 157*: 98–110.

Shannon MC (1998) Adaptation of plants to salinity. *Adv Agron 60*: 75–120.

Shrivastava P, Kumar R (2015) Soil salinity: a serious environmental issue and plant growth promoting bacteria as one of the tools for its alleviation. *Saudi J Biol Sci 22*(2): 123–131. http://doi.org/10.1016/j.sjbs.2014.12.001

Siddiqui MH, Al-Khaishany MY, Al-Qutami MA, Al-Whaibi MH, Grover A, Ali HM, (2015) Morphological and physiological characterization of different genotypes of faba bean under heat stress. *Saudi J Biol Sci 22*: 656–663.

Singh B, Usha K (2003) Salicylic acid induced physiological and biochemical changes in wheat seedlings under water stress. *Plant Growth Regul 39*: 137–141.

Sobahan MA, Akter N, Ohno M, Okuma E, Hirai Y, Mori IC, Nakamura Y, Murata Y (2012) Effects of exogenous proline and glycinebetaine on the salt tolerance of rice cultivars. *Biosci Biotechnol Biochem 76*: 1568–1570.

Statista (2018) *Worldwide; USDA Foreign Agricultural Service; 2008 to 2018*. US Department of Agriculture. https://www.statista.com/statistics/255977/total-global-rice-consumption/. Accessed on 15 July 2019.

Storozhenko S, De Pauw P, Van Montagu M, Kushnir S (1998) The heat shock element is a functional component of the Arabidopsis APX1 gene promoter. *Plant Physiol 116*: 1005–1014.

Subhan D, Zafar-ul-Hye M, Fahad S, Saud S, Martin B, Tereza H, Rahul D (2020) Drought Stress Alleviation by ACC Deaminase Producing *Achromobacter xylosoxidans* and *Enterobacter cloacae*, with and without Timber Waste Biochar in Maize. *Sustain 12*(6286). doi:10.3390/su12156286

Sun W, Van Montagu M, Verbruggen N (2002) Small heat shock proteins and stress tolerance in plants. *Biochimica et Biophysica Acta (BBA)-Gene Structure and Expre 1577*(1): 1–9.

Suzuki N, Mittler R (2006) Reactive oxygen species and temperature stresses: a delicate balance between signaling and destruction. *Physiol Plant 126*: 45–51.

Tariq M, Ahmad S, Fahad S, Abbas G, Hussain S, Fatima Z, Nasim W, Mubeen M, Rehman MH, Khan MA, Adnan M (2018) The impact of climate warming and crop management on phenology of sunflower-based cropping systems in Punjab, Pakistan. *Agri Forest Met 15*(256):270–282.

Tariq M, Khan Z, Arif M, Ali K, Waqas M, Naveed, K, Adnan M (2013) Effect of nitrogen application timings on the seed yield of Brassica cultivars and associated weeds. *Pak J Weed Sci Res 19*(4): 493–502.

Tripathi BN, Bhatt I, Dietz KJ (2009) Peroxiredoxins: a less studied component of hydrogen peroxide detoxification in photosynthetic organisms. *Protoplasma 235*(1–4): 235–243.

Verbruggen N, Hermans C (2008) Proline accumulation in plants: a review. *Amino Acids 35*: 753–759.

Vinocur B, Altman A (2005) Recent advances in engineering plant tolerance to abiotic stress: achievements and limitation. *Curr Opin Biotech 16*: 123–132.

Wahid F, Fahad S, Subhan D, Adnan M,, Zhen Y, Saud S, Manzer HS, Martin B, Tereza H, Rahul D (2020) Sustainable management with mycorrhizae and phosphate solubilizing bacteria for enhanced phosphorus uptake in calcareous soils. *Agri 10*(334). doi:10.3390/agriculture10080334

Wajid N, Ashfaq A, Asad A, Muhammad T, Muhammad A, Muhammad S, Khawar J, Ghulam MS, Syeda RS, Hafiz MH, Muhammad IAR, Muhammad ZH, Muhammad Habib R, Veysel T, Fahad S, Suad S, Aziz K, Shahzad A (2017) Radiation efficiency and nitrogen fertilizer impacts on sunflower crop in contrasting environments of Punjab. *Pakistan Environ Sci Pollut Res 25*: 1822–1836. https://doi.org/10.1007/s11356-017-0592-z

Woodward AW, Bartel B (2005) Auxin: regulation, action, and interaction. *Ann Bot 95*: 707–735. doi:10.1093/aob/mci083.

Wu C, Kehui C, She T, Ganghua L, Shaohua W, Fahad S, Lixiao N, Jianliang H, Shaobing P, Yanfeng D (2020) Intensified pollination and fertilization ameliorate heat injury in rice (*Oryza sativa* L.) during the flowering stage. *Field Crops Res 252*: 107795.

Wu C, Tang S, Li G, Wang S, Fahad S, Ding Y (2019) Roles of phytohormone changes in the grain yield of rice plants exposed to heat: a review. *Peer J 7*:e7792 doi:10.7717/peerj.7792

Yang J, Zhang J, Liu K, Wang Z, Liu L (2007) Involvement of polyamines in the drought resistance of rice. *J Exp Bot 58*: 1545–1555.

Yang Z, Zhang Z, Zhang T, Fahad S, Cui K, Nie L, Peng S, Huang J (2017) The effect of season-long temperature increases on rice cultivars grown in the central and southern regions of China. *Front Plant Sci 8*: 1908. https://doi.org/10.3389/fpls.2017.01908

Yensen NP (2008) Halophyte uses for the twenty-first century. In: Khan MA, Weber DJ (eds.), *Ecophysiology of high salinity tolerant plants*, pp 367–396.

Yildizli A, Çevik S, Ünyayar S (2018) Effects of exogenous myo-inositol on leaf water status and oxidative stress of *Capsicum annuum* under drought stress. *Acta Physiol Plant 40*(6): 122.

Yoshimura K, Miyao K, Gaber A, Takeda T, Kanaboshi H, Miyasaka H, Shigeoka S (2004) Enhancement of stress tolerance in transgenic tobacco plants overexpressing Chlamydomonas glutathione peroxidase in chloroplasts or cytosol. *Plant J 37*: 21–33.

Zafar-ul-Hye M, Ahzeeb-ul-Hassan MT, Muhammad A, Fahad S, Martin B, Tereza D, Rahul D, Subhan D (2020b) Potential role of compost mixed biochar with rhizobacteria in mitigating lead toxicity in spinach. *Scientific Rep 10*: 12159. https://doi.org/10.1038/s41598-020-69183-9.

Zafar-ul-Hye M, Muhammad N, Subhan D, Fahad S, Rahul D, Mazhar A, Ashfaq AR, Martin B, Jiˇrí H, Zahid HT, Muhammad N (2020a) Alleviation of cadmium adverse effects by improving nutrients uptake in bitter gourd through cadmium tolerant rhizobacteria. *Environ 7*(54). doi:10.3390/environments7080054

Zahida Z, Hafiz FB, Zulfiqar AS, Ghulam MS, Fahad S, Muhammad RA, Hafiz MH, Wajid N, Muhammad S (2017) Effect of water management and silicon on germination, growth, phosphorus and arsenic uptake in rice. *Ecotoxicol Environ Saf 144*: 11–18.

Zain NAM, Ismail MR, Mahmood M, Puteh A, Ibrahim MH (2014) Alleviation of water stress effects on MR220 rice by application of periodical water stress and potassium fertilization. *Molecules 19*: 1795–1819.

Zeng L, Shannon MC (2000) Effects of salinity on grain and yield components of rice at different seeding densities. *Agron J 92*: 418–423.

Zheng L, Shannon MC, Lesch SM (2001) Timing of salinity stress affecting rice growth and yield components. *Agric Water Manag 48*: 191–206.

Zonouri M, Javadi T, Ghaderi N (2014) Effect of foliar spraying of ascorbic acid on cell membrane stability, lipid peroxidation, total soluble protein, ascorbate peroxidase and leaf ascorbic acid under drought stress in grapes. *Inter J Advan Bio Biomed Res 2*: 349–354.

2

Biological Nitrogen Fixation in a Changing Climate

Abdel Rahman Mohammad Al-Tawaha
Al Hussein Bin Talal University, Jordan

Sonia Sheoran, Pradeep Sharma, Garima Singroh
ICAR-Indian Institute of Wheat and Barley Research, India

Yousef M Abu-Zaitoon
Al Hussein Bin Talal University, Jordan

Laila Trioui
University Hassan II of Casablanca

Mohammad Ali Shariati
Orel State University Named After I.S. Turgenev, Russia

Huma Naz
Mohammad Ali Nazeer Fatima Degree College, India

Abdel Razzaq Al-Tawaha
Universiti Putra Malaysia, Malaysia

Ali M. Qaisi
University of Jordan Amman, Jordan

Amanullah, Imran
The University of Agriculture Peshawar, Pakistan

Abdur Rauf
University of Swabi, Pakistan

Shah Khalid
The University of Agriculture Peshawar, Pakistan

Mohd Abas Shah
ICAR-Central Potato Research Station, India

Devarajan Thangadurai
Thangadurai Karnatak University, India

Jeyabalan Sangeetha
Central University of Kerala, India

Shah Fahad
The University of Haripur, Pakistan

Khalid Fandi
Al Hussein Bin Talal University, Jordan

CONTENTS

2.1 Introduction

Mineral nutrients such as nitrogen and phosphorus are considered to be a very important factor that limits plant biomass and productivity in many ecosystems (Turk and Tawaha 2001, 2002a, 2002b, 2004; Tawaha and Turk 2002a, 2002b, 2004; Tawaha et al. 2003; Turk et al. 2003a, 2003b, 2003c; Nikus et al. 2004a, 2004b; Abebe et al. 2005; Abera et al. 2005; Al-Tawaha et al. 2005, 2010; Elser et al., 2007; Al-Kiyyam et al. 2008; Al-Ajlouni et al. 2009; Qadeem et al. 2015; Khan et al. 2015; Al-Juthery et al. 2018). The reduced bioavailability of nitrogen and the high dependence of plant growth on molecular nitrogen lead to massive use of Haber-Bosch synthetic fertilizers (Dobermann 2007; Westhoff 2009). Although the green revolution could not have occurred without the use of chemical fertilizers, which has contributed to many environmental hazards. It has been estimated that the excessive use of fertilizers produces approximately 300 million tons of CO_2 per year (Robertson and Vitousek 2009). In addition to being used inefficiently by plants, molecular nitrogen is a very mobile element causing serious water and soil pollution. This, as a result, affects agricultural sustainability in addition to the incidence of many diseases. Cancer, methemoglobinemia, and severe respiratory illness, are among the recorded health problems (Saikia and Vanita 2007). In contrast, the biological nitrogen fixation system could significantly reduce both environmental and economic costs. In this system, fixed N_2 would be assembled directly and efficiently to plants via symbiotic association with certain bacteria. However, a naturally symbiotic relationship is restricted to some plants. Recently, new hopes have been raised for the assembly of a symbiotic relationship in crop plants using developed biotechnology techniques. Only some prokaryotic microorganisms named diazotrophs are able to perform biological nitrogen fixation through the activity of the nitrogenase complex enzyme (Lam et al. 1996; Franche et al. 2009). Among various types of interaction between plants and diazotrophs, nodulation remains the most efficient and highly specific process.

Diazotrophic bacteria include rhizobia that interact with legumes (Oldroyd and Downie 2008; Desbrosses and Stougaard 2011) and *Parasponia*, a non-legume plant species in addition to *Frankia* species that associate with actinorhizal plants. Cyanobacteria, in particular *Nostoc* species, colonize various plant organs either intra- or extracellularly. *Azospirillum*, *Azoarcus* as well as *Herbaspirillum* colonize roots of a wide spectrum of plant species, including cereals (Santi et al. 2013). Most terrestrial plants can form symbiotic associations with diazotrophs. However, nitrogen fixation is limited to legumes (Florence et al. 2016). Recent research on inoculated and synthetic engineered cereal crops was the study of extensive research in this chapter.

Ammonium produced by nitrogen-fixing bacteria in the symbiotic association is highly variable, depending on the plant itself as well as the growth stage, in addition to the associated bacteria and variable environmental conditions that will be investigated in detail in this chapter.

2.2 Expanding Biological Nitrogen Fixation to Non-Legumes

The major goal of research on biological nitrogen fixation (BNF) has been to extend this property to the world's most important crops, including cereals. Wheat, maize, and rice do not have the ability to do nitrogen fixation naturally. This goal can be achieved by implementing several approaches, such as using free-living N_2 fixers, the nodule-independent approach, (inoculant usage), the development of new fixing plant microbe associations, and transfer of the nitrogen-fixing ability to non-fixing organisms.

2.2.1 Free-Living N_2 Fixers

Free-living nitrogen-fixing microorganisms substantially affect the incorporation of reduced N_2 in the biosphere. Several attempts to commercialize free-living fixers have been made. Azotobacterin and Azotobacter-based inoculants have been used mainly in Russia with little success. Currently, Dimargon, and Colombia, as commercialized Azotobacter inoculants, have been reported to reduce N_2 application for rice (Moreno-Sarmiento et al. 2007). Studies have yet to find any evidence of crop improvements as a result of using this mechanism. However, the agricultural contribution of this form of BNF is very limited due to the shortage of carbon fuel in the soil necessary to support the high energy demanded by the nitrogen-fixing process. Moreover, the sensitivity of labile-nitrogenase enzyme to free oxygen is another challenge. As in the application of N_2 fertilizers, the production of ammonia is not directly transferred from N_2-fixers to plants (Mushtaq et al. 2015).

2.2.2 The Nodule-Independent Approach

Biological nitrogen fixation through the nodule-independent approach can be established by manipulating both partners involved in the rhizosphere where plant–microbe interaction occurs. A wide variety of appropriate plant and microbe signals, genes and receptors are needed to achieve successful colonization (Vanbleu and Vanderleyden 2007). Flavonoids as universal plant signals, as well as bacterial polysaccharides, mainly exopolysaccharides and lipopolysaccharides, have been reported to affect root colonization. For instance, flavonoids stimulate the colonization of wheat by *Azospirillumbrasilense* (Webster et al. 1998) as well as the regulation of *Herbaspirillumseropedicae*gene expected to have a role in this process (Tadra-Sfeir et al. 2011). Additionally, the modification of the rhamnose level, a type of polysaccharide, in *Azospirillumbrasilense* as a result of mutation in the *Tn5* gene reduces maize root colonization (Jofre et al. 2004). A substantial level of nitrogen content has been obtained through the inoculation of non-leguminous plants with nitrogen-fixing microorganisms, mainly Azospirillum (Verma et al. 2010; Bhattacharyya and Jha 2012). It has been reported that wheat and maize inoculated with *Pseudomonas protegens* can greatly improve the biomass and nitrogen content of targeted plant reproductive and vegetative tissues. Even manipulating successful nitrogen fixation through nodule-independent association is less complex than nodule-dependent association, meaning that lower nitrogen levels are transported to the plant partner (Florence et al. 2016). The amount of fixed N_2 available to host in this type of association was enhanced by implementing the Azospirillum mutant. A point mutation in glutamine synthetase, an enzyme involved in assimilating fixed ammonium, increased the ammonium level exported from Azospirillum to wheat (Van Dommelen et al. 2009; Basir et al. 2015). In contrast to Azospirillum, some species of cyanobacteria, including *Nostoc* and *Anabaena*, could excrete natural ammonium to hosts more efficiently. Under nitrogen-limitation conditions, more vegetative cells become specialized as heterocyts for nitrogen fixation. The walls of these cells are less permeable to O_2 and glutamine synthetase activity is largely reduced in the symbiotic associations. Changes in the heterocyts number and enzymatic activity have been reported to occur as a result of plant signals (Meeks 2009). Finding out these signals would help in the production of green manures.

2.2.3 Diazotrophic Endophytes

Studies have demonstrated the biological fixing of nitrogen in certain non-legumes, including rice and sugarcane, through specific types of association with diazotrophic endophytes (Hurek and Reinhold-Hurek 2003; Khan et al. 2015; Boddey et al. 2003). In this association, diazotrophs such as *Azoarcus*, *Burkholderia*, *Gluconacetobacter*, and *Herbaspirillum*, were reported to infect roots as well as aerial plant parts without any health or ecological threat. Several successful complex inoculants for both rice and sugarcane have been developed (Silva et al. 2012; Yanni and Dazzo 2010).

2.2.4 Actinorhizal Symbioses

In the actinorhizal system, the actinomycete *Frankia* establishes symbiotic relationships with a wide spectrum of species from across the plant kingdom (Benson and Silvester 1993; Franche and Bogusz 2011). In contrast to the rhizobial system, the specificity of actinorhizal symbioses seems to be low. Actinorhizal plants are well adapted to a wide range of climatic conditions in addition to extremist pH, and saline environments. Other pioneered actinorhizal plants colonize contaminated soil, as well as flood land (Carole et al. 2013). These exceptional characteristics, in addition to the ability of forming mycorrhizal association, make actinorhizal plants ideal for stabilizing desert and coastal lands in addition to extending nitrogen fixation to non-legumes (Dawson 2008).

2.2.5 Parasponia-Rhizobium

In contrast to the above analysis, Parasponia are pioneered tropical trees which have the capability to grow on disturbed and nitrogen-poor soils. The host is the only non-legume reported to be associated with rhizobia (Trinick 1973). *P. andersonii*, for instance, has recently been shown to be nodulated by species from across four different genera of rhizobia. Microscopy studies suggested that Parasponia–rhizobial symbiotic association seem to be less complex than that of legumes (Op den Camp et al. 2012).

2.2.6 Importance of Biological Nitrogen Fixation

Since the first successful horizontal transfer of *nif* cluster genes (encode nitrogenase complex enzyme) in 1970, interests in engineering major crop plants have been renewed (Dixon and Postgate 1971). The *nif* genes present in diazotrophs offer them the ability to utilize atmospheric nitrogen by nitrogen-fixing symbiotic bacteria encoding enzymes and a number of regulatory proteins. The nitrogenase complex is the primary enzyme encoded by the *nif* genes, which regulate the conversion of atmospheric nitrogen (N_2) to other forms of nitrogen which plants can use. The nitrogenase complex is encoded by about 20 different *nif* genes controlling different mechanisms of nitrogen fixation. Necessary techniques to express *nif* genes in tobacco as well as other plants have been developed. Among 15 necessary *nif* genes required to induce active nitrogenase enzyme, the number could be reduced to three due to the primitive nature of expression process in chloroplasts (Rubio and Ludden 2008). However, achieving an efficient nitrogen-fixing plant system now appears to be more complex than was thought to be case about two decades ago (Dixon et al. 1997). This process should occur under highly energetic, but anoxic conditions. This energy requirement is well supplied in plastids whereas the second obstacle could be achieved by separating nitrogen fixation from photosynthesis, either temporally or spatially (Yamamoto et al. 2011).

2.3 Symbiotic Nitrogen Fixation in Cereals

The three most important cereal crops those contribute in more than 50% of the total calories consumed around the world are rice, wheat and maize. Previous studies and reviews on BNF have shown that the contribution of nitrogen fixation in cereal production is very limited (Perrine-Walker et al. 2007; Schmid and Hartmann 2007; Peng et al. 2009; Bulgarelli et al. 2015; Santi et al. 2013; Dent and Cockin 2017; Zhang et al. 2017; Yang etal. 2017; Pii et al. 2015; Yoneyama et al. 2017; Van Deynze et al. 2018;

Rosenblueth et al. 2018; Ma etxal. 2019). On the other hand, Symbiotic nitrogen fixation, which is limited to legumes, is now being explored in cereal crops in an attempt to improve plant growth. Santi et al. (2013) found that inoculation with *A. Brasilense* considerably enhanced the yields of foliage, grain and branching of root hairs in cereal, associative nitrogen fixation can supply 20–25% of total nitrogen needs. The prerequisite for the application of these bacteria as nitrogen fertilizers is the most suitable combination of diazotrphic PGPR strain and plant cultivars. In maize, Peng et al. (2019) disrupted a*acds* gene from a major PGPR, *Rahnellaaquatilis* HX2. Although this action reduced other activities, such as IAA production and biocontrol activity, but it significantly increased nitrogen fixation in the maize rhizosphere in salt-stressed conditions. Several species of Rhizobium, *Acetobacter, Klebsiella, Pseudomonas, Herbaspirillum, Gluconacetobacter*, and *Burkholderia*, have been reported as endophytes. *Rhizobium* strains isolated for roots are involved in developing endopyhtic associations in rice (Zhang et al. 2014) wheat and maize etc. are reported. It has also been shown that seed and stem isolates of endophytic diazotrophs from cereals are able to fix nitrogen. *Lysinibacillussphaericus*, a diazotroph, was reported by Shabanamol et al. (2018) as a N-fixing, true endophyte from the rice crop, promoting both plant growth and biocontrol activities. In a study, Fox et al. (2016) reported a significant gain in nitrogen content and biomass accumulation in wheat and maize when inoculated with engineered strain of nitrogen-fixing bacterium *Psuedomonasprotegens* Pf-5 X940.

2.4 The Association of Diazotrophs with Non-Legumes

Various diazotrophs have been reported to be present in association with non-leguminous plants (Cavaglieri et al. 2009; Elbeltagy and Ando 2008; Ahemad and Kibret 2014; Hadi et al. 2014; Santi et al. 2013; Sambukumar et al. 2015). Diazotrophic bacteria have the potential to expand various kinds of root associations with diverse plant species and are measured as possible BNF contributors in cereal crops. They can be separated into three groups: (i) free living viz., *Azotobacter, Beijerinckia*, and *Clostridium*, have established possible nitrogen fixation in cereals; (ii) symbiotic, chiefly bacteria living inside plant root nodules (*Rhizobium*); and (iii) an endophytic relationship with other organisms such as *Azospirillium*. Over the years, researchers have isolated and identified several isolates of nitrogen-fixing bacteria from plants. Sood et al. (2019) isolated ten bacteria from the rhizosphere of wheat which significantly improved the BNF in wheat and saved at least 18 kilograms of N_2 and 10 kilograms of P_2 on a per hectare basis. Under nitrogen deficit, Zeffa et al. (2019) demonstrated that intensified plant growth, enhanced biochemical character and raised nitrogen use efficiency (NUE)in maize plants inoculated *with Azospirillumbrasilense* Ab-V5. Banik et al. (2019) identified an Azotobacter strain Avi2 (AzA) from the rhizosphere of indigenous rice (*O. sativa* L.) which led to a significant improvement in yield. This strain was found to be useful in improving the photosynthetic rate during the grain-filling stages. In maize, Ke et al. (2019) and Adnan et al. (2014) reported that endophytic *Pseudomonas stutzeri* A1501 inoculated to the maize roots considerably distorted the constitution of diazotrophic society. It became dominant in the rhizosphere and enhanced the population of indigenous diazotrophs serving improved nitrogen fixation and growth in maize. Similarly, many species of cyanobacteria viz., *Nostoc, Anabaena, Tolypothrix, Aulosira, Cylindrospermum, Scytonema*, fix atmospheric nitrogen. Jing et al. (2019) reported that a greater abundance of diazotrophs (heterocystous cyanobacteria *Nostoc*) in hybrid rice cultivar (IIY) than in the inbred rice cultivar (W23) promoted BNF.

2.5 Engineering Symbiotic Nitrogen Fixation in Cereals

Advancement in biotechnological approaches are being explored which could help to fix biological nitrogen in cereals. One of the major approaches to engineer nitrogen fixation in non-legumes is to direct transfer nitrogenase-encoding bacterial *nif* genes into crop plants. A major challenge in implementing this approach is the regulatory complexity of nitrogenase biosynthesis as large cassette of genes is involved and sensitivity of the nitrogenase to oxygen. To address this, Yang et al. (2018) demonstrated an approach

Khan MA, Basir A, Adnan M, Saleem N, Khan A, Shah SRA, ... Ali K (2015) Effect of tillage, organic and inorganic nitrogen on maize yield. *Am Eurasian J Agric Environ Sci 15*(12):2489–2494.

Ke X, Feng S, Wang J et al (2019) Effect of inoculation with nitrogen-fixing bacterium *Pseudomonas stutzeri* A1501 on maize plant growth and the microbiome indigenous to the rhizosphere. *Syst Appl Microbiol 42*(2): 248–260.

Lam HM, Coschigano KT, Oliveira IC et al (1996) The molecular-genetics of nitrogen assimilation into amino acids in higher plants. *Annual Rev Plant Phys Plant Mol Biol 47*:569–593.

Ma J, Bei Q, Wang X et al (2019) Impacts of Mo application on biological nitrogen fixation and diazotrophic communities in a flooded rice-soil system. *Sci Total Environ 649*:686–694.

Meeks JC (2009) Physiological adaptations in nitrogen-fixing Nostocplant symbiotic associations. *Microbiol Monogr 8*:181–205.

Moreno-Sarmiento N, Moreno-Rodríguez L, Uribe-Vélez D (2007) Biofertilizantes para la agricultura en Colombia. Biofertilizantes en Iberoamérica: una visión técnica, científica y empresarial. Imprenta Denad Internacional, Montevideo 38–45.

Mushtaq S, Ahmed N, Khan IA, Shah B, Khan A, Ali M, ... Ahmed S (2015) Study on the efficacy of ladybird beetle as a biological control agent against aphids (*Chaitophorus* spp.). *J Ento Zool Stud 3*(6):117–119.

Nikus O, Al-Tawaha M, Turk MA (2004a) Effect of manure supplemented with phosphate fertilizer on the fodder yield and quality of two sorghum cultivars (*Sorghum bicolor* L.). *Biosci Res 1*:1–7.

Nikus O, Turk MA, Al-Tawaha AM (2004b) Yield response of sorghum (*Sorghum bicolor* L.) to manure supplemented with phosphate fertilizer under semi-arid Mediterranean conditions. *Int J Agric Biol 6*:889–893.

Oldroyd G, Downie JA (2008) Coordinating nodule morphogenesis with rhizobial infection in legumes. *Ann Rev Plant Biol 59*:519–546.

Op den Camp R, Polone E, Fedorova E et al (2012) Non legume Parasponia andersonii deploys a broad rhizobium host range strategy resulting in largely variable symbiotic effectiveness. *Mol Plant–Microbe Interac 25*:954–963.

Peng G, Zhang W, Luo H et al (2009). *Enterobacteroryzae sp.* nov. a nitrogen-fixing bacterium isolated from the wild rice species Oryzalatifolia. *Int J Syst Evol Microbiol 59*(7): 1650–1655.

Peng J, Wu D, Liang Y et al (2019) Disruption of *acdS* gene reduces plant growth promotion activity and maize saline stress resistance by *Rahnellaaquatilis* HX2. *J Basic Microbiol* 1–10.

Perrine-Walker FM, Prayitno J, Rolfe BG et al (2007) Infection process and the interaction of rice roots with rhizobia. *J Exp Botany 58*:3343–3350.

Pii Y, Mimmo T, Tomasi N et al (2015) Microbial interactions in the rhizosphere: beneficial influences of plant growth-promoting rhizobacteria on nutrient acquisition process. A review. *Biol Fert Soils 51*(4):403–415.

Qadeem W, Shahid S, Maid Z, Shah SRA, Junaid K, Said F, ... Rahman NS (2015) Efficacy of botanical and microbial extracts against Angoumois grain moth *Sitotroga cerealella* (Olivier) (Lepidoptera: Gelechiidae) under laboratory conditions. *J Ento Zool Stud 3*(5):451–454.

Robertson G, Vitousek P (2009) Nitrogen in agriculture: Balancing the cost of an essential resource. *Ann Rev Environ Res 34*:97–125.

Rosenblueth M, Ormeño-Orrillo E, López-López A et al (2018) Nitrogen fixation in cereals. *Front Microbiol 9*.

Rubio LM, Ludden PW (2008) Biosynthesis of the iron-molybdenum cofactor of nitrogenase. *Ann Rev Microbiol 62*:93–111.

Saikia S, Vanita J (2007) Biological nitrogen fixation with non-legumes: An achievable target or a dogma? *Current Sci 92*(3):317–322.

Sambukumar S, Affeldt K, Friedrich D et al (2015) Exploring the Rhizobial interactions with non-nodulating legume *Gleditsiatriacanthos*. In: *23rd North italic>American conference on symbiotic nitrogen fixation*. Ixtapa, Mexico, pp 6–10.

Santi C, Bogusz D, Franche C (2013) Biological nitrogen fixation in non-legume plants. *Ann Bot 111*:743–767.

Schmid M, Hartmann A (2007) Molecular phylogeny and ecology of root associated diazotrophic α-and β-proteobacteria. In *Associative and endophytic nitrogen-fixing bacteria and cyanobacterial associations*, Springer, Dordrecht, pp 21–40.

Shabanamol S, Divya K, George T (2018) Characterization and in planta nitrogen fixation of plant growth promoting endo phyticdiazotrophic Lysini bacillus sphaericus isolated from rice (*Oryza sativa*). *Physiol Mol Plant Pathol 102*:46–54.

Silva M, Antonio C, Oliveira P et al (2012) Survival of endophytic bacteria in polymer-based inoculants and efficiency of their application to sugarcane. *Plant Soil 356*:231–243.

Sood G, Kaushal R, Panwar G et al (2019) Effect of indigenous plant growth-promoting rhizobacteria on wheat (*Triticum Aestivum* L.) productivity and soil nutrients. *Commun Soil Sci Plant Anal 50*(2):141–152.

Tadra-Sfeir MZ, Souza EM, Faoro H et al (2011) Naringenin regulates expression of genes involved in cell wall synthesis in Herbaspirillumseropedicae. *Appl Environ Microbiol 77*:2180–2183.

Tawaha AM, Turk MA (2002a) Lentil (*Lens culinaris* Medic.) productivity as influenced by rate and method of phosphate placement in a Mediterranean environment. *Acta Agron Hungarica 50*(2):197–201.

Tawaha AM, Turk MA (2002b) Lentil (*Lens culinaris* Medic.) productivity as influenced by rate and method of phosphate placement in a Mediterranean environment. *Acta Agron Hungarica 50*(2):197–201.

Tawaha AM, Singh VP, Turk MA et al (2003) A review on growth, yield components and yield of barley as influenced by genotypes, herbicides and fertilizer application. *Res Crop 4*(1):1–9.

Tawaha AM, Turk MA (2004) Field Pea Seeding Management for Semi-arid Mediterranean Conditions. *J Agron Crop Sci 190*:86–92.

Trinick M (1973) Symbiosis between *Rhizobium* and the non-legume, Tremaaspera. *Nature 244*:459–460.

Turk MA, Tawaha AM (2001) Common vetch (*Vicia sativa* L.) productivity as influenced by rate and method of phosphate fertilization in a Mediterranean environment. *Agric Mediterranea 131*:108–111

Turk MA, Tawaha, AM (2002a) Onion (*Allium cepa* L.) as influenced by rate and method of phosphorus placement. *Crop Res 23*(1):105–107.

Turk MA, Tawaha, AM (2002b) Impact of seeding rate, seeding date and method of phosphorous application in faba bean (*Viciafaba* L. minor) in the absence of moisture stress. *Biotechnol Agric Soc Environ 6*(3):171–178

Turk MA, Tawaha, AM, Shatnawi M (2003a) Lentil (*Lens culinaris* Medik) Response to plant density, sowing date, phosphorus fertilization and Ethephon application in the absence of moisture stress. *J Agron Crop Sci 189*(1):1–6.

Turk MA, Tawaha, AM, Samara N (2003b) Effects of seeding rate and date, and phosphorus application on growth and yield of narbon vetch (*Vicia narbonensis*). *Agronomie 23*:1–4.

Turk MA, Hameed KM, Aqeel AM, Tawaha AM (2003c) Nutritional status of durum wheat grown in soil supplemented with olive mill by-products. *Agrochimica 47*(5–6):209–219.

Turk M, Tawaha AR (2004) *Effect of variable sowing ratios and sowing rates of bitter vetch on the herbage yield of oat-bitter vetch mixed cropping*. In *Proceedings 7th International Oat Conference/Pirjo Peltonen-Sainio and Mari Topi-Hulmi*. MTT.

Van Deynze A, Zamora P, Delaux P et al (2018) Nitrogen fixation in a landrace of maize is supported by a mucilage-associated diazotrophic microbiota. *PLoS Biol 16*(8):e2006352.

Van Dommelen A, Croonenborghs A, Spaepen S et al (2009) Wheat growth promotion through inoculation with an ammonium-excreting mutant of *Azospirillum brasiliense*. *Biol Fert Soils 45*:449–553.

Vanbleu E, Vanderleyden J (2007) Molecular genetics of rhizosphere and plant-root colonization. In: Elmerich C, Newton WE (eds) *Associative and endophytic nitrogen-fixing bacteria and cyanobacterial associations*. Kluwer Academic Publishers, Dordrecht, pp 85–112.

Verma JP, Yadav J, Tiwari KN et al (2010) Impact of plant growth promoting rhizobacteria on crop production. *Int J Agric Res 5*:954–983.

Webster G, Jain V, Davey MR et al (1998) The flavonoid naringenin stimulates the intercellular colonization of wheat roots by Azorhizobiumcaulinodans. *Plant Cell Environ 21*:373–383.

Westhoff P (2009) The economics of biological nitrogen fixation in the global economy. In: Emerich DW, Krishnan HB (eds) *Nitrogen fixation in crop production*. Agronomy Monograph No. 52. American Society of Agronomy, Madison, WI, pp 309–328.

Yamamoto H, Kurumiya S, Ohashi R et al (2011) Functional evaluation of a nitrogenase-like protochlorophyllide reductase encoded by the chloroplast DNA of Physcomitrella patens in the cyanobacterium Leptolyngbyaboryana. *Plant Cell Physiol 52*:1983–1993.

Yanni Y, Dazzo F (2010) Enhancement of rice production using endophytic strains of Rhizobium leguminosarumbv. trifolii in extensive field inoculation trials within the Egypt Nile delta. *Plant Soil 336*:129–142.

Yoneyama T, Terakado-Tonooka J, Minamisawa K (2017) Exploration of bacterial N2-fixation systems in association with soil-grown sugarcane, sweet potato, and paddy rice: a review and synthesis. *Soil Sci Plant Nutr 63*(6):578–590.

Zeffa D, Perini L, Silva M et al (2019) *Azospirillumbrasilense* promotes increases in growth and nitrogen use efficiency of maize genotypes. *PloS one 14*(4):e0215332.

Zhang X, Zhang R, Gao J et al (2017) Thirty-one years of rice-rice-green manure rotations shape the rhizosphere microbial community and enrich beneficial bacteria. *Soil Biol Biochem 104*:208–217.

3

Organic Agriculture and Its Promotion

Sikander Khan Tanveer
PARC- National Agricultural Research Center (NARC), Pakistan

Muhammad Ayub Khan and Muhammad Asim
Plants Sciences Division PARC, Pakistan

CONTENTS

3.1 Introduction

In 1999, the total area under organic agriculture was 11.0 million hectares; by 2016, this had increased up to 57.8 m.ha. In 2016, there were 178 countries which had organic activities and share of organic land to total agricultural land was 1.2%. The area under organic agriculture in different areas of the world varied from region to region. According to FiBL, IFOAM – Organics International report (2018), the area under organic agriculture in Oceania, Europe, Latin America, North America, Asia, North America and Africa is about 27.3, 13.5, 7.1, 4.9, 3.1 and 1.8 million hectares, respectively, while the value of organic products sold globally in 2016 was about 90 billion US dollars and even double-digit growth rates were recorded in many advanced markets for organic products. In 2016, a growth rate of even 20% and more of the organic retail sales value was noticed for Ireland and France. In Switzerland in the same year, where the market has been evolving over several years with high growth rates, it grew by 8.4%.

The production of organic agriculture is keeping pace and the total organic area has increased to almost 58 million ha, which is being managed by 2.7 million producers. In particular, for some crops such as citrus, fruit, dry pulses and grapes area increased at the rate of 15% and even more reached in 2016. Overall, 57.8 million hectares of increased soil fertility, farm and field diversity and billions of farm animals were raised under animal welfare standards, which are an important contribution to the Sustainable Development Goals (SDGs) of the United Nations (FiBL, IFOAM – Organics International report 2018a).

In 1999, there were a total of 200,000 organic producers in the world, which increased to 2.7 million by 2016. In 1999, the area under wild collection/non-agricultural area was 4.1 million hectares, which increased up to 39.9 million hectares up to 2016. In 2016, the organic market worth was 89.7 billion US dollars (more than 80 billion euros), while it was 17.9 billion US dollars in 2000 and the per capita consumption in 2016 was 12.1 US dollars (11.3 euros). In 2017, 87 countries of the world had organic regulations, while in 2017, there were 1003 affiliates from 127 countries and maximum affiliates (111) were from India, followed by Germany (88), United States (63) and China (56), respectively. In terms of continents, the highest percentage (some 40%) of the world's organic producers were in Asia, 27% in Africa and 17% in Latin America. In 2016, a quarter of the world's organic agricultural land (i.e. 14.3 million hectares) and more than 87% (i.e. 2.4 million) of the producers were in developing countries. With a total of almost 10.6 million hectares, arable land constitutes 18% of the organic agricultural land. During the 2015–16 growing season, 107,980 metric tons of organic fiber was produced globally by 219,947 farmers on 302,562 hectares of land (FiBL, IFOAM – Organics International report 2018).

In 2016, the world's largest single organic market was the United States (comprising 47% of the global market). At the same time, the European Union had an organic market of 30.7 billion euros (37%) and China had a market of 5.9 billion euros (6 %). The highest per capita consumption of organic produce, more than 200 euros, was found in the cases of both Switzerland and Denmark (FiBL, IFOAM – Organics International report 2018b).

According to the FiBL Survey report (2018a), in 2016, Australia had the maximum area (27.15 m.ha) under organic agriculture, followed by Argentina (3.01 m.ha), China (2.28 m.ha), the USA (2.03 m.ha), Spain (2.02 m.ha), Italy (1.80 m.ha), Uruguay (1.66 m.ha), France (1.54 m.ha), India (1.49 m.ha) and Germany (1.25 m.ha). The distribution of all organic areas in 2016 included 40% wild collection, agricultural land and crops 59% and others 0.4%, while the distribution of organic producers by region varied: Latin America (17%), Africa (27%), Asia (40%), Europe (14%), Oceania (1%), and North America (1%). In 2016, the maximum organic producers (835,000) in the world were in India, followed by Uganda (210352), Mexico (210,000), Ethiopia (203,602), the Philippines (165,994), Tanzania (148,610), Peru (91,771), Turkey (67,879), Italy (64,210) and Paraguay (58,258), respectively. Similarly, in 2016, the top ten countries having the largest number of organic markets included USA having maximum number (i.e. 38,938) of organic markets, followed by Germany (9478), France (6736), China (5900), Canada (3002), Italy (2644), UK (2460), Switzerland (2298), Sweden (1944) and Spain (1686), respectively. While in 2016, Switzerland had the highest per capita consumption (274 euros) of organic products, followed by Demark, Sweden, Luxembourg, Austria, Liechtenstein, the USA, Germany, France and Canada having per capita consumption of 227, 197, 188, 177, 171, 121, 116, 101 and 83, respectively (FiBL, IFOAM – Organics International report 2018).

In 2016, the distribution data of organic arable cropland by region shows that 23% of the area was in Asia, 57% in Europe, 11% in North America, 5% in Africa and 4% in Latin America (FiBL Survey 2018b), while the use of organic arable cropland by crop group in 2016 shows that 5% of the area was under textile crops, 5% under dry pulses, 12% under oil seeds crops, 26% under green fodder, 38% under cereals and 14% under other different crops. While in 2016, data regarding the distribution of organic permanent cropland by region shows that Asia, Latin America, Africa, Europe, North America and Oceania had 17%, 22%, 23%, 33%, 2%, and 3% areas under permanent organic cropland, respectively (FiBL, IFOAM – Organics International report 2018).

In 2016, the use of permanent cropland by crop group shows that the area under different crops, i.e., cocoa, coconut, fruits (both tropical and subtropical), grapes, nuts, olives, coffee and other crops were 8%, 8%, 8%, 8%, 13%, 16%, 20%, and 19%, respectively (Saleem et al. 2015).

According to 2016 data, the distribution of organic wild collection and beekeeping areas by region shows that Europe had the highest area (42%), followed by Africa (31%), Asia (16%), and Latin America (11%), while the smallest area (0.2%) was in North America. In 2016, in the world there were 2.1 million organic beehives, accounting for 2.4% of the world's beehives. Similarly, in 2016, a production volume of over 400,000 metric tons of organic aquaculture was reported; of this total organic aquaculture, 22% was in Europe, while Asia had the highest share (77%) of aquaculture and this was mostly produced in China (FiBL, IFOAM – Organics International report 2018).

In 2016, at least 4.1 million hectares of cereals were under organic management. Comparing the organic figure with FAO's figure for the total world's harvested cereal area of 718 million hectares in 2016 (FAO STAT), 0.6% of the total cereal area was under organic management. The data of distribution of global organic area by cereal type in 2016, shows that 36% wheat, 12% oat, 11% maize (grain), 10% rice, 9% barley, 4% rye, 3% Andean grains and 6% other crops were grown under organic condition.

According to FAO, in 2016, almost 91,000 hectares of citrus fruits were being grown organically worldwide and this area was 1 percent of the world's total citrus area of 9.4 million hectares (FAO STAT). Data regarding the distribution of different tropical and subtropical fruits production area under organic condition in 2016 shows that 18% (avocados), 16% (bananas), 11% (dates), 6% (mangoes), 5% (figs), 3% (guava) and 41% (other fruits) were being produced under organic conditions.

Data regarding the production of different oil seeds crops indicates that in 2016, 43% (soybeans), 9% (sesame), 8% (sunflower seed), 7% (peanuts), 7% (rape and turnip rape), 3% (linseed or oil flax) and similarly 23% (other oil seed crops), were being grown under organic conditions (FiBL Survey 2018a).

In 2016, the total area under organic vegetables production (more than 437,000 hectares) was 0.7% of the total area of vegetables being grown in the world (62 million hectares) and the four most important countries in the world for growing organic vegetables of the world were China, India, Nigeria and Viet Nam.

Over the course of the past two decades, organic food and drink sales have increased from less than 15 billion US dollars to almost 90 billion US dollars. Similarly, the production of organic cow's milk has almost doubled since 2007. In 2016, this was 4.4 million metric tons and was more than 2.8% of the European Union's milk production from dairy cows in 2016 (FiBL, IFOAM – Organics International report, 2018).

3.2 History

During the past two decades organic foods have been the highest-growing market segments within the global food industry (Willer et al. 2009; Adnan et al. 2015) and studies have shown that organic farming is eco-friendly, and both more sustainable and more efficient in terms of energy consumption (Lotter 2003; Mader et al. 2002), but it is knowledge-intensive rather than resource-based because in organic environments, instead of synthetic fertilizers and antibiotics etc., farmers make use of manures and composts. In the past, before the introduction of synthetic fertilizers and chemical pesticides, organic sources of fertilizers and similarly physical methods were in common use. In organic farming modern equipment, certified seed, soil and water management, the latest crop varieties and modern innovations in the handling and feeding of livestock are all used. Organic farming systems range from strict closed cycle systems that go beyond organic certification guidelines by limiting external inputs as much as possible to move standard systems that simply follow organic certification guidelines. In 1928, the first standards for biodynamically grown food were developed under the name Demeter (Schmid 2007; Gomiero et al. 2011). Lord Walter Northbourne wrote '*Look to the Land*', in which he became the first to use the term organic farming (Northbourne 1940). Influenced by the German and the British work on organic farming, organic thinking spread to France under the name 'agriculture biologique' (Vogt 2007). Calud Aubert, who is the author of the popular book, *L'Agriculture Biologique* (Aubert 1970), helped to found the French association Nature et Progrès and in France, this resulted in the foundation of the International Federation of Organic Agriculture Movement (IFOAM) in Versailles in 1972. With the passage of time, from its small roots, IFOAM grew quickly, becoming an important global network, for the promotion of organic farming practices around the world and in the setting of basic global standards for organic practices. In 1973, the Research Institute of Organic Agriculture (known by its German initial FiBL) was founded, and this continues today to be the world's largest organic agriculture research establishment (Niggli 2007). The United States Department of Agriculture (USDA) did not acknowledge organic farming as a noteworthy form of agriculture untill the controversial publication of the USDA's Report and Recommendation on Organic farming (USDA 1980).

In 1981 the national meeting of the American Society of Agronomy, Crop Science Society of America, and the Soil Science Society of America included a significant discussion on organic agriculture and which was recorded in a 1984 proceedings report entitled *Organic Farming: Current Technology and its Role in a Sustainable Agriculture* (Bezdicek et al. 1984). In 1989, a National Research Council report entitled *Alternative Agriculture* attested that the wider adoption of alternative farming systems such as organic farms 'Would result in even greater economic benefits to farmers and environmental gains for the nation' (National Research Council 1989, p. 6). The organic farming movement in the USA received a further boost with the passage of the Organic Foods Production Act in 1990. In 1990, Nicolas Lampkin published his book *Organic Farming* (Lampkin 1990).

In the start organic food was either sold directly from farmers, or through health food stores and small cooperatives. Until the early 1980's organic food had a global market share of less than 0.1% (Aschemann et al. 2007). The main reasons given for consuming organic foods included health, environment, taste and animal welfare (Lockie et al. 2006). The Food and Agriculture Organization created its *Codex alimentarius* in 1961 and its guidelines for the production, processing, labeling and marketing of organically produced foods became effective in 2001 (Courville 2006). Both IFOAM and the Codex Alimentarius Commission serve as global standards for certification programs worldwide. Although several European countries had existing certification programs, such as France's Agriculture Biologique label and Germany's Bioland label, the European Union developed its own regulatory framework to oversee organic practices across the whole of Europe, beginning with the passage of the EU Council Regulation 2092/91 in 1991 (Schmid 2007). These standards prohibit the use of synthetic pesticides and fertilizers, the use of growth hormones and antibiotics in livestock (with some exceptions for the sake of humane treatment), and the use of genetically modified organisms (GMOs). New regulations became effective in 2009 and these placed greater emphasis on environmental protection, animal welfare, and the preservation of biodiversity (European Commission 2012). These new regulations permit an organic product to contain 5% non-organic ingredients, a limit of 0.9% of unintentionally present genetically modified materials, and also require organically certified food to use the EU organic label. The US Congress passed the Organic Foods Production Act in 1999, charging the USDA with developing national standards for organic production. The USDA National Organic Program standards went into effect in 2002, setting standards for all food, domestic and imported, that is labeled 'Organic' and specifically prohibiting the use of genetic engineering methods, ionizing radiation, and sewage sludge for fertilization. The USDA National Organic Program was established; this requires organic foods to be certified by a USDA-accredited third-party certifier marked by the USDA organic foods label.

Another system, which is an alternative to governmental or third-party certification, called the Participatory Guarantee Systems (PGS) has been growing in popularity. This system is designed for small farmers and similarly for short supply chains, in which stakeholders verify producers and practices and hold them to be standards for a chosen certifier. This system is less expensive and more locally focused, making it more accessible for small farmers in both developing and developed countries (IFOAM 2012a). The PGS is furthermore seen as returning to the local quality assurance systems, built around social networks and trust, on which the organic movement was established (Katto-Aandrighetto 2010). Organic farms try to build up the interconnections between components to achieve a self-emerging, integrated agro-ecosystem akin to an organism or a natural ecosystem (Kristiansen and Merifield 2006b). In 1980, IFOAM published a set of basic principles for organic agriculture, and their list has been updated from time to time. The application of synthetic pesticides, fertilizers, antibiotics and additives is virtually banned for risk of negative health effects (IFOAM 2012b).

3.3 IFOAM Principles of Organic Agriculture through Time

3.3.1 1980 IFOAM Basic Principles

- To work as much as possible within a closed system, and draw on local resources.
- To maintain the long-term fertility of soils.
- To avoid all forms of pollution that may result from agricultural techniques.
- To produce foodstuffs of high quality and in sufficient quantities.
- To reduce the use of fossil energy in agricultural practice to a minimum.
- To give livestock conditions of life that confirms to their physiological needs and to humanitarian principles.
- To make it possible for agricultural producers to earn a living through their work and develop their potentialities as human beings.

3.3.2 Current IFOAM Principles of Organic Agriculture

- Principles of Health. According to this, organic agriculture should sustain and enhance the health of soil, plants, animals and humans as one indivisible system.
- Principles of Ecology. According to this, Organic agriculture should be based on living ecological systems and cycles, work with them, emulate them and help sustain them.
- Principles of Fairness. According to this, organic agriculture should build relationships that ensure fairness with regard to the common environment and life opportunities.
- Principles of Care. Organic agriculture should be managed in a precautionary and responsible manner to protect the health and well-being of current and future generations and the environment.

The utilization of organic amendments can also promote an abundant and diverse social microbial population, which may lead to competition against soil pathogens (Letourneau and Van Bruggen 2006). According to Shenan (2008), increasing biological diversity, in flora, fauna and soil microbiology, may increase the stability of the system, reducing outbreaks of pests and diseases. In organic farming, the composting of wastes and returning them to the field is a central practice in organic farming. Mixed crop and livestock operations are also common in organic agriculture (Lampkin 1990). By recycling all materials and growing leguminous nitrogen-fixing crops and green manures, organic farms minimize the need for nutrients. To replace nutrients lost with the export of crops organic forms use biologically fixed nitrogen from green manures and cover crops, supplemented with compost, animal manure or rock minerals like rock phosphate for phosphorous and greensand for potassium (Und Niemsdorff and Kristiansen 2006).

Organic farms have lower nitrate leaching, nitrous oxide and ammonia emissions. According to Drinkwater and Snap (2007), the minimizing of nutrient losses both keeps nutrients in the farm and also limits atmospheric and surface water pollution. The use of organic (carbon-coupled) nutrient sources such as manure, compost and plant residues play a particularly important role in dropping of nutrient losses from organically managed forms.

Organic farms are more environmentally friendly because these have good soil quality, more diverse in forms of both flora and fauna and similarly organic farms have higher water-holding capacity as compared to inorganic farms. In the case of organic farming, there is no risk of synthetic pollution of surface and groundwater and similarly no nitrate and phosphorus and similarly greenhouse gas emissions (Mondelaers et al. 2009).

Han et al. (2009) conducted a study, by using a model of riverine nitrogen export for 18 Lake Michigan Basin watersheds based on nitrogen budgets at five-yearly intervals over the period 1974 to 1992. Their findings indicate that, if farmers choose organic practices, and decrease their use of fertilizers, nitrogen export levels by river could decline to below present-day levels.

Several studies have been reported, regarding energy consumption both per hectare and similarly per unit of output (tons) for different crops from organic and conventional farming systems in Germany, Italy, Sweden and Switzerland (Alfoldi et al. 2002). According to the results, in the case of organic winter wheat, citrus, olive and milk, energy consumption per unit was less in organic farming than in conventional production; in the case of potato and apple production, however, energy consumption per unit was higher in the case of organic farming. However, in comparison with conventional farming systems, organic farming has benefits in terms of energy efficiency, environmental viability and stability.

The use of organic fertilizers diminishes nutrient leaching, while it also eliminates high energy requirements needed for the production of synthetic fertilizers (Drinkwater & Snap 2007).

According to Kristiansen and Merifield (2006a) and Reganold et al. (2011), policies can be supportive in the promotion of organic agriculture in terms of research and development and similarly in the case of the removal of, or changes in, agricultural subsidies to support organic farms and similarly the development of regulations to limit the overuse of synthetic pesticides and fertilizers (Kristiansen and Merifield 2006a; Reganold et al. 2011; Arif et al. 2015; Ali et al. 2016).

Different results have been reported regarding the production of different crops under organic and inorganic conditions. For example, de Ponti et al. (2012) reported the results of 362 studies which were carried out to find out the performances of crops under organic and inorganic conditions. According to them, organic yields were 21% lower in developed countries and similarly 20% lower globally, as compared to inorganic yields. Similarly Seufert et al. (2012) reported that the organic farming system had 20% lower yields in developed countries and 25% lower yields when data from both developed and developing countries were compared.

According to de Ponti et al. (2012), yields of organic rice, soybean, corn and grass clover were 6%, 8%, 11% and 11% lower than the yields of these different crops under inorganic conditions. Similar results have been reported by Seufert et al. (2012). According to them, yields of organic fruits, soybean (under rainfed conditions), and oil seed (also under rainfed conditions) produced under organic conditions were, respectively, 3% , 5% and 11% lower than the crops produced under inorganic conditions.

According to Baker et al. (2002), organic foods use around one-third of the pesticides employed in inorganic food production. Regarding the nutrition levels of organic foods as compared to inorganic foods, there have been a range of sometimes conflicting results. Some studies have shown that organic foods are more nutritious than inorganic foods, while others have reported that there are no nutritional differences between organic and inorganic foods.

Production trends also show that the initial three years can be economically more challenging for the organic producers. With the passage of time, however, organic farming becomes financially more beneficial than the conventional inorganic production.

3.4 Organic Legislation Worldwide

According to the current situation, up to 2017, 87 countries had organic standards. 18 countries were in the process of drafting legislation. Some countries have regulations, but these countries do not enforce these regulations. Some countries have adopted legislation and or still in the process of finalizing its implementation. Some countries have adopted legislation but are not providing the resources necessary for its implementation (FiBL, IFOAM–Organics International report 2018).

3.5 Important Requirements of Major Markets

The most important organic markets in the world are the European Union, the United States, Canada and Japan. All of these markets have strict regulations for the importation of organic products. In the European Union, the United States and Japan, products may only be imported if the certifying agent have been approved by the respective authority and the approval of certification requires compliance or equivalency with the requirements of the importing countries, which can be achieved through

a. Bilateral agreements between the exporting country and the target import country.
b. Direct acceptance of the certifying agency by the target import country.

3.6 Conclusion

Organic farming offers many sustainability benefits in terms of environmental viability, productivity, stability and energy efficiency. The use of organic manure and compost is not only environmentally friendly, but also improves the water-holding capacity. Similarly, uses of organic fertilizers reduce nutrient leaching losses. However, challenges still exist in terms of the implementation of meaningful organic standards and the development of organic standards worldwide can be helpful in maintaining the minimum organic practices. Many areas or regions, and especially the developing countries, still face a lot of barriers in the promotion of organic farming. As organic farming is basically knowledge-intensive and complex, all stakeholders, including governments, nongovernmental organizations and international agencies, can combine to play a key role in the promotion of large-scale organic farming. The implementation of proper regulations, certification, marketing systems and truthful labelling can be supportive in the promotion of organic farming.

REFERENCES

Ali A, Imtiyaz, Adnan M, Arshad M, Inayat UR, Muhammad YJ, Saleem N, Rahman, Z (2016) Effect of zinc activities on shoot, root biomass and phosphorus uptake in wheat gynotypes. *American-Eurasian J Agric Environ Sci 16*(1): 204–208.

Alfoldi T, Fliessbach A, Geier U et al. (2002) Organic agriculture, environment and food security. Rome Italy Food and Agriculture Organization Chapter 2. http://www.fao.org/docrep/005/y4137e/y4137e00.htm. Accessed 8 August 2013.

Aschemann J, Hamm U, Naspetti S, Zanoli R (2007) The organic market. In: Kristiansen P, Taji A, Reganold J (ed), *Organic agriculture: a global perspective*. Comstock Publishing Associates, Ithaca, New York, pp 123–150.

Adnan M, Basir A, Arif M, Shah SRA, Khan M, Jamal Y (2015) Impact of grazing on wheat yield and associated weeds. *Pak J Weed Sci Res 21*(3): 351–358.

Arif M, Shah T, Ilyas M, Ahmad W, Mian AA, Jadoon MA, Adnan M (2015). Effect of organic manures and their levels on weeds density and maize yield. *Pakistan Pak J Weed Sci Res 21*(4): 517–522.

Aubert C, Le Courrier du Livre, Badgley C, Moghtader J, Quintero E et al. (1970) Organic agriculture and the global food supply. *Renew Agr Food Syst 22*: 86–108.

Baker BP, Benbrook CM, Groth III E, Benbrook KL (2002) Pesticide residues in conventional integrated pest management (IPM) – grown and organic foods Insights from three U S data sets. *Food Addit Contam 19*: 427–446.

Courville S (2006) Organic standards and certification. In: Kristiansen P, Taji A, Reganold J (ed), *Organic Agriculture, a global perspective*. Comstock Publishing Associates, Ithaca, New York, pp 201–209.

de Ponti T, Rijk B, van Ittersun MK (2012) The crop yield gap between organic and conventional agriculture. *Agri Sys 108*: 1–9.

Drinkwater LE, Snap SS (2007) Nutrients in agroecosystems: rethinking the management paradigm. *Adv Agron 92*: 163–186.

European Commission (2012) Agriculture and Rural development, Organic farming. http://ec.europa.Eu/agriculture/organic/. Accessed 08 August 2013.

FiBL survey (2018a) Based on national data sources and data 2018 from certifier * Global market. Ecovia Intelligence (formerly Organic Monitor).

FiBL Survey (2018b) Based on information from the private sectors certifiers and governments.

Gomiero T, Pimentel D, Paoletti MG (2011) Environmental impact of different agricultural management practices: Conventional Vs organic agriculture. *Crit Rev Plant Sci 30*: 95–124.

Han H, Allan JD, Scavia D (2009) Influence of climate and human activities on the relationship between watershed nitrogen input and river export. *Environ Sci Technol 43*: 1916–1922.

IFOAM (2012a) Participatory guarantee systems. International Federation of Organic Agriculture Movements. http://www.ifoam.org/en/value-chain/participatory-guarantee-systems-pgs. Accessed 8 August 2013.

IFOAM (2012b) The principles of organic agriculture. International Federation of Organic Agriculture Movements. http://www.ifoam.org/en/organic-landmarks/principles-organic-agriculture. Accessed 8 August 2013.

Katto-Aandrighetto J (2010) Overview of participatory guarantee system world-wide. In: Weidmann G, Kilcher L, Garibay S (eds), *The world of organic agriculture: statistics and emerging trends*. International Federation of Organic Agriculture Movements, Bonn, Germany, pp 85–86.

Kristiansen P, Merifield C (2006a) *Overview of organic agriculture: a global perspective*. Comstock Publishing Associates, Ithaca, NY, pp 1–23.

Kristiansen P, Merifield C (2006b) Overview of organic agriculture. In: Kristiansen P, Taji A, Reganold J (eds), *Organic agriculture: a global perspective*. Comstock, Ithaca, NY, pp 421–441.

Lampkin N (1990) *Organic farming*. Farming Press, Alexandria Bay, NY.

Letourneau D, Van Bruggen A (2006) Crop protection in organic agriculture. In: Kristiansen P, Taji A, Reganold J (eds), *Organic agriculture: a global perspective*. Comstock Publishing Associates, Ithaca, NY, pp 93–121.

Lockie S, Halpin D, Pearson D (2006) Understanding the market for organic food. In: Kristiansen P, Taji A, Reganold J (eds), *Organic agriculture: a global perspective*. Comstock Publishig Associates, Ithaca, NY, pp 245–258.

Lotter DW (2003) Organic agriculture. *J Sustain Agric 21*: 59–128.

Mader P, Fliessbach A, Dubois D et al. (2002). Soil fertility and biodiversity in organic farming. *Science 296*: 1694–1697.
Mondelaers K, Aertsens J, Van Huylenbroeck G (2009) A meta-analysis of differences in environmental impacts between organic and conventional farming. *British Food J 111*: 1098–1119.
National Research Council (1989) *Alternative agriculture*. The National Academies Press, Washington, DC.
Niggli U (2007) FiBL and organic research in Switzerland. In: Lockeretz W (ed), *Organic farming: an international history*. CAB International, Oxford, UK, pp 242–252.
Northbourne L (1940) *Look to the Land*. Dent, London.
Reganold JP, Jackson-Smith D, Batle SS et al. (2011) Transforming US Agriculture. *Science 332*: 670–671.
Saleem N, Adnan M, Khan NA, Zaheer S, Jalal F, Amin M, Jamal Y (2015). Dual purpose canola: grazing and grains options. *Pak J Weed Sci Res 21*(2): 295–304.
Schmid O (2007) Development of standards for organic farming. In: Lockeretz W (ed), *Organic farming: an international history*. CAB International, Oxfordshire, UK, pp 152–174.
Shenan C(2008) Biotic interactions, ecological knowledge and agriculture. *Philosophical Trans Royal Soc B 363*: 717–739.
Seufert V, Ramankutty N, Foley JA (2012) Comparing the yields of organic and conventional agriculture. *Nat 485*: 282–292.
USDA (1980) *Report and recommendations on organic farming*. United States Department of Agriculture; United States Government Printing Office, Washington, DC.
Und Niemsdorff PF, Kristiansen P (2006) Crop management in organic agriculture. In: Kristiansen P, Taji A, Reganold J (eds), *Organic agriculture: a Global Perspective*. Comstock Publishing Associates, Ithaca, NY, pp 53–82.
Vogt G (2007) The origin of organic farming. In: Lockeretz W (ed), *Organic farming: an international history*, CAB International, Oxfordshire, UK, pp 9–29.
Willer H, Rohwedder M, Wynen E (2009) Current statistics. In: Weidmann G, Kilcher L, Garibay S (eds), *The world of organic agriculture: statistics and Emerging Trends*, IFOAM FiBL and ITC, Bonn, Frick and Geneva, pp 25–58.

4

Soil Salinity Management and Plant Growth Under Climate Change

Muhammad Akmal, Khalid Saifullah Khan, Qaiser Hussain, Hafeez Ullah Rafa, and Muhammad Subtain Abbas
PMAS-Arid Agriculture University Rawalpindi, Pakistan

CONTENTS

4.1 Introduction

Climate change is brought about by increases in the temperature along with disruptions in precipitation patterns. Both of these play a direct role in agriculture and carbon losses from the soil is increased, thereby contributing to an increase in the concentration of carbon dioxide in the atmosphere. The disturbances in rainfall patterns, i.e more quantity/intensity in one region/season as compared to another have increased the chances of desertification/land degradation. In recent years, overall climate change has increased the incidence of severe weather events such as drought, floods, and heatwaves (Adnan et al. 2018a, 2018b, 2019; Akram et al. 2018a, 2018b; Aziz et al. 2017a, 2017b; Habibur et al. 2017; Hafiz et al. 2016, 2018, 2019; Kamam et al. 2017; Muhammad et al. 2019; Sajjad et al. 2019; Saud et al. 2013, 2014, 2016, 2017, 2020; Shah et al. 2013; Qamar-uz et al. 2017; Wajid et al. 2017; Yang et al. 2017; Zahida et al. 2017; Depeng et al. 2018; Hussain et al. 2020; Hafiz et al. 2020a, 2020b; Shafi et al. 2020; Wahid et al. 2020; Subhan et al. 2020; Zafar-ul-Hye et al. 2020a, 2020b; Adnan et al. 2020; Ilyas et al. 2020; Saleem et al. 2020a, 2020b, 2020c; Rehman et al. 2020; Farhat et al. 2020; Wu et al. 2020; Mubeen et al. 2020; Farhana et al. 2020; Jan et al. 2019; Wu et al. 2019; Ahmad et al. 2019; Baseer et al. 2019; Tariq et al. 2018). As a result of the melting of ice/glaciers, changes in the course of rivers, and increases in sea levels, many coastal areas have experienced erosion or even submersion. Climate change has led to disturbances in the balance of water across the globe, leading to droughts in some areas and widespread area or seasonal floods on the other. Accordingly, changes in land usage are becoming increasingly common (Fahad and Bano 2012; Fahad et al. 2013, 2014a, 2014b, 2015a, 2015b, 2016a, 2016b, 2016c, 2016d, 2017, 2018, 2019a, 2019b; Hesham and Fahad 2020).

4.2 Causes of Climate Change

There are two major causes of climate change: (1) Human-induced; (2) Natural causes.

Human-influenced activities leading to climate change have included fossil fuel usage, agricultural, urban, industrial activities, and deforestation. By contrast, natural causes that accelerates the production of greenhouse gases (GHGs) include: (a) Earth orbital changes; (b) sunlight intensity; (c) water cycle in ocean and atmosphere, and so on. Together, these contribute to greenhouse gas emissions and disturb the climate on Earth. Among the gases that contribute to climate change are carbon dioxide, nitrous oxide, and methane. These individual gases make a diverse impact on climate change. Any factor which can limit or control the emission of these gases will significantly decrease the levels of global warming.

4.3 Climate Change Impact on Soil

Climate change has a considerable impact on land degradation. The increase/change in temperature and rainfall, and the rise in sea level has accelerated this process in the form of soil erosion and salinization of soil (Fahad and Bano 2012). In some parts of the world salinity has been a threat to agriculture for over 3000 years, yet in recent times this threat has accelerated. Furthermore, there has been a huge increase in human population (Fahad and Bano 2012), meaning that more food is required. As an agriculturalist, one is faced with two options. The first option is the vertical expansion of agriculture (an increase in the crop yield from area under cultivation). The second option is the horizontal expansion of agriculture (increasing the area under plough, especially in salt-hit areas). Keeping all of this in focus the management of salt-affected soils can be a workable option for ensuring food security. Salt-affected soils can be found around the globe in more than 100 countries, covering an area of about 9.32 10^6 ha. Their contribution in terms of irrigated and cultivated areas is about 25 and 60%, respectively. Of this total, 38% are saline and 62% are saline-sodic (Tanji 1990). The total saline area is about 351.5 m.ha, and sodic area is about 581.0 m.ha, respectively. In different continents of world i.e America, Africa, Asia, Europe, Australia, saline soil area are 77.6, 53.5, 194.9, 7.8 and 17.4 m.ha and sodic soil area are 69.2 27.0 121.9, 22.9, 340.0 m.ha respectively (Rengasamy, 2006). Khan (1998) reported that the area of salt-affected soil in Pakistan is about 6.6710 6 ha. Maximum area 3.04 m.ha is in Sindh followed by Punjab 1.234 m.ha, Baluchistan 0.12 m.ha and Khyberpakhtoon khawa 0.11 m.ha.

4.4 Categories of Salt-Affected Soils

Salt-affected soils are classified as follows (Figure 4.1).

In such soils, physical properties, such as soil structure, soil aeration and hydraulic properties, have deteriorated, meaning that they are not suitable for plant growth.

4.5 Plant Growth under Salinity Stress

Plants grown in a saline environment can be characterized by fewer/smaller no of leaves and a pattern of stunted growth. There are two ways in which salinity can effect plants: (1) Osmotic effect; (2) Specific Ion toxicity/ Nutritional disorder (Läuchli and Epstein 1990). The osmotic effect resembles water stress, in which plant roots are unable to extract water from the saline rhizosphere environment. Later on, as the growth progresses there is an excessive uptake of a single ion, e.g. Na, by the plants at the cost of other ions such as calcium (Ca) and potassium (K). Similarly, boron (B) and chlorine (Cl) toxicity can also occur. It is reported that crop salt tolerance within a specie may vary according to age or from one growth stage to the next (Bernstein and Hayward, 1958). Initially, plant leaves faces growth reduction just a few minutes after its exposure to salts in the rhizosphere. Instantly, there is a reduction in the ability of plants to absorb water. After a steady state in the root zone the plant leaves regain their growth rate. Finally,

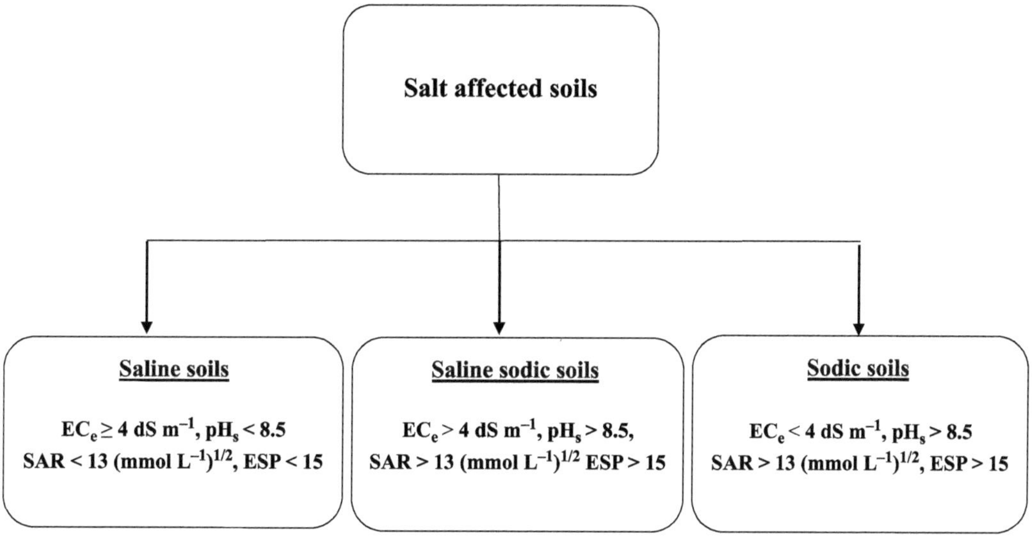

FIGURE 4.1 Classification of salt affected soils.

salt toxicity in plants occurred, slowly (within days, weeks, months) due to excessive uptake of salts in the older leaves. This causes leaf death, which in turn decreases the total photosynthetic leaf area of the plant. If the leaf death rate of older leaves is greater than the development of new leaves, under such conditions there is a fall in photosynthesis and the plant is unable to survive and produce seeds. Most annual crops can tolerate salt stress during the germination stage and are sensitive during emergence/early vegetative development (Maas and Grattan 1999). Salinity has an impact on both vegetative and reproductive development and it often reduces shoot growth more than root growth. In rice and wheat, it impacts maturity, time of flowering and the sterility. Crop response to high salt concentration is also affected by climatic conditions. In hot and dry climates, for example, crops are more sensitive to salinity than is the case in cool and humid climatic conditions, although this can vary from crop to crop. For example, both cotton and tomatoes are more sensitive to salinity at higher temperatures than lower temperatures. By contrast, barley, cotton, onion, and radish are more tolerant to salinity in higher humidity. In addition, the actual crop performance is associated with the interactive effect of salinity with flooding of water. In reduced soil conditions, the diffusion of oxygen, nitrate, sulfate, iron and manganese is decreased, leading to physiological dysfunction of plants. The true success of plants in saline environments is associated with composition and type of cation, anions, i.e. Na/Ca, Na/K, Ca/Mg, and Cl/NO_3, water stress, plant pathogenicity, soil physical conditions in rhizosphere.

4.6 Management of Salt-Affected Soils

Many strategies have been proposed for the rehabilitation of salt-affected soils. The objective of these approaches is to increase the crop yield per unit area along with more area under plough. In this way farmers' living standards and lifestyle will be improved. There are different methods of land reclamation. (1) Physical methods include subsoiling, deep ploughing, sanding, and so on. They enhance the permeability of soil, making good soil tilth. (2) Biological methods include the raising of crops on problem soil, and also at certain vegetative phases their incorporation into the soil. This practice increases the level of organic matter in the soil. Among other additions, farm wastes/manures, chicken manure, compost, pressmud organic polymers proved effective for improvement soil properties during the reclamation of salt-affected soils. (3) Chemical methods include the incorporation of gypsum, sulfuric acid, sulfur, and hydrochloric acid, which are useful for the reclamation of calcareous saline sodic soil. While gypsum application can be customized for saline, saline sodic soil to sequester carbon along with improvements in the physical properties of soil.

4.6.1 Management of Reclaimed Soils

Reclaimed soils need careful management in order to sustain agricultural productions. General measures for controlling the buildup in salt in reclaimed soils include the following: (1) Every effort should be made to have a net downward balance of water which in turn will favor the disappearance of salts from the root zone. This is practiced during reclamation of saline soils. (2) Keeping in view the climatic conditions at site, water usage should be made for the growth of crops. (3) Mulching and green manuring should be made at regular intervals during crop growth to check the evaporation of water along with the improvement of soil physical properties (Murtaza et al. 2009; Rahman et al. 2016; Adnan et al. 2016) during the reclamation of salt-affected soil. In order to be reclaimed, saline sodic and sodic soils need chemical amendments as described above.

4.6.1.1 Management Practices for Salt-Affected Soils

Management practices for salt-affected soils include the following:

4.6.1.1.1 Crop Selection/Raising of Salt Tolerant Crops

Salt tolerance vary among different crops, e.g. barley, cotton, sugar beet, sorghum, wheat, and rice can tolerate up to EC_e 8.0, 7.7, 7.0, 6.6, 6.0, 3.0 dSm^{-1} respectively (Food and Agriculture Organization 1994).

4.6.1.1.2 Irrigation Practices

The leaching of salts through the rhizosphere can be accomplished by the flooding or heavy irrigation of good-quality water. This practice will ensure a favorable environment for seed germination and root development and other plant processes. During early growth stages, in particular, more frequent irrigation is found useful for overall plant growth in a saline environment.

4.6.1.1.3 Soil Tillage Operations

Soil tillage practices are of particular significance as crops are sensitive to salt stress during early growth stages. Ridge sowing of different field crops (wheat, cotton) is found to be successful in avoiding salt stress at the germination and seedling stages. The clogging of soil pores is observed in sodic soil during puddling, which in turn impedes the water flow through the soil. The tilling of such soil under dry conditions favors clod formation. Subsoiling and deep ploughing helps to loosen the soils, thereby favoring a deeper root zone and better water movement during reclamation.

4.6.1.1.4 Fertilizer Use in Salt Affected Soils

Salt-affected soils are characterized by low levels of organic matter and nutrient availability, and also an imbalance in nutrition. Fertilizers, except Phosphorus containing fertilizers, are applied at higher rates (15-30%) when compared with the practices in normal soil. Green manuring is beneficial to rectify nitrogen deficiency through an improvement of the physical properties of the soil. Ammonium sulfate and ammonium nitrate should be the preferred nitrogen source over urea in these soils. Similarly nitrophos, single super phosphate is beneficial when compared with other phosphatic fertilizers e.g Di ammonium phosphate.

4.7 Saline Agriculture

Saline agriculture is the economical use of salt-affected land and water on a sustained basis. It involves the integrated use of all available sources, i.e land, plants, animals with a special genetic approach and agronomic manipulations. The use of brackish water with certain management strategies, salt-tolerant crops/varieties, grasses, trees, shrubs, and fish culture are among its component parts. Slight, moderately and highly salt-affected soils should be allocated for salt-tolerant crops, trees and shrubs, and fish culture, respectively.

4.7.1 Workable Strategies to Manage Salt-Affected Soils under Climate Impact

1. The screening of plant varieties bearing in mind salt stress, drought, heat, cold tolerance is need of the day.
2. Agronomic interventions that promote conservation agriculture. The crop leftover should be managed on site and burning should be avoided.
3. The use of biotechnology for the identification and incorporation of tolerant genes (salt, heat, moisture, etc.) in transgenic plants.
4. Greenhouse gas emissions should be reduced through a policy of carbon sequestration. The cultivation of crops and tree plantation on marginal lands should be encouraged for this purpose.
5. Early weather warning information system be installed so that the end user i.e farmer may have time (10–15 days) to manage the risk factor.
6. Animals are the largest producers of methane. Any intervention that improved the livestock housing and feeding can be beneficial in terms of the reduction of greenhouse gases/climatic change.
7. A crop insurance scheme, and climate literacy, should be popularized among general masses and farmers.

4.8 Conclusion

Salinization is becoming an increasingly significant aspect of present-day agriculture. Fertile cultivated areas have become unviable because of this menace. Yet this threat can also be an opportunity under the climatic change scenario. As we have seen, salt-affected soils are generally low in organic matter and carbon. They have the potential to sequester carbon from the atmosphere through crop growth, and through tree plantation during a policy of reclamation. These soils provide an opportunity for the horizontal expansion of agriculture, thereby ensuring the supply of food for an ever-increasing population. Among different amendments, gypsum is the cheapest source of Ca-amendment that can sequester carbon, and improve both soil properties and crop yields. In this way it has the potential to manage and cut down emissions of CO_2 in the environment. The rapid reclamation of calcareous salt-affected soils can be done by acids and acid formers, albeit at a higher cost. Similarly, this saline agriculture is another viable option for the wise use of marginal lands. In this respect there is need for attention rather than leaving this resource fallow. With an increase in the level of public awareness and farmer education, great benefits can be obtained from uncultivated salt-affected soil.

REFERENCES

Adnan M, Fahad S, Muhammad Z, Shahen S, Ishaq AM, Subhan D, Zafar-ul-Hye M, Martin LB, Raja MMN, Beena S, Saud S, Imran A, Zhen Y, Martin B, Jiri H, Rahul D (2020) Coupling Phosphate-Solubilizing Bacteria with Phosphorus Supplements Improve Maize Phosphorus Acquisition and Growth under Lime Induced Salinity Stress. *Plants 9*(900). doi:10.3390/plants9070900.

Adnan M, Fahad S, Khan IA, Saeed M, Ihsan MZ, Saud S, Riaz M, Wang D, Wu C (2019) Integration of poultry manure and phosphate solubilizing bacteria improved availability of Ca bound P in calcareous soils. *Biotech 9*(10):368.

Adnan M, Shah Z, Saleem N, Basir A, Ullah H, Ibrahim M, Shah SRA (2016). Isolation and evaluation of summer legumes rhizobia as PGPR. *Pure and Appl Bio 5*(1):127–133.

Adnan M, Shah Z, Sharif M, Rahman H (2018b) Liming induces carbon dioxide (CO_2) emission in PSB inoculated alkaline soil supplemented with different phosphorus sources. *Environ Sci Poll Res 25*(10): 9501–9509.

Adnan M, Zahir S, Fahad S, Arif M, Mukhtar A, Imtiaz AK, Ishaq AM, Abdul B, Hidayat U, Muhammad A, Inayat-Ur R, Saud S, Muhammad ZI, Yousaf J, Amanullah, Hafiz MH, Wajid N (2018a) Phosphate-solubilizing bacteria nullify the antagonistic effect of soil calcification on bioavailability of phosphorus in alkaline soils. *Sci Rep 8*: 4339. https://doi.org/10.1038/s41598-018-22653-7.

Ahmad S, Kamran M, Ding R, Meng X, Wang H, Ahmad I, Fahad S, Han Q (2019) Exogenous melatonin confers drought stress by promoting plant growth, photosynthetic capacity and antioxidant defense system of maize seedlings. *Peer J 7*: e7793. http://doi.org/10.7717/peerj.7793.

Akram R, Turan V, Hammad HM, Ahmad S, Hussain S, Hasnain A, Maqbool MM, Rehmani MIA, Rasool A, Masood N, Mahmood F, Mubeen M, Sultana SR, Fahad S, Amanet K, Saleem M, Abbas Y, Akhtar HM, Waseem F, Murtaza R, Amin A, Zahoor SA, ul Din MS, Nasim W (2018a) Fate of organic and inorganic pollutants in paddy soils. In: Hashmi MZ, Varma A (eds.), *Environmental pollution of paddy soils, soil biology*. Springer International Publishing AG, Cham, Switzerland, pp 197–214.

Akram R, Turan V, Wahid A, Ijaz M, Shahid MA, Kaleem S, Hafeez A, Maqbool MM, Chaudhary HJ, Munis, MFH, Mubeen M, Sadiq N, Murtaza R, Kazmi DH, Ali S, Khan N, Sultana SR, Fahad S, Amin A, Nasim W (2018b) Paddy Land Pollutants and Their Role in Climate Change. In: Hashmi MZ, Varma A (eds) *Environmental Pollution of Paddy Soils, Soil Biology*. Springer International Publishing AG, Cham, Switzerland, pp 113–124.

Aziz K, Daniel KYT, Fazal M, Muhammad ZA, Farooq S, Fan W, Fahad S, Ruiyang Z (2017a) Nitrogen nutrition in cotton and control strategies for greenhouse gas emissions: a review. *Environ Sci Pollut Res 24*: 23471–23487. https://doi.org/10.1007/s11356-017-0131-y.

Aziz K, Daniel KYT, Muhammad ZA, Honghai L, Shahbaz AT, Mir A, Fahad S (2017b) Nitrogen fertility and abiotic stresses management in cotton crop: a review. *Environ Sci Pollut Res 24*:14551–14566. https://doi.org/10.1007/s11356-017-8920-x.

Baseer M, Adnan M, Fazal M, Fahad S, Muhammad S, Fazli W, Muhammad A, Amanullah Jr, Depeng W, Saud S, Muhammad N, Muhammad Z, Fazli S, Beena S, Mian AR, Ishaq AM (2019) Substituting urea by organic wastes for improving maize yield in alkaline soil. *J Plant Nutrition*. doi.org/10.1080/01904167.2019.1659344.

Bernstein L, Hayward H E (1958) Physiology of salt tolerance. *Ann Rev Plant Physiol 9*:25–46.

Depeng W, Fahad S, Saud S, Muhammad K, Aziz K, Mohammad NK, Hafiz MH, Wajid N (2018) Morphological acclimation to agronomic manipulation in leaf dispersion and orientation to promote "Ideotype" breeding: evidence from 3D visual modeling of "super" rice (*Oryza sativa* L.). *Plant Physiol Biochem*. https://doi.org/10.1016/j.plaphy.2018.11.010.

Fahad S, Bajwa AA, Nazir U, Anjum SA, Farooq A, Zohaib A, Sadia S, Nasim W, Adkins S, Saud S, Ihsan MZ, Alharby H, Wu C, Wang D, Huang J (2017) Crop production under drought and heat stress: plant responses and management options. *Front Plant Sci 8*: 1147. doi:10.3389/fpls.2017.01147.

Fahad S, Bano A (2012) Effect of salicylic acid on physiological and biochemical characterization of maize grown in saline area. *Pak J Bot 44*: 1433–1438.

Fahad S, Chen Y, Saud S, Wang K, Xiong D, Chen C, Wu C, Shah F, Nie L, Huang J (2013) Ultraviolet radiation effect on photosynthetic pigments, biochemical attributes, antioxidant enzyme activity and hormonal contents of wheat. *J Food Agri Environ 11*(3&4): 1635–1641.

Fahad S, Hussain S, Bano A, Saud S, Hassan S, Shan D, Khan FA, Khan F, Chen Y, Wu C, Tabassum MA, Chun MX, Afzal M, Jan A, Jan MT, Huang J (2014a) Potential role of phytohormones and plant growth-promoting rhizobacteria in abiotic stresses: consequences for changing environment. *Environ Sci Pollut Res 22*(7): 4907–4921. https://doi.org/10.1007/s11356-014-3754-2.

Fahad S, Hussain S, Matloob A, Khan FA, Khaliq A, Saud S, Hassan S, Shan D, Khan F, Ullah N, Faiq M, Khan MR, Tareen AK, Khan A, Ullah A, Ullah N, Huang J (2014b) Phytohormones and plant responses to salinity stress: A review. *Plant Growth Regul 75*(2): 391–404. https://doi.org/10.1007/s10725-014-0013-y.

Fahad S, Hussain S, Saud S, Hassan S, Chauhan BS, Khan F et al. (2016a) Responses of rapid viscoanalyzer profile and other rice grain qualities to exogenously applied plant growth regulators under high day and high night temperatures. *PLoS One 11*(7): e0159590. https://doi.org/10.1371/journal.pone.0159590.

Fahad S, Hussain S, Saud S, Hassan S, Ihsan Z, Shah AN, Wu C, Yousaf M, Nasim W, Alharby H, Alghabari F, Huang J (2016c) Exogenously applied plant growth regulators enhance the morphophysiological growth and yield of rice under high temperature. *Front Plant Sci 7*: 1250. https://doi.org/10.3389/fpls.2016.01250.

Fahad S, Hussain S, Saud S, Hassan S, Tanveer M, Ihsan MZ, Shah AN, Ullah A, Nasrullah KF, Ullah S, Alharby HNW, Wu C, Huang J (2016d) A combined application of biochar and phosphorus alleviates heat-induced adversities on physiological, agronomical and quality attributes of rice. *Plant Physiol Biochem 103*:191–198.

Fahad S, Hussain S, Saud S, Khan F, Hassan S, Jr A, Nasim W, Arif M, Wang F, Huang J (2016b) Exogenously applied plant growth regulators affect heat-stressed rice pollens. *J Agron Crop Sci 202*: 139–150.

Fahad S, Hussain S, Saud S, Tanveer M, Bajwa AA, Hassan S, Shah AN, Ullah A, Wu C, Khan FA, Shah F, Ullah S, Chen Y, Huang J (2015a) A biochar application protects rice pollen from high-temperature stress. *Plant Physiol Biochem 96*: 281–287.

Fahad S, Muhammad ZI, Abdul K, Ihsanullah D, Saud S, Saleh A, Wajid N, Muhammad A, Imtiaz AK, Chao W, Depeng W, Jianliang H (2018) Consequences of high temperature under changing climate optima for rice pollen characteristics-concepts and perspectives. *Archives Agron Soil Sci.* DOI:10.1080/03650340.2018.1443213.

Fahad S, Nie L, Chen Y, Wu C, Xiong D, Saud S, Hongyan L, Cui K, Huang J (2015b) Crop plant hormones and environmental stress. *Sustain Agric Rev 15*:371–400.

Fahad S, Adnan M, Hassan S, Saud S, Hussain S, Wu C, Wang D, Hakeem KR, Alharby HF, Turan V, Khan MA, Huang J (2019b) Rice responses and tolerance to high temperature. In: Hasanuzzaman M, Fujita M, Nahar, K and Biswas JK (eds) *Advances in rice research for abiotic stress tolerance*. Woodhead Publ. Ltd, Abington Hall Abington, Cambridge CB1 6AH, Cambs, England, pp 201–224.

Fahad S, Rehman A, Shahzad B, Tanveer M, Saud S, Kamran M, Ihtisham M, Khan SU, Turan V, Rahman MHU (2019a) Rice responses and tolerance to metal/metalloid toxicity. In: Hasanuzzaman M, Fujita M, Nahar K, Biswas JK (eds) *Advances in rice research for abiotic stress tolerance*. Woodhead Publ Ltd, Cambs, England, pp 299–312.

FAO 1994. Basic texts of the Food and Agriculture Organization of the United Nations.

Farhana G, Ishfaq A, Muhammad A, Dawood J, Fahad S, Xiuling L, Depeng W, Muhammad F, Muhammad F, Syed AS (2020) Use of crop growth model to simulate the impact of climate change on yield of various wheat cultivars under different agro-environmental conditions in Khyber Pakhtunkhwa, Pakistan. *Arabian J Geosci. 13*: 112 https://doi.org/10.1007/s12517-020-5118-1.

Farhat A, Hafiz MH, Wajid I, Aitazaz AF, Hafiz FB, Zahida Z, Fahad S, Wajid F, Artemi C (2020) A review of soil carbon dynamics resulting from agricultural practices. *J Environ Manage 268*(2020):110319.

Habibur R, Ashfaq A, Aftab W, Manzoor H, Fahd R, Wajid I, Md Aminul I, Vakhtang S, Muhammad A, Asmat U, Abdul W, Syeda RS, Shah S, Shahbaz K, Fahad S, Manzoor H, Saddam H, Wajid N (2017) Application of CSM-CROPGRO-Cotton model for cultivars and optimum planting dates: Evaluation in changing semi-arid climate. *Field Crops Res* http://dx.doi.org/10.1016/j.fcr.2017.07.007.

Hafiz MH, Abdul K, Farhat A, Wajid F, Fahad S, Muhammad A, Ghulam MS, Wajid N, Muhammad M, Hafiz FB (2020b) Comparative effects of organic and inorganic fertilizers on soil organic carbon and wheat productivity under Arid Region, *Commun Soil Sci Plant Anal.* DOI:10.1080/00103624.2020.1763385.

Hafiz MH, Farhat A, Ashfaq A, Hafiz FB, Wajid F, Carol Jo W, Fahad S, Gerrit H (2020a) Predicting Kernel growth of Maize under controlled water and Nitrogen applications. *Int J Plant Prod.* https://doi.org/10.1007/s42106-020-00110-8.

Hafiz MH, Farhat A, Shafqat S, Fahad S, Artemi C, Wajid F, Chaves CB, Wajid N, Muhammad M, Hafiz FB (2018) Offsetting land degradation through nitrogen and water management during maize cultivation under arid conditions. *Land Degrad Dev* 1–10. 10.1002/ldr.2933.

Hafiz MH, Muhammad A, Farhat A, Hafiz FB, Saeed AQ, Muhammad M, Fahad S, Muhammad A (2019) Environmental factors affecting the frequency of road traffic accidents: a case study of sub-urban area of Pakistan. *Environ Sci Pollut Res* https://doi.org/10.1007/s11356-019-04752-8.

Hafiz MH, Wajid F, Farhat A, Fahad S, Shafqat S, Wajid N, Hafiz FB (2016) Maize plant nitrogen uptake dynamics at limited irrigation water and nitrogen. *Environ Sci Pollut Res 24*(3):2549–2557. https://doi.org/10.1007/s11356-016-8031-0.

Hesham FA and Fahad S (2020) Melatonin application enhances biochar efficiency for drought tolerance in maize varieties: Modifications in physio-biochemical machinery. *Agron J* 1–22.

Hussain MA, Fahad S, Rahat S, Muhammad FJ, Muhammad M, Qasid A, Ali A, Husain A, Nooral A, Babatope SA, Changbao S, Liya G, Ibrar A, Zhanmei J, Juncai H (2020) Multifunctional role of brassinosteroid and its analogues in plants. *Plant Growth Regul* https://doi.org/10.1007/s10725-020-00647-8.

Ilyas M, Mohammad N, Nadeem K, Ali H, Aamir HK, Kashif H, Fahad S, Aziz K, Abid U (2020) Drought tolerance strategies in plants: a mechanistic approach. *J Plant Growth Regulation.* https://doi.org/10.1007/s00344-020-10174-5.

Jan M, Muhammad Anwar, Adnan NS, Muhammad Y, Javaid I, Xiuling L, Depeng W, Fahad S (2019) Modulation in growth, gas exchange, and antioxidant activities of salt-stressed rice (*Oryza sativa* L.) genotypes by zinc fertilization. *Arabian J Geosci. 12*:775. https://doi.org/10.1007/s12517-019-4939-2.

Kamarn M, Wenwen C, Irshad A, Xiangping M, Xudong Z, Wennan S, Junzhi C, Shakeel A, Fahad S, Qingfang H, Tiening L (2017) Effect of paclobutrazol, a potential growth regulator on stalk mechanical strength, lignin accumulation and its relation with lodging resistance of maize. *Plant Growth Regul 84*:317–332. https://doi.org/10.1007/s10725-017-0342-8.

Khan GS (1998) *Soil Salinity/Sodicity Status in Pakistan.* Soil Survey of Pakistan, Lahore, Pakistan.

Läuchli A, Epstein E (1990) Plant responses to saline and sodic conditions. In Tanji KK (ed) *Agricultural salinity assessment and management.* ASCE manuals and reports on engineering practice No, 71. pp. 113–137. ASCE, New York.

Maas EV, Grattan SR (1999) Crop yields as affected by salinity. In RW Skaggs and J van Schilfgaarde (eds), *Agricultural Drainage.* Agron. ASA, CSSA, SSA, Madison, WI, pp. 55–108.

Mubeen M, Ashfaq A, Hafiz MH, Muhammad A, Hafiz UF, Mazhar S, Muhammad Sami ul Din, Asad A, Amjed A, Fahad S, Wajid N (2020) Evaluating the climate change impact on water use efficiency of cotton-wheat in semi-arid conditions using DSSAT model. *J Water Clim Change.* doi:10.2166/wcc.2019.179/622035/jwc2019179.pdf.

Muhammad Z, Abdul MK, Abdul MS, Kenneth BM, Muhammad S, Shahen S, Ibadullah J, Fahad S (2019) Performance of Aeluropus lagopoides (mangrove grass) ecotypes, a potential turfgrass, under high saline conditions. *Environ Sci Pollut Res.* https://doi.org/10.1007/s11356-019-04838-3.

Murtaza G, Ghafoor A, Owens G, Qadir M, Kahloon U Z (2009) Environmental and economic benefits of saline sodic soil reclamation using low quality water and soil amendments in conjugation with a rice wheat cropping system. *J Agron Crop Sci 1995*: 124–136.

Qamar-uz Z, Zubair A, Muhammad Y, Muhammad ZI, Abdul K, Fahad S, Safder B, Ramzani PMA, Muhammad N (2017) Zinc biofortification in rice: leveraging agriculture to moderate hidden hunger in developing countries. *Arch Agron Soil Sci 64*:147–161. https://doi.org/10.1080/03650340.2017.1338343.

Rahman IU, Ali S, Adnan M, Ibrahim M, Saleem N, Khan IA (2016) Effect of seed priming on growth parameters of Okra (*Abelmoschus esculentus* L.). *Pure and Appl Bio 5*(1):165–171.

Rehman M, Fahad S, Saleem MH, Hafeez M, Muhammad Habib ur Rahman, Liu F, Deng G (2020) Red light optimized physiological traits and enhanced the growth of ramie (*Boehmeria nivea* L.). *Photosynthetica 58*(4):922–931.

Rengasamy P (2006) World salinization with emphasis on Australia. *J Experiment Botany 57*(5): 1017–1023.

Sajjad H, Muhammad M, Ashfaq A, Waseem A, Hafiz MH, Mazhar A, Nasir M, Asad A, Hafiz UF, Syeda RS, Fahad S, Depeng W, Wajid N (2019) Using GIS tools to detect the land use/land cover changes during forty years in Lodhran district of Pakistan. *Environ Sci Pollut Res.* https://doi.org/10.1007/s11356-019-06072-3.

Saleem MH, Fahad S, Adnan M, Mohsin A, Muhammad SR, Muhammad K, Qurban A, Inas AH, Parashuram B, Mubassir A, Reem MH (2020a) Foliar application of gibberellic acid endorsed phytoextraction of copper and alleviates oxidative stress in jute (Corchorus capsularis L.) plant grown in highly copper-contaminated soil of China. *Environ Sci Pollution Res.* https://doi.org/10.1007/s11356-020-09764-3.

Saleem MH, Fahad S, Shahid UK, Mairaj D, Abid U, Ayman ELS, Akbar H, Analía L, Lijun L (2020c) Copper-induced oxidative stress, initiation of antioxidants and phytoremediation potential of flax (*Linum usitatissimum* L.) seedlings grown under the mixing of two different soils of China. *Environ Sci Poll Res.* https://doi.org/10.1007/s11356-019-07264-7.

Saleem MH, Rehman M, Fahad S, Tung SA, Iqbal N, Hassan A, Ayub A, Wahid MA, Shaukat S, Liu L, Deng G (2020b) Leaf gas exchange, oxidative stress, and physiological attributes of rapeseed (Brassica napus L.) grown under different light-emitting diodes. *Photosynthetica 58*(3):836–845.

Saud S, Chen Y, Fahad S, Hussain S, Na L, Xin L, Alhussien SA (2016) Silicate application increases the photosynthesis and its associated metabolic activities in Kentucky bluegrass under drought stress and post-drought recovery. *Environ Sci Pollut Res 23* (17):17647–17655. https://doi.org/10.1007/s11356-016-6957-x.

Saud S, Chen Y, Long B, Fahad S, Sadiq A (2013) The different impact on the growth of cool season turf grass under the various conditions on salinity and drought stress. *Int J Agric Sci Res 3*: 77–84.

Saud S, Fahad S, Cui G, Chen Y, Anwar S (2020) Determining nitrogen isotopes discrimination under drought stress on enzymatic activities, nitrogen isotope abundance and water contents of Kentucky bluegrass. *Sci Rep 10*:6415. https://doi.org/10.1038/s41598-020-63548-w.

Saud S, Fahad S, Yajun C, Ihsan MZ, Hammad HM, Nasim W, Arif M and Alharby H (2017) Effects of Nitrogen Supply on Water Stress and Recovery Mechanisms in Kentucky Bluegrass Plants. *Front Plant Sci 8*:983. doi:10.3389/fpls.2017.00983.

Saud S, Li X, Chen Y, Zhang L, Fahad S, Hussain S, Sadiq A, Chen Y (2014) Silicon application increases drought tolerance of Kentucky bluegrass by improving plant water relations and morph physiological functions. *Sci World J 2014*:1–10. https://doi.org/10.1155/2014/ 368694.

Shafi MI, Adnan M, Fahad S, Fazli W, Ahsan K, Zhen Y, Subhan D, Zafar-ul-Hye M, Martin B, Rahul D (2020) Application of Single Superphosphate with Humic Acid Improves the Growth, Yield and Phosphorus Uptake of Wheat (*Triticum aestivum* L.) in Calcareous Soil. *Agron 10*: 1224. doi:10.3390/agronomy 10091224.

Shah F, Lixiao N, Kehui C, Tariq S, Wei W, Chang C, Liyang Z, Farhan A, Fahad S, Huang J (2013) Rice grain yield and component responses to near 2°C of warming. *Field Crop Res 157*:98–110.

Subhan D, Zafar-ul-Hye M, Fahad S, Saud S, Martin B, Tereza H, Rahul D (2020) Drought Stress Alleviation by ACC Deaminase Producing *Achromobacter xylosoxidans* and *Enterobacter cloacae*, with and without Timber Waste Biochar in Maize. *Sustain 12*(6286). doi:10.3390/su12156286.

Tanji KK (ed). (1990) Agricultural salinity assessment and management. ASCE manuals and reports on engineering practice No. 71. Am. Soc. Civil Eng. New York.

Tariq M, Ahmad S, Fahad S, Abbas G, Hussain S, Fatima Z, Nasim W, Mubeen M, ur Rehman MH, Khan MA, Adnan M. (2018). The impact of climate warming and crop management on phenology of sunflower-based cropping systems in Punjab, Pakistan. *Agri Forest Met 15*(256):270–282.

Wahid F, Fahad S, Subhan D, Adnan M, Zhen Y, Saud S, Manzer HS, Martin B, Tereza H, Rahul D (2020) Sustainable management with mycorrhizae and phosphate solubilizing bacteria for enhanced phosphorus uptake in calcareous soils. *Agri 10*(334). doi:10.3390/agriculture10080334.

Wajid N, Ashfaq A, Asad A, Muhammad T, Muhammad A, Muhammad S, Khawar J, Ghulam MS, Syeda RS, Hafiz MH, Muhammad IAR, Muhammad ZH, Muhammad Habib ur R, Veysel T, Fahad S, Suad S, Aziz K, Shahzad A (2017) Radiation efficiency and nitrogen fertilizer impacts on sunflower crop in contrasting environments of Punjab. *Pakistan Environ Sci Pollut Res 25*:1822–1836. https://doi.org/10.1007/s11356-017-0592-z.

Wu C, Kehui C, She T, Ganghua L, Shaohua W, Fahad S, Lixiao N, Jianliang H, Shaobing P, Yanfeng D (2020) Intensified pollination and fertilization ameliorate heat injury in rice (*Oryza sativa* L.) during the flowering stage. *Field Crops Res 252*: 107795.

Wu C, Tang S, Li G, Wang S, Fahad S, Ding Y (2019) Roles of phytohormone changes in the grain yield of rice plants exposed to heat: a review. *Peer J 7*:e7792. doi:10.7717/peerj.7792.

Yang Z, Zhang Z, Zhang T, Fahad S, Cui K, Nie L, Peng S, Huang J (2017) The effect of season-long temperature increases on rice cultivars grown in the central and southern regions of China. *Front Plant Sci 8*:1908. https://doi.org/10.3389/fpls.2017.01908.

Zafar-ul-Hye M, Muhammad T, ahzeeb-ul-Hassan, Muhammad A, Fahad S, Martin B, Tereza D, Rahul D, Subhan D (2020b) Potential role of compost mixed biochar with rhizobacteria in mitigating lead toxicity in spinach. *Scientific Rep 10*:12159. https://doi.org/10.1038/s41598-020-69183-9.

Zafar-ul-Hye M, Muhammad N, Subhan D, Fahad S, Rahul D, Mazhar A, Ashfaq AR, Martin B, Jiří H, Zahid HT, Muhammad N (2020a) Alleviation of cadmium adverse effects by improving nutrients uptake in bitter gourd through cadmium tolerant rhizobacteria. *Environ 7*(54). doi:10.3390/environments7080054.

Zahida Z, Hafiz FB, Zulfiqar AS, Ghulam MS, Fahad S, Muhammad RA, Hafiz MH, Wajid N, Muhammad S (2017) Effect of water management and silicon on germination, growth, phosphorus and arsenic uptake in rice. *Ecotoxicol Environ Saf 144*: 11–18.

5

The Application of Biochar for the Mitigation of Abiotic Stress-Induced Damage

Imran, Amanullah, Shah Khalid, and Muhammad Arif
The University of Agriculture, Pakistan

Shah Fahad
The University of Haripur, Pakistan

Abdel Rahman Mohammad Al-Tawaha
Al Hussein Bin Talal University, Jordan

CONTENTS

5.1 Introduction

Abiotic stress is the negative impact of non-living factors on the living organisms in a specific environment (Rolf et al., 2004; Khan et al., 2015; Akhtar et al., 2015a; Imran, 2018a,b,c; Imran and Khan, 2015c,d,e; Imran et al., 2015a,b,c). The non-living variable must influence the environment beyond its normal range of variation to have a significant adverse effect on the population performance or individual physiology of the organism (Ping et al., 2007; Akhtar et al., 2015b; Khan et al., 2016; Zhan et al., 2011). Abiotic stress is essentially unavoidable. Abiotic stress effects on plants are especially dependent, if not solely dependent, on environmental factors, and are thus particularly constraining. Abiotic stress is the most harmful factor concerning the growth and productivity of crops worldwide. Mittler (2006) stated that abiotic stressors are most harmful when they occur together, in combinations of abiotic stress factors. These findings were in relation with those of Roelofs et al. (2008), who revealed that salinity and drought stress combine reduced grain yield of maize.

Abiotic stress comes in many forms. The most common of the stressors are the easiest to identify, but there are many other, less recognizable abiotic stress factors which have a continuous effect on environments (Wang et al., 2007; Amonette and Joseph, 2009; Armor et al., 2005; Arshad et al., 2008; Ayaz et al., 2000; Babar et al., 2016a). The most basic abiotic stressors are high winds, extreme temperatures, drought, flood and other natural disasters, such as tornadoes and wildfires. Lesser-known stressors generally occur on a smaller scale. They include: poor edaphic conditions like rock content andpH levels, high radiation, compaction, contamination, and other, highly specific conditions like rapid rehydration during seed germination (Bohnert and Jensen, 1996; Bos et al., 2005; Bruinsma, 1963; Brussaard et al., 2007; Capell and Doerffling, 1993; Chimenti et al., 2006). Abiotic stress, as a natural part of every ecosystem, will affect plants in a variety of ways. Although these effects may be either beneficial or detrimental, the location of the area is crucial to determining the extent of the impact that abiotic stress will have. The higher the latitude of the area affected, the greater the impact of abiotic stress will be on that area. So, a taiga or boreal forest is at the mercy of whatever abiotic stress factors may come along, while tropical zones are much less susceptible to such stressors (Cramer et al., 2011; Babar et al., 2015a, 2016a; Badridze et al., 2009; Bates et al., 1973). A plant's first line of defense against abiotic stress is in its roots. If the soil holding the plant is healthy and biologically diverse, the plant will have a higher chance of surviving stressful conditions (Conde, 2011; Cramer et al., 2011; Imran, 2018d,e,f; Imran and Khan, 2015a, 2017; Duan et al., 2007; Farahbakhsh, 2012). The plant responses to stress are dependent on the tissue or organ affected by the stress. For example, transcriptional responses to stress are tissue- or cell-specific in roots and are quite different, depending on the stress involved (Maestre et al., 2007). Various publications report a generally positive effect of biochar soil amendment on field crops under stress condition. Major et al. reported that a single application of 20 t ha^{-1} biochar to a Colombian savanna soil resulted in an increase in maize yield by 28 to 140% as compared with the unamended control under the saline stress condition. Biochar act synergistically to improve crop performance under stress conditions. Specific mechanisms underlying the contribution of biochar to plant response are poorly understood (Flowers, 2004; Gasco and Paz-Ferreiro, 2012; Goncalves-Alvim & Fernandez, 2001; González-Villagra et al., 2018). Regional conditions, including climate, soil chemistry and soil conditions, all influence biochar agronomic benefits. Evidence is mounting that the presence of biochar in soil has significant effects on soil microorganisms (Lehmann et al., 2011; Imran et al., 2015a, 2016, 2017, 2018). In the majority of the studies assessed in that review, microbial biomass increased in biochar-amended soils which act a biocontrol agent and mitigate abiotic stress induce damages (Heans, 1984; Imran, 2018a; Imran and Khan, 2015c,d,e).

5.2 Biochar Amendments and Salinity Stress

One of the primary responses of biochar application to tolerate abiotic stress such as high salinity is the disruption of the Na+/K+ ratio in the cytoplasm of the plant cell (Jackson, 1962; Jagatheeswari and Ranganathan, 2012; Joseph and Jini, 2011; Kamara et al., 2014). High concentrations of Na+, for example, can decrease the capacity for the plant to take up water and also alter enzyme and transporter functions. Soil salinization, the accumulation of water-soluble salts to levels that negatively impact plant production, is a global phenomenon that affects approximately 831 million hectares of land (Martinez-Beltran and Manzur, 2005; Kammann et al., 2011; Khan et al., 2010; Imran et al., 2019; Imran et al., 2015; Lashari et al., 2013; Lehmann and Joseph, 2009). More specifically, the phenomenon is estimated to threaten around 19.5% of the world's irrigated agricultural land and 2.1% of the world's non-irrigated (dry-land) agricultural lands (Lei et al., 2011; Maestre et al., 2007; Martinez-Beltran and Manzur, 2005; Imran et al., 2015, 2017; Imran, 2017, 2018a). High soil salinity content can be harmful to plants because water-soluble salts can alter osmotic potential gradients and may consequently inhibit many cellular functions (Raghothama, 1999) For example, high soil salinity content can inhibit the process of photosynthesis by limiting a plant's water uptake; high levels of water-soluble salts in the soil can decrease the osmotic potential of the soil and consequently decrease the difference in water potential between the soil and the plant's roots, thereby limiting electron flow from H_2O to P680 in Photosystem II's reaction center (Rubio et al., 2001; Lehmann et al., 2011; Lei et al., 2011; Lu and Vonshak, 2002).

Biochar application has been proposed to increase tolerance to abiotic stresses in crop plants. Biochar application enhances enzyme activation and may increase the levels of stress-inducing chemicals in plants so that the plant begins preparing defence mechanisms (Paz-Ferreiro et al., 2014; Peng et al., 2008; Ping et al., 2007; Prado et al., 2000; Raghothama, 1999). Thus, when abiotic stress occurs, the plant has already prepared defense mechanisms that can be activated faster and increase tolerance. Over generations, many plants have mutated and built different mechanisms to counter salinity effects (Pant et al., 2015a; Rajalakshmi et al., 2015; Rashid et al., 2003; Imran, 2015a, 2018a; Iqbal et al., 2017). A good combatant of salinity in plants is the biochar amendment. Biochar is known for regulating plant growth and development and dealing with stress conditions (Palm & Van Volkenburgh, 2012; Palta and Farag, 2006; Pant et al., 2015a). With biochar application under stress conditions, plants have a range of very different adaptations, even from a plant living in the same area. When a group of different plant species was prompted by a variety of different stress signals, such as drought or cold, each plant responded in a unique fashion. Hardly any of the responses were similar, even though the plants had become accustomed to identical home environments (Neto et al., 2004; Rolf et al., 2004; Rubio et al., 2001; Savvides, 2015; Shahida et al., 2016; Shamsi and Kobraee, 2013; Shamsuddin et al., 2015; Shazma et al., 2016).

5.3 Biochar Amendments and Drought Stress

Drought stress is one of the main causes of crop loss in the agricultural world (Figure 5.1). This is due to the necessary role of water in so many fundamental processes in plant growth. It has become especially important in recent years to find a way to combat drought stress (Sidari et al., 2008; Karim and Imran, 2019; Tank and Saraf, 2010; Tiwari et al., 2011; Tombesi et al., 2018; Ulfat et al., 2016). In future, a decrease in precipitation and a subsequent increase in drought are extremely likely due to an increase in global warming (Zargar et al., 2017; Van Hoom et al., 2001; Wang et al., 2007; Wang and Xu, 2013; Wang and Jin, 2007; Wolfe, 2007). Plants have come up with many mechanisms and adaptations to try and deal with drought stress (Figure 5.1). One of the leading ways that plants combat drought stress is by increasing the rates of water infiltration and soil porosity (Xu et al., 2008; Yeo et al., 1990; Younis et al., 2015; Zakaria et al., 2012; Zargar et al., 2017). Biochar application can enhance soil porosity and increasing water infiltration in various types of soil. It has been reported that biochar application regulating hormone secretion maintains stomatal opening and closing. The hormone which is regulating by biochar application is abscisic acid (ABA). The synthesis of this hormone causes the ABA to bind to receptors. This binding then effects the opening of ion channels, thereby decreasing turgor pressure in the stomata and causing them to close. Recent studies showed that how ABA levels increased in drought-stressed plants (Van Hoom et al., 2001; Wang et al., 2007; Wang and Xu, 2013; Wang and Jin, 2007). They showed that when plants were placed in a stressful situation they produced more ABA to try and conserve any water they had in their leaves (Goncalves-Alvim and Fernandez, 2001; Rolf et al., 2004;

FIGURE 5.1 Biochar could mitigate drought stress and enhance the growth and productivity of soybean.

Rubio et al., 2001; Savvides, 2015; Shahida et al., 2016; Shamsi and Kobraee, 2013). Biochar application can also regulate aquaporins (AQPs), another extremely important factor dealing with drought stress and the uptake of water. AQPs are integral membrane proteins that make up channels. The main job of these channels is the transportation of water and other necessary solutes. AQPs are both transcriptionally and post-transcriptionally regulated by many different factors such as ABA, GA3, pH and Ca2+ and the specific levels of AQPs in certain parts of the plant, such as roots or leaves, helps to draw as much water into the plant as possible (Rubio et al., 2001; Savvides, 2015; Shahida et al., 2016; Shamsi and Kobraee, 2013; Shamsuddin et al., 2015).

5.4 Biochar Amendments and Soil Microbial Stress

Evidence is mounting that biochar in soil has significant effects on soil microorganisms, as reviewed recently by Lehmann et al. (2011). In the majority of the studies assessed in that review, the level of microbial biomass increased in biochar-amended soils (Palm and Van Volkenburgh, 2012; Palta and Farag, 2006; Pant et al., 2015b; Paz-Ferreiro et al., 2014; Peng et al., 2008; Ping et al., 2007; Prado et al., 2000; Raghothama, 1999). Biochar addition also caused significant changes in microbial community composition and enzyme activities in both bulk soil and the rhizosphere (Lehmann et al., 2011; Lei et al., 2011; Lu and Vonshak, 2002). For instance, biochar amendment was generally characterized by an increase in the relative abundance of members of the *Actinobacteria* and *Bacteriodetes* phyla. While little is understood regarding the mechanisms by which biochar affects microbial abundance and community structure, it is well known that soil microorganisms can have a tremendous impact on plant productivity. Therefore, biochar-induced changes in soil microorganisms may certainly play a role in 'The Biochar Effect' (Lehmann et al., 2011; Lei et al., 2011; Lu and Vonshak, 2002).

5.5 Biochar Amendments and Soil Fertility Stress (Nutrients Stress)

Currently, very little biochar is utilized in agriculture, in part because its agronomic value in terms of crop response and soil health benefits have yet to be quantified, and because the mechanisms by which it improves soil fertility are poorly understood (Bohnert and Jensen, 1996; Bos et al., 2005; Bruinsma, 1963; Brussaard et al., 2007; Capell and Doerffling, 1993; Chimenti et al., 2006). The positive effects of biochar on crop productivity under conditions of extensive agriculture are frequently attributed to the direct effects of biochar-supplied nutrients and to several other indirect effects, including increased water and nutrient retention, improvements in soil pH, increased soil cation exchange capacity, effects on P and S transformations and turnover, the neutralization of phytotoxic compounds in the soil, improved soil physical properties, the promotion of mycorrhizal fungi, and the alteration of soil microbial populations and functions (Izhar et al., 2019; Imran and Khan, 2015; Imran et al., 2014, 2015; Jackson, 1962; Jagatheeswari and Ranganathan, 2012; Joseph and Jini, 2011; Kamara et al., 2014). Yet the biochar effect is also evident under conditions of intensive production, where many of these parameters are unlimited (Gasco and Paz-Ferreiro, 2012; Goncalves-Alvim and Fernandez, 2001). Biochar addition to soil alters microbial populations in the rhizosphere, albeit via mechanisms not yet understood, and may cause a shift towards beneficial microorganism populations that promote plant growth and resistance to biotic stresses. In addition to some scant evidence for biochar-induced plant protection against soil-borne diseases, the induction of systemic resistance towards several foliar pathogens in three crop systems has been demonstrated (Khan et al., 2010; Imran et al., 2019; Imran et al., 2015; Lashari et al., 2013; Lehmann and Joseph, 2009). There are indications that biochar induces responses along both systemic-acquired resistance (SAR) and induced systemic resistance (ISR) pathways, resulting in a broad spectrum controlling capacity in the canopy (Lei et al., 2011; Lu and Vonshak, 2002). This review examines the effects of biochar soil amendment on the different soil–plant–microbe interactions that may play a role in plant health. The improvement of plant responses to disease can be among the benefits gained from applying biochar to soil (Paz-Ferreiro et al., 2014; Peng et al., 2008; Ping et al., 2007; Prado et al., 2000; Raghothama, 1999; Rajalakshmi et al., 2015; Rashid et al., 2003; Imran, 2015a, 2018a; Iqbal et al., 2017; Raza, 2012; Rengasamy, 2002; Roelofs et al., 2008).

5.6 Biochar Amendments and Plant Growth

Generally, biochar has a positive effect on field crops grown under greenhouse and field conditions. Early studies reported that charcoal added to soil increased the yield of moong, soybean and pea (Paz-Ferreiro et al., 2014; Peng et al., 2008; Ping et al., 2007; Prado et al., 2000; Raghothama, 1999; Rajalakshmi et al., 2015; Rashid et al., 2003; Imran, 2015a, 2018a; Iqbal et al., 2017; Raza, 2012; Rengasamy, 2002; Roelofs et al., 2008) and of soybean. Shoot and root biomass of birch and pine were greater in charcoal-amended soil (Rashid et al., 2003; Imran, 2015a, 2018a; Iqbal et al., 2017). Similarly, five years following the soil application of charcoal, biomass production of sugi trees (*Cryptomeria japonica*) was increased substantially (Sidari et al., 2008; Karim and Imran, 2019). A single application of 20 t ha^{-1} biochar to a Colombian savanna soil resulted in an increase in maize yield by between 28% and 140% as compared with the unamended control in the 2nd to 4th years after application (Peng et al., 2008; Ping et al., 2007; Prado et al., 2000). With the addition of biochar (at 90 g kg^{-1}) to a tropical, low-fertility Ferralsol, the proportion of N fixed by bean plants (*Phaseolus vulgaris*) increased from 50% (without biochar) to 72%, and biomass production and bean yield improved significantly (Van Hoorn et al., 2001; Wang et al., 2007; Wang and Xu, 2013; Wang and Jin, 2007; Wolfe, 2007; Xu et al., 2008; Yeo et al., 1990; Younis et al., 2015). On the same type of soil, total N recovery in soil, crop residues, and grains was significantly higher on compost (16.5%), charcoal (18.1%), and charcoal plus compost treatments (17.4%) in comparison to mineral-fertilized plots (10.9%). Biochar soil application resulted in higher upland rice (Oryza sativa) grain yields at sites in northern Laos with low P availability, and improved the response to N and NP chemical fertilizer treatments (Munns et al., 2006; Munns, 2002; Munns and Tester, 2008; Nazarli et al., 2011; Neto et al., 2004). Large volume applications of biochar (30 and 60 t ha^{-1}) in the Mediterranean basin increased durum wheat (Triticum durum) biomass and yield by up to 30%, an effect which was sustained over two consecutive seasons (Nazarli et al., 2011; Neto et al., 2004). Overall, these results demonstrate the potential of biochar application to improve plant productivity. The means by which biochar improves crop response can be attributed to direct effects via biochar-supplied nutrients (Munns et al., 2006; Munns, 2002; Munns and Tester, 2008), and to several other indirect effects, including: increased nutrient retention (Rajalakshmi et al., 2015; Rashid et al., 2003; Imran, 2015a, 2018a; Iqbal et al., 2017; Raza, 2012; Rengasamy, 2002; Roelofs et al., 2008); improvements in soil pH; increased soil cation exchange capacity (Rashid et al., 2003; Imran, 2015a, 2018a; Iqbal et al., 2017; Raza, 2012; Rengasamy, 2002; Roelofs et al., 2008); effects on P and S transformations and turnover (Imran, 2015a; Iqbal et al., 2017; Raza, 2012; Rengasamy, 2002; Roelofs et al., 2008); the neutralization of phytotoxic compounds in the soil (Rajalakshmi et al., 2015; Raza, 2012; Rengasamy, 2002; Roelofs et al., 2008); improved soil physical properties, including water retention (Rajalakshmi et al., 2015; Rashid et al., 2003; Imran, 2015a, 2018a; Iqbal et al., 2017; Raza, 2012; Rengasamy, 2002; Roelofs et al., 2008); the promotion of mycorrhizal fungi (Rajalakshmi et al., 2015; Rashid et al., 2003; Imran, 2015a, 2018a; Iqbal et al., 2017; Raza, 2012; Rengasamy, 2002; Roelofs et al., 2008); and the alteration of soil microbial populations and functions (Rubio et al., 2001; Shahida et al., 2016; Shamsi and Kobraee, 2013; Shamsuddin et al., 2015). Many of these effects are interrelated and may act synergistically to improve crop performance (Rubio et al., 2001; Savvides et al., 2015; Shahida et al., 2016; Shamsi and Kobraee, 2013; Shamsuddin et al., 2015). Specific mechanisms underlying the contribution of biochar to plant response are poorly understood. Regional conditions, including climate, soil chemistry and soil conditions, all influence biochar agronomic benefits. In addition, dissimilar biomass feedstocks and pyrolysis conditions create biochars with different physical and chemical properties (Savvides et al., 2015; Shahida et al., 2016; Shamsi and Kobraee, 2013; Shamsuddin et al., 2015), resulting in different impacts on plant response (Shahida et al., 2016; Shamsi and Kobraee, 2013; Shamsuddin et al., 2015). Given that the biochar–soil–plant–water–environment interaction is highly complex, it is difficult to isolate those factors which actually play an instrumental role in the 'Biochar Effect'. To reduce the number of potential factors involved, Savvides et al. (2015) tested whether biochar addition could impact plant growth when nutritional and soil physical aspects of biochar amendment were eliminated. This was achieved by examining the impact of a nutrient-poor, wood-derived biochar on tomato.

5.7 Biochar Application and Phosphate Starvation in Plants

Phosphorus (P) is an essential macronutrient required for plant growth and development, but most of the world's soil is deficient in this important plant nutrient (Palta and Farag, 2006; Pant et al., 2015b; Paz-Ferreiro et al., 2014; Peng et al., 2008; Ping et al., 2007; Prado et al., 2000). Plants can utilize P mainly in the form of soluble inorganic phosphate (Pi), but are subjected to abiotic stress of P-limitation when there is insufficient soluble PO_4 in the soil. Phosphorus forms insoluble complexes with Ca and Mg in alkaline soils and Al and Fe in acidic soils that make it unavailable for plant roots (Imran, 2015a, 2018a; Adnan et al., 2016; Iqbal et al., 2017; Raza, 2012; Adnan et al., 2016; Rengasamy, 2002; Roelofs et al., 2008; Rolf et al., 2004; Rubio et al., 2001; Savvides et al., 2015; Arshad et al., 2016a). When there is limited bioavailable P in the soil, plants show extensive abiotic stress phenotypes such as short primary roots and more lateral roots and root hairs to make more surface available for Pi absorption, the exudation of organic acids and phosphatase to release Pi from complex P-containing molecules and make it available for growing plants organs (Imran, 2015a, 2018a; Iqbal et al., 2017; Raza, 2012; Rengasamy, 2002; Roelofs et al., 2008; Rolf et al., 2004; Rubio et al., 2001; Savvides et al., 2015; Arshad et al., 2016b). It has been shown that biochar application is a master regulator of P-starvation response in plants (González-Villagra et al., 2018). Biochar-amended soil has been shown to regulate extensive secondary metabolites during phosphorus limitation stress (Bos et al., 2005; Bruinsma, 1963; Brussaard et al., 2007; Capell, and Doerffling, 1993).

5.8 Conclusion

Biochar is an activated carbon soil conditioner that can alleviate the negative impacts of salinity, drought, heat and cold, water-logging and heavy metal stresses. Drought and salt stress have a negative impact on soil fertility and plant growth. The application of biochar ameliorates the negative effects of drought and salt stress on plants. The biochar application increased plant growth, biomass, and yield under either drought or salt stress and also increased photosynthesis, nutrient uptake, and modified gas exchange characteristics in drought- and salt-stressed plants. Under drought stress, biochar increased the water-holding capacity of soil and improved the physical and biological properties of soils. Under salt stress, biochar decreased Na^+ uptake, while simultaneously increasing K^+ uptake by plants. Biochar-mediated increase in the salt tolerance of plants is primarily associated with an improvement in soil properties, improving plant water status, reducing Na^+ uptake, and increasing the uptake of minerals, and the regulation of stomatal conductance and phytohormones. This chapter highlights both the potential of biochar in alleviating abiotic stresses in plants and also the future prospects for the role of biochar under abiotic stresses in plants.

REFERENCES

Adnan M, Khan MA, Saleem N, Hussain Z, Arif M, Alam M, Ullah H (2016) Nitrogen depletion by weeds from organic and inorganic nitrogen sources in wheat crop. *Pak J Weed Sci Res 22*(1):103–110.

Akhtar SS, Andersen MN, Liu F (2015a) Biochar mitigates salinity stress in potato. *J Agr Crop Sci 201*:321–400.

Akhtar SS, Andersen MN, Liu F (2015b) Residual effects of biochar on improving growth, physiology and yield of wheat under salt stress. *Agric Water Manage 158*:61–68.

Amonette JE, Joseph S (2009) Characteristics of biochar: microchemical properties. In: Lehmann J, Joseph S (eds) *Biochar for environmental management: Science and technology*. Earthscan, London: Sterling, pp 33–52.

Armor NB, Hamed KB, Debez A, Grignon Crabbedly C (2005) Physiological and antioxidant responses of perennial halophyte *Crithmum maritimum* to salinity. *Plant Sci 4*:889–899.
Arshad M, Adnan M, Ahmed S, Khan AK, Ali I, Ali M, Khan MA (2016a) Integrated effect of phosphorus and zinc on wheat crop. *Am Eurasian J Agric Environ Sci 16*(3):455–459.
Arshad M, Adnan M, Ali A, Khan AK, Shahwar D (2016b) Integrated effect of phosphorous and zinc on wheat quality and soil properties. *Adv Environ Bio 10*(2):40–45.
Arshad M, Shaharoona B, Mahmood T (2008) Inoculation with *Pseudomonas* spp. containing ACC-deaminase partially eliminates the effects of drought stress on growth, yield, and ripening of pea (*Pisum sativum* L.). *Pedosphere 18*:611–620.
Ayaz FA, Kadioglu A, Turgut TR (2000) Water stress effects on the content of low molecular weight carbohydrates and phenolic acids in *Cienanthe setosa*. *Can J Plant Sci 80*:373–378.
Babar I, Ahmad B, Ullah I, Imran, Khan AA, Anwar S, Ali A, Shahzad K, Khan S (2016a) Effect of phosphorus, sulphur and different irrigation levels on phenological traits of Triticale. *Pure Appl Biol 5*(2):303–310. http://dx.doi.org/10.19045/bspab.2016.50039.
Babar I, Jan MT, Imran, Muhammad Z, Khan AA, Anwar S, Shahzad K (2015a) Growth and phenology of maize as affected by integrated management of compost and fertilizers nitrogen. *J Pure Appl Biol 4*(4).
Babar I, Jan MT, Inamullah, Imran, Khan AA, Muhammad Z, Anwar S, Khan S, Baqa S, Usman A (2015b) Integrated management of compost type and fertilizer-N in maize. *J Pure Appl Biol 4*(4).
Babar I, Jan MT, Muhammad Z, Khan AA, Anwar S, Imran, Shahzad K (2016b) Phenological traits of Maize influenced by integrated management of compost and fertilizer Nitrogen. *J Pure Appl Biol 5*(1):58–63.
Badridze G, Weidner A, Asch F, Borner A (2009) Variation in salt tolerance within Georgian wheat germplasm collection. *Genet Res Crop Evol 56*:1125–1130.
Bates LS, Waldren RP, Teare ID (1973) Rapid determination of free proline for water stress studies. *Plant Soil 39*:205–207.
Bohnert HJ, Jensen RG (1996) Strategies for engineering water stress tolerance in plants. *Trends Biotechnol 14*:89–97.
Bos C, Juillet B, Fouillet H, Turlan L, Dare S, Luengo C, Ntounda R, Benamouzig R, Gausseres N, Tome V, Gaudichon C (2005) Postprandial metabolic utilization of wheat protein in humans. *Am J Clin Nutr 81*:87–94.
Bruinsma J (1963) The quantitative analysis of chlorophyll a and b in plant extract. *Photochem Photobiol 2*:241–249.
Brussaard L, de Ruiter PC, Brown GG (2007) Soil biodiversity for agricultural sustainability. *Agric Ecosyst Environ 121*(3):233–244. doi:10.1016/j.agee.2006.12.013.
Capell B, Doerffling K (1993) Genotype specific difference in chilling tolerance of maize in relation to chilling induce changes in water status and abscisic acid accumulation. *Physiologia Plantarum 88*:638–646.
Chimenti CA, Marcantonio M, Hall JA (2006) Divergent selection for osmotic adjustment results in improved drought tolerance in maize (*Zea mays* L.) in both early growth and flowering phases. *Field Crops Res 95*:305–315.
Conde A (2011) Membrane transport, sensing and signaling in plant adaptation to environmental stress (PDF). *Plant Cell Physiol 52*(9):1583–1602. doi:10.1093/pcp/pcr107.
Cramer GR, Urano K, Delrot S, Pezzotti M, Shinozaki K (2011) Effects of abiotic stress on plants: a systems biology perspective. *BMC Plant Biol 11*:163. doi:10.1186/1471-2229-11-163.
Duan DY, Li WQ, Liu XJ, Ouyang HP, An P (2007) Seed germination and seedling growth of *Suaeda salsa* under salt stress. *Annales Botanici Fennici 44*:161–169.
Farahbakhsh H (2012) Germination and seedling growth in un-primed and primed seeds of fennel as affected by reduced water potential induced by NaCl. *Int Res J Appl Basic Sci 3*:737–744.
Flowers TJ (2004) Improving crop salt tolerance. *J Exp Bot 55*:307–319.
Gasco G, Paz-Ferreiro, Mendez A (2012) Thermal analysis of soil amended with sewage sludge and biochar from sewage sludge pyrolysis. *J Thermal Anal Calorimetry 108*:769–775.
Goncalves-Alvim SJ, Fernandez GW (2001) Biodiversity of galling insects: historical, community and habitat effects in four neotropical savannas. *Biodiver Conserv 10*:79–98. doi:10.1023/a:1016602213305.
González-Villagra J, Rodrigues-Salvador A, Nunes-Nesi A, Cohen JD, Reyes-Díaz MM (2018) Age-related mechanism and its relationship with secondary metabolism and abscisic acid in Aristotelia chilensis plants subjected to drought stress. *Plant Physiol Biochem 124*:136–145. doi:10.1016/j.plaphy.2018.01.010.

Heans DL (1984) Determination of total organic carbon in soils by an improved chromic acid digestion and spectrophotometric procedure. *Commun Soil Sci Plant Anal 15*:1191–1213.

Imran (2015a) Effect of germination on proximate composition of two maize cultivars. *J Bio Agric Healthcare*. ISSN 2224-3208 (Paper) ISSN 2225-093X (Online).

Imran (2015b) Influence of nitrogen levels and decapitation stress on biological potential of rapeseed (*Brassica Napus* L.) under water deficit condition of Swat-Pakistan. *J N Sci Res* ISSN 2224-3186 (Paper) ISSN 2225-0921 (Online).

Imran (2017) Climate change is a real fact confronting to agricultural productivity. *Int J Environ Sci Nat Res 3*(3):555613. doi:10.19080/IJESNR.2017.03.555613.

Imran (2018a) Climate change is threat toward agronomy (base of food, fiber system), and food security. *Food Nutr J*. doi:10.29011/2575-7091.100060.

Imran (2018b) Phosphorous fertilization influenced weeds attributes and phenological characteristics of mungbean cultivars (*vigna radiata* L.) *Russ Agric Sci 44*(3):229–234.

Imran (2018c) Physiological and morphological traits of agronomic crops influenced by climate change. *Mod Concep Dev Agrono* 1(4). MCDA.000524.

Imran A (2018d) Review. Global impact of climate change on water, soil resources and threat towards food security: evidence from Pakistan. *Adv Plants Agric Res 8*(5):350–355. doi:10.15406/apar.2018.08.00349.

Imran (2018e) Ecological environmental variability influence growth and yield potential of rice under Northern climatic scenario. *Russian Agric Sci 44*(1):18–24.

Imran (2018f) Organic matter amendments improve soil health, productivity and profitability of maize and soybean. *Ann Rev Resear 1*(3):555564.

Imran AB, Ali R, Ahmad N, Ahmad Z, Khattak MI, Ali A, Ahmad F, Khan I, Naveed S (2017b) Traditional rice farming accelerate CH_4 & N_2O emissions functioning as a stronger contributors of climate change. *Int J Environ Sci Nat Res*. doi:10.19080/ARTOAJ.2017.09.555765.

Imran A, Bari A, Ali R (2018) Peach sources, phosphorous and beneficial microbes enhance productivity of soybean. *Soy Res 16*(2):39–48.

Imran, Jamal N, Alam A, Khan AA (2017a) Grain yield, yield attributes of wheat and soil physio-chemical characteristics influenced by biochar, compost and inorganic fertilizer application. *Agric Res Tech*. Open Access. doi:10.19080/ARTOAJ.2017.10.555795.

Imran, Khan AA (2015a) Grain yield and phenology of maize cultivars influenced by various phosphorus sources. *Food Sci Qual Manage*. ISSN 2224-6088 (Paper) ISSN 2225-0557.

Imran, Khan AA (2015c) Biochar application and shoot cutting duration (days) influenced growth, yield and yield contributing parameters of *Brassica Napus* L. *J Bio Agric Healthcare*. ISSN 2224-3208 (Paper) ISSN 2225-093X (Online).

Imran, Khan AA (2015b) Influence of compost application and seed rates on production potential of late sown maize on high elevation in Swat-Pakistan. *J Environ Eng Sci*. ISSN 2224-3216 (Paper) ISSN 2225-0948.

Imran, Khan AA (2015d) Phenological characteristics of *Brassica Napus* L. as influenced by biochar application and shoot cutting duration (days). *Civil Environ Res*. ISSN 2224-5790 (Paper) ISSN 2225-0514 (Online).

Imran, Khan AA (2015e) Effect of transplanting dates on yield and yield components of various rice genotypes in hilly area cold climatic region of Khyber Pakhtunlhwa-Pakistan. *J Bio Agric Healthcare*. ISSN 2224-3208 (paper) ISSN 2225-093x (online).

Imran, Khan AA (2017) Canola yield and quality enhanced with sulphur fertilization Russ. *Agricult Sci 43*:113. doi:10.3103/S1068367417020100.

Imran, Khan AA, Ahmad F (2015a) Phenology of Various Rice Genotypes as Affected by Different Transplanting Dates under Cold Climatic Region of Khyber Pakhtunkhwa-Pakistan. *J Environ Eng Sci*. ISSN 2224-3216 (Paper) ISSN 2225-0948.

Imran, Khan AA, Ahmad F, Irfanullah (2015b) Influence of hydrated calcium sulphate ($CaSO_4.2H_2O$) and nitrogen levels on water infiltration rate and maize varieties productivity in rainfed area of Swat, Pakistan. *Chem Environ Res*. ISSN 2224-3224 (Print) ISSN 2225-0956 (Online).

Imran, Khan AA, Ahmad F, Irfanullah (2015c) Nitrogen levels, tillage practices and irrigation timing influenced yield, yield components and oil contents of Canola. *Civ Environ Res*. ISSN 2224-5790 (Paper) ISSN 2225-0514 (Online).

Imran, Khan AA, Akhtar K, Zaheer S, Faisal S, Ali S (2015d) Rice seedling characteristics of various genotypes influenced by different sowing dates in Swat-Pakistan. *J Environ Eng Sci*. ISSN 2224-3216 (Paper) ISSN 2225-0948.

Imran A, Khan AA, Bari A (2019) Production statistics and modern technology of maize cultivation in Khyber Pakhtunkhwa Pakistan. *Plant Sci Arc 2*(2):1–12.

Imran, Khan AA, Inam I, Ahmad F (2016c) Yield and yield attributes of Mungbean (*Vigna radiata* L.) cultivars as affected by phosphorous levels under different tillage systems. *Cogent Food Agric 2*:1151982.

Imran, Khan AA, Inamullah, Luqman (2015e) Weeding stages and their effect on yield and yield components of rice in upper Swat, Pakistan. *Pak J Weed Sci Res 21*(4):555–563.

Imran, Khan AA, Irfanullah, Ahmad F (2014) Production potential of rapeseed (*brassica napus* L.) As influenced by different nitrogen levels and decapitation stress under the rainfed agro-climatic condition of swat Pakistan. *J Glob Innov Agric Soc Sci*. ISSN (Online): 2311-3839; ISSN (Print): 2312-5225.

Imran, Khan AA, Khan IU, Naveed S (2016a) Weeds density and late sown maize productivity influenced by compost application and seed rates under temperate environment. *Pak J Weed Sci Res 22*(1):169–181.

Imran, Khan AA, Ullah I, Zada H, Ahmad F, Shah ST, Usman A, Ullah I (2015f) Yield and yield attributes of rapeseed cultivars as influence by sulfur level under swat valley conditions. *J Pure Appl Biol 4*(3).

Imran, Khan AA, Zada H, Ahmad F, Irfanullah (2015g) Graine yield and yield components of wheat cultivar "Siran 2010" as affected by phosphorous levels under rain fed condition. *J N Sci Res*. ISSN 2224-3186 (Paper) ISSN 2225-0921 (Online).

Imran, Khattak I, Hussain I, Rehman A, Anwar S, Ahmad F, Khan AA, Zada H (2015h) Growth and yield of maize hybrids as effected by different sowing dates in Swat Pakistan. *J Pure Appl Biol 4*(4).

Imran, Hussain I, Khattak I, Rehman A, Ahmad F, Shah ST (2015i) Roots nodulation, yield and yield contributing parameters of mungbean cultivars as influenced by different phosphorous level in swat-Pakistan. *J Pure Appl Biol 4*(4).

Imran, Maula F, Uzair M, Zada H (2015j) Farmers income enhancement through off-season vegetables production under natural environment in Swat-Pakistan. *J Environ Eng Sci*. ISSN 2224-3216 (Paper) ISSN 2225-0948.

Imran, Naveed S, Khan AA, Khattak I (2015k) Impact of phosphorus levels and seed rates on growth and yield of late sown maize on high elevation in Swat, Pakistan. *Pakistan J Agric Res 28*(4):406–413.

Imran, Uzair M, Maula F, Vacirca M, Farfaglia S, Khan MN (2015l) Introduction and promotion of off-season vegetables production under natural environment in hilly area of upper swat-Pakistan. *J Biol Agric Healthcare 5*(11).

Imran, Zada H, Naveed S, Khattak I, Ahmad S (2016b) Variable rates of phosphorous application influenced phenological traits of green gram (*Vignaradiata* L.). *J Agric Res 1*(3).

Iqbal A, Amanullah, Ali A, Iqbal M, Ikramullah, Imran (2017) Integrated use of phosphorus and organic matter improve fodder yield of moth bean *Vigna aconitifolia* (Jacq.) under irrigated and dryland conditions of Pakistan. *Journal of AgriSearch 4*(1):10–15.

Islam M, Anwar S, Bashir S, Khattak WA, Imran, Ali M, Khan N (2015) Growth and Yield Components of Wheat Varieties As Affected by Dual Purpose Practices. *J Pure Applied Biology 4*(4): 491–496.

Izhar A, Khan AA, Imran, Inamullah, Khan A (2019) Humic acids and nitrogen levels optimizing productivity of green gram (*vigna radiate* L.). *Russian Agric Sci 45*(1):43–47.

Jackson M (1962) *Soil chemical analysis*. Constable and Company Limited, London, p. 496.

Jagatheeswari D, Ranganathan P (2012) Influence of Mercuric Chloride on seeds germination, seedling growth and biochemical analysis of green gram (*Vigna radiata* (L.) Wilczek. Var. Vamban-3). *Int J Pharma Biol Arch 3*:291–295.

Joseph B, Jini D (2011) Development of salt stress-tolerant plants by gene manipulation of antioxidant enzymes. *Asian J Agric Res 5*:17–27.

Kamara A, Kamara A, Mansaray MM, Sawyerr PA (2014) Effects of biochar derived from maize stover and rice straw on the germination of their seeds. *Am J Agric Forestry 2*:246–249.

Kammann CI, Linsel S, Gobling JW, Koyro H (2011) Influence of biochar on drought tolerance of *Chenopodium quinoa* wild and on soil-plant relation. *Plant Soil 345*:195–210.

Karim SA, Imran (2019) Reduction in substrate moisture content reduce final yield of wheat. *Biomed J Sci Tec Res* 1–5.

Khan AZ, Imran, Muhammad A, Khalil A, Gul H, Akbar H, Wahab S (2016) Impact of fertilizer priming on seed germination behavior and vigor of maize. *Pure Appl Biol 5*(4):744–751. http://dx.doi.org/10.19045/bspab.2016.50093.

Khan MA, Islam E, Shirazi MU, Mumtaz S, Mujtaba SM, Khan MA, Shereen A, Ashrafand MY, Kaleri GM (2010) Physiological responses of various wheat genotypes to salinity. *Pakistan J Bot 42*:3497–3505.

Khan AA, Khan MN, Shah S, Arshad IUR, Muhammad I, Zeb A (2015) Effect of potash application on growth, yield and yield components of spring maize hybrids. *Pure Appl Biol 4*(2):195.

Lashari MS, Liu Y, Li W, Pan J, Pan G, Zheng J, Zheng G, Zhang X, Yu X (2013) Effect of amendment of biochar manure compost in conjunction with pyroligneous solution on soil quality and wheat yield of a salt stressed cropland from central China Great Plain. *Food Crop Res 144*:113–118.

Lehmann J, Joseph S (2009) Biochar for environmental management: an introduction. In: Lehmann J, Joseph S (eds) *Biochar for environmental management: Science and technology*. Earthscan, London, pp 1–12.

Lehmann J, Rillig MC, Thies J, Masiello CA, Hockaday WC, Crowley D (2011) Biochar effects on soil biota. *Soil Biol Biochem 43*:1812–1836.

Lei G, Shen M, Li Z-G, Zhang B, Duan K-X, Wang N, Cao Y-R, Zhang W-K, Ma B (2011) EIN2 regulates salt stress response and interacts with a MA3 domain-containing protein ECIP1 in Arabidopsis. *Plant, Cell Environ 34*(10):1678–1692. doi:10.1111/j.13653040.2011.02363.x.

Lu C, Vonshak A (2002) Effects of salinity stress on photosystem II function in cyanobacterial Spirulina platensis cells. *Physiol Plant 114*:405–413.

Maestre FT, Cortina J, Bautista S (2007) Mechanisms underlying the interaction between *Pinus halepensis* and the native late-successional shrub *Pistacia lentiscus* in a semi-arid plantation. *Ecography 27*(6):776–786. doi:10.1111/j.0906-7590.2004.03990.x.

Martinez-Beltran J, Manzur CL (2005) Overview of salinity problems in the world and FAO strategies to address the problem. Proceedings of the international salinity forum, Riverside, CA, pp 311–313.

Mittler R (2006) Abiotic stress, the field environment and stress combination. *Trends Plant Sci 11*(1):15–19. doi:10.1016/j.tplants.2005.11.002.

Munns R (2002) Comparative physiology of salt and water stress. *Plant Cell Environ 25*:239–250.

Munns RR, James A, Lauchli A (2006) Approaches to increasing the salt tolerance of wheat and other cereals. *J Exp Bot 57*:1025–1043.

Munns R, Tester M (2008) Mechanisms of salinity tolerance. *Ann Rev Plant Biol 59*:651–681.

Nazarli H, Faraji F, Zardashti MR (2011) Effect of drought stress and polymer on osmotic adjustment and photosynthetic pigments of sunflower. *Carcetriari Agronomica in Maldova 1*:45.

Neto A, De AD, Prisco JT, Enéas-Filho J, de Lacerda CF, Silva JV, da Costa PHA, Gomes-Filho E (2004). Effects of salt stress on plant growth, stomatal response and solute accumulation of different maize genotypes. *Brazilian J Plant Physiol 16*(1):31–38. doi:10.1590/S1677-04202004000100005.

Palm B, Van Volkenburgh E (2012) Serpentine tolerance in *Mimuslus guttatus* does not rely on exclusion of magnesium. *Funct Plant Biol 39*(8): 679–688. doi: 10.1071/FP12059.

Palta JP, Farag K (2006) Methods for enhancing plant health, protecting plants from biotic and abiotic stress related injuries and enhancing the recovery of plants injured as a result of such stresses. United States Patent 7101828.

Pant BD, Burgos A, Pant P, Cuadros-Inostroza A, Willmitzer L, Scheible W-R (2015a) The transcription factor PHR1 regulates lipid remodeling and triacylglycerol accumulation in Arabidopsis thaliana during phosphorus starvation. *J Exp Bot 66*(7):1907–1918. doi:10.1093/jxb/eru535.

Pant BD, Pant P, Erban A, Huhman D, Kopka J, Scheible W-R (2015b) Identification of primary and secondary metabolites with phosphorus status-dependent abundance in Arabidopsis, and of the transcription factor PHR1 as a major regulator of metabolic changes during phosphorus limitation. *Plant Cell Environ 38*(1):172–187. doi:10.1111/pce.12378.

Paz-Ferreiro J, Fu S, Mendez A, Gasco G (2014) Interactive effects of biochar and the earthworm *Pontoscolex corethrurus* on plant productivity and soil enzymes activities. *J Soils Sediments 14*:483–494.

Peng Y, Gao Z, Gao Y, Liu G, Shengand L, Wang D (2008) Ecophysiological characteristics of alfalfa seedlings in response to various mixed salt-alkaline stresses. *J Integr Plant Biol 50*:29–39.

Ping J, Geo K, Ron M (2007) Understanding abiotic stress tolerance mechanisms: recent studies on stress response in rice. *J Integr Plant Biol 49*(6):742–750. doi:10.1111/j.1744-7909.2007.00495.x.

Prado FE, Boero C, Gallarodo M, Gonzalez AJ (2000) Effect of NaCl on germination, growth and soluble sugar content in *Chenopodium quinoa* will seeds. *Bot Bull – Academia Sinica Taipei 14*:27–34.

Raghothama KG (1999) Phosphate acquisition. *Ann Rev Plant Physiol Plant Mol Biol 50*(1):665–693. doi:10.1146/annurev.arplant.50.1.665.

Rajalakshmi AG, Kumar SK, Bharathi CD, Karthika R, Visalakshi KD, Meera R, Mohanapriya S (2015) Effect of biochar in seed germination: *In-vitro* study. *Int J Biosci Nanosci 2*:132–136.

Rashid A, Saleem Q, Nazir A, Kazım HS (2003) Yield potential and stability of nine wheat varieties under water stress conditions. *Int J Agric Biol 5*:7–9.

Raza A (2012) Grain and field annual 2012 gain report. Global Agricultural Information Network. *USDA Foreign Agricultural Services*, 1–3.

Rengasamy P (2002) Transient salinity and subsoil constraints to dryland farming in Australian sodic soils: an overview. *Austr J Exp Agric 42*:351–361.

Roelofs D, Jiwan P, Kwangy I (2008) Functional ecological genomics to demonstrate general and specific responses to abiotic stress. *Func Ecol 22*:8–18. doi:10.1111/j.1365-2435.2007.01312.x.

Rolf D, Vinebrooke K, Mittler J (2004) Impacts of multiple stressors on biodiversity and ecosystem functioning: the role of species co-tolerance. *Oikos 104*(3):451–457. doi:10.1111/j.0030-1299.2004.13255.x.

Rubio V, Linhares F, Solano R, Martín AC, Iglesias J, Leyva A, Paz-Ares J (2001) A conserved MYB transcription factor involved in phosphate starvation signaling both in vascular plants and in unicellular algae. *Genes Dev 15*(16):2122–2133. doi:10.1101/gad.204401.

Savvides A (2015) Chemical priming of plants against multiple abiotic stresses: mission possible? *Trends Plant Sci 21*(4):329–340. doi:10.1016/j.tplants.2015.11.003.

Shahida N, Ibrar M, Khattak I, Khan I, Imran, Khan MI, Khan H (2016) Anthelmintic, antilice, insecticidal, cytotoxic and phytotoxic potential of ethanolic extracts of two wild medicinal plants *Iphiona Grantioides* and *Plucheaarguta*. *J Woulfenia 11*(23):13–25.

Shamsi K, Kobraee S (2013) Biochemical and physiological responses of three wheat cultivars (*Triticum aestivum* L.) to salinity stress. *Ann Biol Res 4*:180–185.

Shamsuddin B, Khan AZ, Inamullah, Imran, Khan AA, Anwar S, Iqbal B, Khan S, Usman A (2015) Influence of farm yard manure and phosphorus application on yield and yield components of wheat. *J Pure Appl Biol 4*(4).

Shazma A, Israeel BI, Khan AA, Imran, Shah WA, Islam M, Khattak WA, Ikramullah, Akram W, Abbas W (2016) Nitrogen and phosphorus fertilization of improved varieties for enhancing phenological traits of wheat. *Pure Appl Biol 5*(3):511–519. http://dx.doi.org/10.19045/bspab.2016.50065.

Sidari M, Mallamaci C, Muscolo A (2008) Drought, salinity and heat differently affect seed germination of *Pinus pinea*. *J Forest Res 13*:326–330.

Tank N, Saraf M. 2010. Salinity-resistant plant growth promoting rhizobacteria ameliorates sodium chloride stress on tomato plants. *J Plant Inter 5*:51–58.

Tiwari S, Singh P, Tiwari R, Meena KK, Yandigeri M, Singh DP, Arora DK (2011) Salt-tolerant rhizobacteria-mediated induced tolerance in wheat (*Triticum aestivum*) and chemical diversity in rhizosphere enhance plant growth. *Biol Fertile Soils 47*:907–916.

Tombesi S, Frioni T, Poni S, Palliotti A (2018) Effect of water stress "memory" on plant behavior during subsequent drought stress. *Environ Exp Bot 150*:106–114. doi:10.1016/j.envexpbot.2018.03.009.

Ulfat S, Muhammad Ibrar L, Naveed S, Imran, Inayat K (2016) Ethnobotanical study of subtropical hills of Darazinda, Takht-e-Suleman range F.R D.I. Khan, Pakistan. *Pure Appl Biol 5*(1):149–164. http://dx.doi.org/10.19045/bspab.2016.50020.

Van Hoorn JW, Nader K, Hamdy A, Mastrorilli M (2001) Effect of salinity on yield and nitrogen uptake of four grain legumes and on biological nitrogen contribution from the soil. *Agric Water Manage 51*:87–98.

Wang H, Jin J (2007) Effects of zinc deficiency and drought on plant growth and metabolism of reactive oxygen species in maize (*Zea mays* L.). *Agric Sci China 6*:988–995.

Wang W, Vinocur B, Altman A (2007) Plant responses to drought, salinity and extreme temperatures towards genetic engineering for stress tolerance. *Planta 218*(1):1–14. doi:10.1007/s00425-003-1105-5.

Wang G, Xu Z (2013) The effects of biochar on germination and growth of wheat in different saline-alkali soil. *Asian Agric Res 5*:116–119.

Wolfe A (2007) "Patterns of biodiversity." Ohio State University, doi:10.1016/j.agee.2006.12.013.

Xu D, Huang J, Guo S, Yang X, Bao Y, Tang H, Zhang H (2008) Over-expression of a TFIII-type zinc finger protein gene ZFP252 enhances drought and salt tolerance in rice (*Oryza sativa* L.). *Feder Eur Biochem Soc Lett 582*:1037–1043.

Yeo AR, Yeo ME, Flowers SA, Flowers TJ (1990) Screening of rice (*Oryza sativa* L.) genotypes for physiological characters contributing to salinity resistance, and their relationship to overall performance. *Theor Appl Gen 79*:377–384.

Younis U, Athar M, Malik SA, RazaShah MH, Mahmood S (2015) Biochar impact on physiological and biochemical attributes of spinach *Spinacia oleracea* (L.) in nickel contaminated soil. *Global J Environ Sci Manage 1*:245–254.

Zakaria M, Daniel S, Murphy V, Abbott LK (2012). Biochars influence seed germination and early growth of seedlings. *Plant Soil 353*:273–287.

Zargar SM, Nagar P, Deshmukh R, Nazir M, Wani AA, Masoodi KZ, Agrawal GK, Rakwal R (2017) Aquaporins as potential drought tolerance inducing proteins: Towards instigating stress tolerance. *J Proteomics 169*:233–238. doi:10.1016/j.jprot.2017.04.010.

Zhan HX, Chang ZJ, Wei AL, Zhang XJ, Li X (2011) Impact of drought to wheat physiological index. *J Shanxi Agric Sci 39*:1049–1051.

6

Heavy Metals Stress and Plants Defense Responses

Adnan Rasheed
Jiangxi Agriculture University, PR China

Muhammad Umair Hassan
Jiangxi Agricultural University, China

Shah Fahad
The University of Haripur, Pakistan

Muhammad Aamer
Jiangxi Agricultural University, China

Maria Batool
Huazhong Agricultural University, China

Muhammad Ilyas
University of Poonch, Pakistan

Fang Shang and Ziming Wu
Jiangxi Agriculture University, PR China

Huijie Li
Jiangxi Agriculture University, PR China; Jiangxi Agricultural University, China

CONTENTS

6.1 Introduction

Heavy metals stress is a serious constraint for crop production and a continuous increase in their concentration owing to human activities, which also poses a serious threat to humans (Rasheed et al. 2020a, 2020b). World population is estimated to increase up to 9.1 billion up to the year 2050; thus, it is a big challenge to satisfy global hunger through sustainable agricultural production (Arif et al. 2016). The anthropogenic activities such as industrialization, mining, intensive agriculture due to rapidly increasing population and urbanization not only reduce the available resources on Earth but also cause the adulteration of key components of plants (Hassan et al. 2019). Human-induced disturbances in biogeochemical cycles and the rapid accumulation of HMs have key significance regarding nutritional, ecological, and environmental issues (Mukta et al. 2019). Thus, the management of HMs should be a primary concern in maintaining the stability of the environment, which is linked to the health of organisms. HMs are non-biodegradable inorganic chemicals that have atomic mass over 20, and their mass is larger than 5 $g \cdot cm^{-3}$, which induces mutational changes in humans and plants and contaminating water resources, food chain, irrigation, and surrounding atmosphere (Chen et al. 2019; Ciriaková 2009; Flora et al. 2008; Rascio and Navari-Izzo 2011). Metals are usually categorized into two groups, such as essential metals, Zn, Cu, Fe, Ni, and Mo (Shanmugam et al. 2011), and the second category is non-essential metals, Cr, Cd, Sb, Pb, Co, As Ag, Hg and Se, which are not compulsory for plants for their normal growth as well as development (Aamer et al. 2018; Schutzendubel and Polle 2002; Tangahu et al. 2011).

Plants take up HMs occurs both underground, via absorption through the root membrane, and the above-ground parts through the phloem to the aerial parts (Patra et al. 2004; Tao et al. 2018). The synthesis of enzymes, proteins, growth, development, and several metabolic processes are carried out by these HMs, and, therefore, their quantity is one of the main concerns, as an excess can induce a reduction in plant growth and cause other deleterious changes (Zengin and Munzuroglu 2005). At toxic levels, HMs hinder normal growth and plant functioning in many different ways, including the displacement of amino acids, which emerges from the creation of bonds of HMs with sulfhydryl groups (Hall 2002), disturbing the function of important metals in biomolecules like enzymes (Ali et al. 2013), affecting cytoplasmic membrane integrity (Farid et al. 2013), which leads to a decline in essential processes or events like respiration, photosynthesis and several enzymatic functions. The HMs also lead to the creation of ROS, like hydroxyl free radicles ($OH\cdot^-$), superoxide free radicles ($O_2\cdot^-$) or hydrogen peroxide ($H_2\,0_2$), which induce oxidative stress via the disturbance of the equilibrium among antioxidant and pro-oxidant homeostasis in plant cells (Hossain et al. 2012; Sytar et al. 2013; Zengin and Munzuroglu 2005) and this redox status is one of the main reasons for HMs' toxicity (Tiwari and Lata 2018).

Oxidative stress leads to multiple disorders, including the oxidation of DNA protein and lipids and a denaturing of the cell membrane and cell structure, which eventually leads to programmed cell death (Flora et al. 2008; Rascio and Navari-Izzo 2011; Sharma et al. 2012). The plants use a variety of defense strategies, both natural and artificial, to detoxify the HMs. The plants reduce the uptake and bind the HM with Phyto-chelation and activate many antioxidants to detoxify the HMs (Shahid et al. 2015). The first step towards plants' response to heavy metals stress is an avoidance mechanism, which includes preventing the entry of HMs, via excluding it from uptake through roots (Dalvi and Bhalerao 2013; Viehweger 2014). This process is activated by immobilization via mycorrhizal association, metals sequestration, and the release of organic acids (Dalvi and Bhalerao 2013). The second step towards metals detoxification is only adopted by plants when the first step fails, and metals enter into the cell. This step includes metals sequestration (Dalvi and Bhalerao 2013) and compartmentalization into vacuoles (Patra et al. 2004), and the tying of metals ions into the cell wall, and the chelation of metal ions by the release of organic acids, like ascorbic acid which play a key role in converting metals into non-available forms (John et al. 2009; Prasad 2004). In the end, if all defense techniques fail and plant become overloaded with HMs then the activation of antioxidant defense response is continued (Manara 2012a). A large number of antioxidants

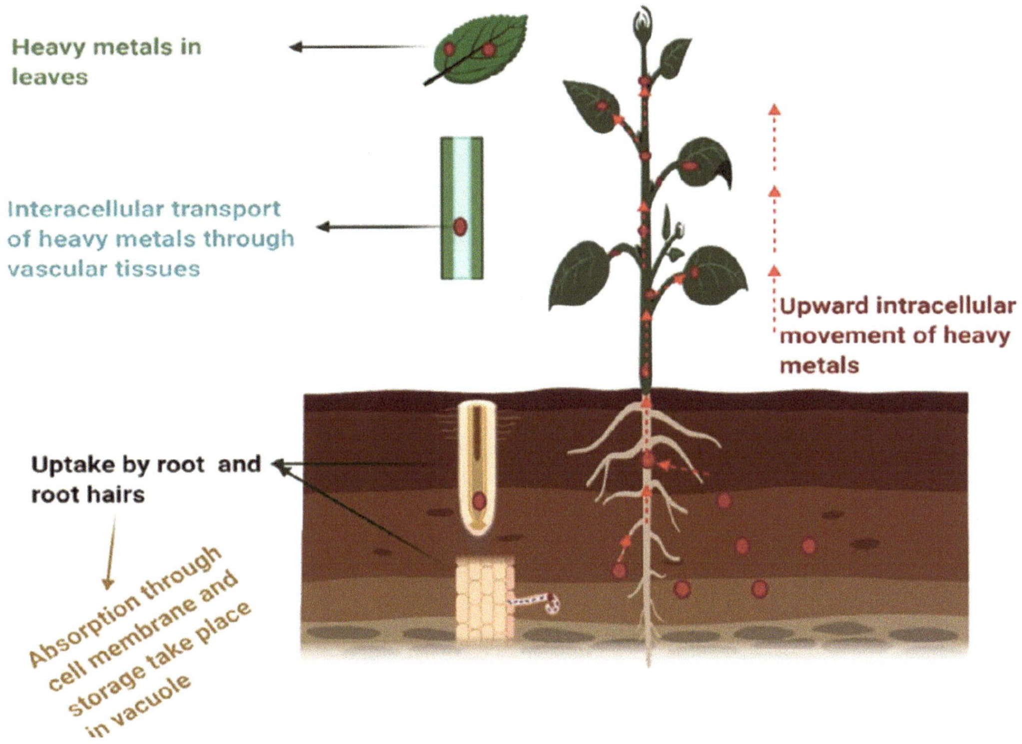

FIGURE 6.1 Graphical representation of absorption as well as movement of HMs in plants.

comprising superoxide dismutase, peroxidase as well as catalase, which may competently change the superoxide radicals into hydrogen peroxide and, consequently, water and oxygen whereas low molecular weight non-enzymatic antioxidants containing Proline directly reclaim the ROS (Singh et al. 2015b).

This chapter deals with a brief study of past trends and ongoing developments regarding research on HMs' stress and tolerance in plants, and highlights the role of numerous defense mechanisms adopted by plants to counter HMs' stress. Moreover, we also focused on the role of PCs, Proline, HTs, AMF, anti-oxidant defense systems, amino acids, miRNA and salicylic acid (SA) to improve HMs tolerance in plants. Additionally, some genes have a key role in enhancing plant tolerance to HMs and important suggestion to improve the HMs tolerance also discussed. Absorption as well as translocation of HMs in plant in shown in Figure 6.1.

6.2 Toxic Effects of HMs Stress on Plants

Bioactive metals are generally classified into two main groups on the basis of their physicochemical features. The first group consists of redox-active metals like iron, manganese, chromium, copper and the second group consists of non-redox metals like aluminum, cadmium, nickel, mercury and zinc (Engwa et al. 2019). Redox metals cause direct injury to plants by Harberwesis and Fenton reactions which lead to the production of ROS, resulting in a disruption of homeostasis processes, the breaking of the double helix strand, proteins, and damage to photosynthesis and ultimately triggered cell death (Flora et al. 2008; Schutzendubel and Polle 2002). On the other hand, non-redox metals have different types of damaging mechanisms to plants which includes attachment to the sulphydryl group of protein and the depletion of glutathione (Valko et al. 2005). The main criteria for choosing HMs for their study depends on the way in which they work in plants, since they are either redox or non-redox active metals. On the basis of this criteria, the effects of three metals, Mn, Cu and Cr, which are redox active metals and three metals, Al, Zn and Ni, which are non-redox metals, are described in detail in this chapter. Heavy metals stress in plants, and their effects, are shown in Figure 6.2.

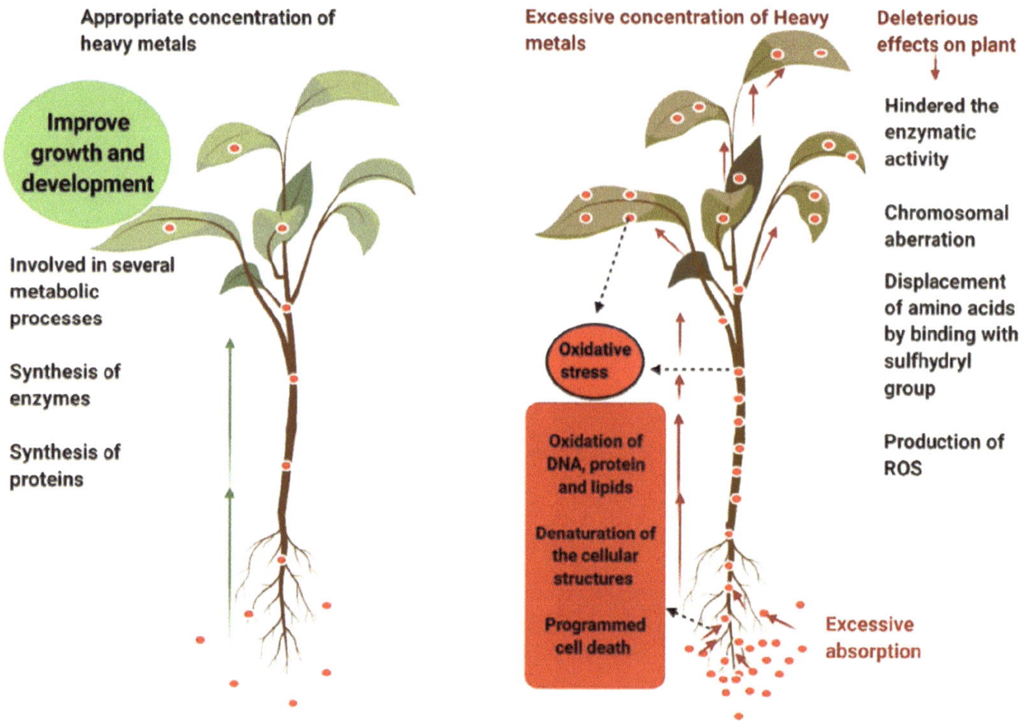

FIGURE 6.2 Heavy metals effects on plants at various levels.

6.3 Effects of Redox Active HMs

6.3.1 Chromium (Cr) Toxicity Stress in Plants

Cr is the most common metal which is adding to soil (Sharma et al. 2020a), groundwater and sediments owing to the industrial sector. Therefore, it poses severe challenges to global environmental, food production systems, plants, animals and humans beings (Anjum et al. 2017; Farid et al. 2013). In plants, Cr usually occurs in its trivalent form Cr^{3+} and in its less toxic (Sharma et al. 2020a) and hexavalent form Cr^{6+} of Cr (Sihag and Joshi 2018). Cr has many forms, but Cr (II) is more toxic to plants (Mukta et al. 2019). Cr is toxic to plants because of its great oxidizing and solubility power, which increases its accessibility to plants and it severely reduced the growth of plant growth as well as production (Ranieri et al. 2013). Cr uptake in plants occurs through the uptake of water as well as nutrients from soil by plant roots. Thus, it causes drastic alterations in several physiological and biochemical processes of plants (Ranieri and Gikas 2014). The variable factors, including the soil pH, soluble salts, salinity and concentration of Cr, affect Cr uptake (Anjum et al. 2017). In plants, Cr-mediated oxidative stress leads to the production of ROS (Anjum et al. 2011) at higher concentrations, which causes oxidative damage and higher ROS leads to noticeable alterations in cell structure, the oxidation of lipids and proteins, the inhibition of important enzymatic functions, damage to DNA and RNA and, ultimately, cell death (Adrees et al. 2015a; Anjum et al. 2017). The toxic effects of Cr regarding germination, growth, development, photosynthesis and yield are given below.

Seed germination is a major physiological process, which is affected by Cr toxicity. The treatment with 200 $\mu M Cr^{6+}$ was shown to reduce to 25% of seed germination of *Echinochloa colona* L (Rout et al. 2000). The toxic effects of Cr on sunflower (*helianthus annus* L) growth traits were studied by Fozia et al. (2008), who showed that Cr toxicity induces injury to plants and caused a decrease in germination and shoot length and root length (Fozia et al. 2008). Sihag and Joshi (2018) studied the effects of diverse Cr

(VI) levels (0.0–4.0 mg) on the performance of sorghum. They found that Cr stress significantly reduced the chlorophyll contents, protein content and fresh and dry weight. Likewise, Anjum et al. (2017) carry out an experiment to investigate the consequences of Cr applied with various concentration 0, 30, 60, 90, 120, and 150 -mol $^{L-1}$, to two maize cultivars and revealed that Cr toxicity stress reduced the cob formation, leaf area and 1000 grain weight and yield. This reduction in root growth was associated with imbalance in Ca^{2+} in cell induced by Cr toxicity, leading to a disruption in the transport of Ca (Zou et al. 2006). The effect of Cr toxicity was studied in rice and significant effects were found in protein content and membrane stability by Mukta et al. (2019). In the *Pisum sativum* plant, the growth in the potassium-containing medium caused an increase in Cr content in various parts in the following order, roots>, stems>, leaves>, and seeds (Stambulska et al. 2018). Cr decreases the root growth of *Amaranthus viridis* L. The consequence of various trace metals on three Veronica species were studied by Živković et al. (2012), and a highly significant positive correlation was noticed among the quantities of Cr and Fe in tissues. Seeds of cowpea (*Vigna sinensis* L) were treated with various levels of Cr^{6+}·, and Cr stress resulted in a decrease in the total amount of sugar and germination traits (Nath et al. 2008). The studies cited above demonstrated that Cr toxicity has drastic effects on different plant characters; however, detailed studies are required to evaluate the response of different plant species under Cr toxicity conditions, along with changing environmental conditions.

6.3.2 Copper (Cu) Toxicity Stress in Plants

Copper (Cu) is one of the oldest and well-known metals and it is 25th most important metal and trace element which is present in a very small quantity in tissues, but it is essential for life (Adrees et al. 2015b). The solubility of Cu depends on the dissolved organic matter (Bravin et al. 2012) and soil pH and it is quickly available to plants at pH below 6 (Adrees et al. 2015a). Plant require varying quantities of Cu for their healthy growth and development (Mantovi et al. 2003). Cu exists in Cu^{2+} which is readily taken up by plants (Maksymiec 1998). Cu is toxic at higher concentration for both plants and animals (Michaud et al. 2007) and it is much more toxic to plants as compared to other metals (Dresler et al. 2014). The poor seed germination, low yield, stunted growth, and an alteration in the uptake of other metals are the most widely observed symptoms of Cu toxicity as studied by Adrees et al. (2015a). Seed germination is most common and broadly studies the symptoms of Cu stress. In an experiment, wheat seed germination decreased by about 40% compared to control at 100 ppm Cu levels after 14 days of germination (Singh et al. 2007).

Cu stress affects the growth of plants. This is one of the common effects of Cu stress and a lot of studies have been conducted into this issue. The effects of Cu on mungbean (*Vigna radiate*) was studied by Verma et al. (2011), who found a significant decline in radicle length of mungbean at different Cu concentrations of 20 and 50 uM. In an experiment Sheldon and Menzies (2004) found that Cu toxicity reduces the root growth in Rhod grass. In spinach, a positive negative association between shoot and root elongation and an increase in Cu levels was observed (Sharma et al. 2020a). A decrease in photosynthesis is one of the symptoms of Cu stress and this is related to the decrease in chlorophyll content and it is reported in many plants. Alaoui-Sossé et al. (2004) studied the consequences of Cu stress on photosynthesis of cucumber leaves. Younger seedlings displayed a decline in leaf area and photosynthesis (Alaoui-Sossé et al. 2004), while old seedlings showed a decrease in photosynthesis. In rapeseed (*Brassica napus*), the contents of chlorophyll were decreased when plants were treated with 6 μmol·dm^{-3} of Cu (Peško and Kráľová 2013). Cu stress ultimately affects the yield of plant species. There was a considerable decrease in the grain yield of rice with soil levels of Cu (over 100 mg kg^{-1} of soil) (Xu et al. 2005). In an experiment, eggplant seedlings were treated with Cu toxicity stress, and significant changes were observed in morphological traits (Ozkay et al. 2014). Karimi et al. (2012) applied different concentrations of Cu (0, 50, 100 and 150 uM) to *Astragalus*. They observed an increase in activity of peroxidase, catalase and superoxide dismutase with an increase in Cu concentration. On the other hand, chlorophyll content decreased with increasing concentration. The Cu toxicity also induced the photo inhibition and disruption of the repair cycle of PSII action center and also resultantly the production of biomass (Pätsikkä et al. 1998). The studies cited above demonstrated that Cu toxicity has drastic effects on different plant characters; however, detailed studies are required to evaluate the response of different plant species under Cr toxicity condition along with changing environmental conditions.

6.3.3 Manganese (Mn) Toxicity Stress in Plants

Manganese (Mn) is the second most important trace metal in the Earth's crust after Fe and it is extensively dispersed in soil as well as other biological ingredients (Geszvain et al. 2012). Mn exists in soil in many states, including Mn (II), Mn (III) and Mn (VII) etc. (Schmidt and Husted 2019). Of these, Mn (II) is more soluble and readily available from soil to plants. Redox conditions and soil pH strongly influence Mn solubility and availability (Sparrow and Uren 2014). Mn becomes insoluble at higher pH and under low pH Mn is easily converted or reduced to its soluble and divalent form (Watmough et al. 2007). Mn plays a key role in the variety of processes, like photosynthesis, respiration, metabolic processes, protein synthesis and it is also an essential part for synthesis of secondary metabolites such as acyl lipids and lignin's (Lidon et al. 2004). Despite its beneficial effects, Mn is a heavy metal and also cause harmful effects in plants. When concentration of Mn reaches to 150 mg per kg in above-ground parts than Mn toxicity occurs and it usually occurs in acidic soils (Mukhopadhyay and Sharma 1991). The effects of Mn stress on different plant stages are discussed below. Mn toxicity inhibits essential enzymatic functions, induce oxidative stress, reducing photosynthesis and restricting the uptake of minerals and nutrients like phosphorus and potassium from soils (Ducic and Polle 2005). Mn stress inhabited uptake as well as the concentration of mineral elements like Ca and Fe (Xianghua et al. 2007) In general, a high level of Mn had direct cytotoxic influences in the outer root cap and meristematic cells, such as extensive cytoplasmic injuries and plasma membrane ruptures. Many reports have suggested that excessive Mn could cause oxidative stress (Ren and Liu 2007).

The most visible symptoms of Mn toxicity are chlorosis in young leaves, necrosis on mature leaves, and, finally, plant death. Mn toxicity symptoms vary from species to species. Likewise, Fernando and Lynch (2015) reported chlorosis and necrosis in leaves of common bean (*Phaseolus valguris*) and Rosas et al. (2007) in clover (*Trifolium repens*). Brown spots are noticed in cow pea (*Vigna unguiculata*) by Fecht-Christoffers et al. (2003) and barley (*Hordeum vulgare*) (Führs et al. 2010). Silva et al. (2017) studied the association among surplus Mn and water-logging tolerance in maize. Leaves and plant biomass were evaluated at the beginning of treatment and after 7, 14 and 21 days of treatment. They have studied various parameters, such as chlorophyll content, biomass and antioxidant metabolism. Mn was transported from roots to leaves, resulting in a decrease in the chlorophyll content and increased ROS and lipid peroxidation. Likewise, Dimkpa et al. (2018) studied the effect of Mn nanoparticles on nutritional acquisition in wheat. Wheat was exposed to different Mn types in soil, and nano-Mn was repeated in foliar spray. Nano-Mn particles induced other effects on nutritional composition, including a decrease in shoots, P and K. The soybean (*Glycine max*) plant was exposed to excess Mn stress and demonstrated a decline in stomatal conductance and CO_2 assimilation rate, which resulted in a decrease in shoot biomass.

6.4 Effects of Non-Redox Active HMs

6.4.1 Nickel (Ni) Toxicity Stress in Plants

Ni ranked 22nd among heavy metals (Sreekanth et al. 2013). Ni is an essential heavy metal that assists urease enzyme in the conversion of urea into ammonia and carbon dioxide (Gajewska and Skłodowska 2009). Ni plays a key role in plant germination and physiology and yield (Sreekanth et al. 2013). Barley crops cannot complete their cycle of life without Ni (Brown et al. 1990). Ni when supplied in excess to soil via irrigation, sewage sludge, and mining leads to Ni toxicity (Brown et al. 1990). The natural concentration of Ni in soils is 22 mg Kg^{-1} (McIlveen and Negusanti 1994). Ni is present in soils, in many forms and most soluble and available form is Ni^{2+} (Lešková et al. 2020). The toxic symptoms of Ni include the inhibition of enzymatic activity, chlorophyll synthesis, Calvin cycle which decreases photosynthetic rate in plants (Gajewska et al. 2012). Plant water status and the anti-oxidant system of plants is also affected by Ni toxicity (Amjad et al. 2020). Ni toxicity induces the production of a large number of ROS in many plants (Gill and Tuteja 2010), which causes damage to DNA, protein, enzyme and chloroplast (Gopal and Nautiyal 2012). The effect of Ni toxicity on cell wall-related processes was studied in an experiment. The results showed that Ni toxicity induces root growth inhibition by surpassing cells elongation. Nickel toxicity significantly caused chlorosis in common beans plants at concentration of 100 mg L^{-1} after 24 hours of treatment (Campanharo et al. 2010). The growth and biochemical variables of wheat were studied under different toxic levels of Ni (0, 25, 50 ug per litter). Lipid peroxide was found higher at treatment of 25 and 50 ug, at the same time plant height and chlorophyll contents were also decreased (Parlak 2016). Rice plants were exposed to a higher concentration of Ni toxicity and a significant decline in root and shoot length and fresh and dry weights of rice plants was observed, which caused a decline in pigments of photosynthesis (Zahoor et al. 2016b; Rizwan et al. 2017).

Ni toxicity induces the deficiency of other metals like Fe and Zn and restricts the uptake of Cr and Cd (Myśliwa-Kurdziel et al. 2004). A decline in protein and carbohydrate contents due to Ni toxicity was reported in *Myplis snavelus*, and sunflower (Pillay et al. 1996), maize (Baccouch et al. 2001) and soybean (El-Shintinawy and El-Ansary 2000). Ni toxicity disrupts the uptake of macro and micro nutrients and hinders their movement from root to shoot and grain (Ameen et al. 2019); it also effects the assimilation of nutrients (Chen et al. 2009). Two maize hybrids were evaluated to study the effect of Ni toxicity on growth, physiology and nutrients composition. The results showed that Ni concentration had a significant correlation with growth of hybrids (Amjad et al. 2020). Ni stress reduced the concentration of N in roots and leaves of mungbean as well as chickpea (Yusuf et al. 2011). The protein content as well as carbohydrates reduced with Ni toxicity levels (0, 200, 400, and 800 mg kg^{-1}) in *Myplis snavelus* and maize were studied by (Baccouch et al. 2001).

6.4.2 Zinc (Zn) Toxicity Stress in Plants

Zn is an important micronutrient mandatory for normal plants growth as well as development (Cabot et al. 2019; Shahzad et al. 2014), which is taken by plants from soil via the root system. Zinc toxicity occurs when level of zinc exceeds 200 ppm. Zinc exists in the divalent form Zn^{2+}, which is the most prevalent form found in soil and taken by plants (Sharma et al. 2013). The typical symptoms of Zn stress includes small leaf size, chlorosis, necrosis, the inhibition of whole plant growth, and a deficiency of plant essential nutrients such as P, Fe, Mn and Cu. Zinc is readily available to plants at low pH (Rizwan et al. 2019). In an experiment seedlings of wheat were exposed to Zn toxicity under hydroponic experiment. After four days of treatment, seed germination rate, radicle length and plume length were reduced and simultaneously chlorophyll content also decreased (Wang et al. 2010). Zn toxicity severely affected the shoot length, number of leaves, roots and fresh weight of tomato at high levels (Al Khateeb and Al-Qwasemeh 2014). In another study, the effects of many doses of Zn stress on biomass and chlorophyll contents of sorghum bicolor and *Chenopodium album* was studied. Leaf morphology has drastically changed by Zn toxicity in *Populus alba* and barley (Todeschini et al. 2011). The results indicated that at the end of Zn treatment, plant height, chlorophyll content and biomass production were all decreased (Xianghua et al. 2007). In two lettuce (*Lactuca sativa*) varieties the interaction of phosphate and Zn was studied. The results showed that due to an increase in the concentration of Zn, a reduction in dynamics of Pi transporter in lettuce varieties was observed (Bouain et al. 2014). The effect of Zn on rice plants were studied in an experiment and rice seedlings were treated with four Zn levels (0.5, 5, 50 and 100 uM) and concluded that Zn decreased ascorbate peroxidase and guaiacol peroxidase (Asadi et al. 2015). The chlorophyll content and lipid peroxidation response to Zn stress in soil was studied and it was noted that superoxide dismutase and peroxidase activity was increased by increasing Zn stress (Cui and Zhao 2011). Zinc toxicity caused the dysfunctioning of photosynthesis in many crops, such as mungbean and naked pumpkin (Lalelou et al. 2013; Vassilev et al. 2011).

6.4.3 Aluminum (Al) Toxicity Stress in Plants

Al is the third most widespread element in the Earth's crust after oxygen and silicon and it has no proper biological function in plants (Bojórquez-Quintal et al. 2017). Al stimulate plant growth, helps in nutrients uptake and proves resistant to biotic and abiotic stress in some plant species, rice and tea plant (Bojórquez-Quintal et al. 2017). Al is solubilized at a pH below 5.0 and becomes available to plants as Al^{3+} form, which is a toxic from of Al (Kinraide 1997). Al is the main growth-reducing element in acidic soils (Kochian et al. 2005). Al toxicity effects on plants can be defined into two types: placed in two types,

TABLE 6.1
Effect of toxic level of different heavy metals on different plants traits

HMs	Plants	Trait	Con	Effect	References
Cu	Rhodgrass	Root	0.2 μM	Reduction of root growth	(Sheldon and Menzies 2004)
Cu	Wheat	Yield	100 ppm	Reduced seed germination	(Singh et al. 2007)
Cr	Rice	CC	100 μM	Decline in CC	(Zeng et al. 2012)
Cr	Sorghum	Protein	4.0 mg	Reduced protein content	(Baccouch et al. 2001)
Cr	Maize	Leaf area	150 umol L^{-1}	Decreased leaf area	(Anjum et al. 2017)
Cr	Wheat	FM and DM	300 μM	Reduced FM and DM	(Mathur et al. 2016)
Mn	Rice	Yield	100 ppm	Decrease in yield	(Alam 1985)
Mn	Wheat	Seedling	6.4 mg	Decrease in shoot length nitrogen	(Dimkpa et al. 2018)
Al	Maize	RL (cm)	50 uM	Root growth inhibition	(Stass et al. 2006)
Al	Rice	SL (cm)	90 μg/mL	Reduction in shoot length	(Bidhan and Bhadra 2014)
Al	Wheat	Plant growth	200 μM	Inhibition of plant growth	(Pintro and Taylor 2004)
Zn	Rice	Biomass	200 μM	Reduced biomass	(Gu et al. 2012)
Zn	Wheat	Yield	7 ppm	Decline in yield	(Takkar and Mann 1978)
Ni	Rice	RDW (mg)	40 mg kg^{-1}	Decreased RDW	(Aziz et al. 2015)
Ni	Maize	Growth	40 mg L^{-1}	Reduced growth	(Amjad et al. 2020)
Ni	Wheat	CC	50 ug	Decreased CC	(Parlak 2016)
Cd	Rice	RL	50 mg kg^{-1}	Root length decreased	(Jianguo et al. 2010)
Cd	Wheat	DMY	100 mg kg^{-1}	Reduction in DMY	(Singh et al. 1989)

HMs, heavy metals: CC, chlorophyll content: FM, fresh matter: DM, dry matter: SL, shoot length: RL, root length: RDW, root dry weight: Con, concentration: μM, micrometer: mg, milligram: ug, microgram: ppm, parts per million: kg, kilogram: Ref, references.

short-term response in which Al toxic effects can be observed within a few minutes and long-term effects in which toxic effects can be seen after days of exposure (Kochian et al. 2004; Simões et al. 2012). The severity of Al toxicity depends on plant genotype, age and growth conditions (Bojórquez-Quintal et al. 2017). Al inhibits cell elongation after short exposure (Ahn and Matsumoto 2006) and root growth in many plant species (Awasthi et al. 2017; El-Moneim et al. 2014, 2017), such as the high blue blueberry (Inostroza-Blancheteau et al. 2011). The root length of wheat seedlings grown in hydroponic conditions was decreased after exposure to 0.1 mM of Al toxicity (Szab-Nagy 2015). A decrease in photosynthesis was noticed in highbush blueberry due to Al toxicity (Bidhan and Bhadra 2014; Zahoor et al. 2016a).

Al toxicity reduced the primary root, the number of leaves/seedlings, the seedling fresh and dry weight of rice when exposed to 30, 60 and 90 ug/mL of Al toxicity level (Bidhan and Bhadra 2014). Al toxicity produces ROS, cause damage to membrane and leads to a decline in photosynthesis process and enzymatic function and, finally, causes cell death (Chen et al. 2005; XU et al. 2018). Al induces various harmful cellular alterations, in cell division, and expression of fibrillarin protein (Zhang et al. 2014). Al reduces synthesis of cytokinins, alters the structural and functional features of cell wall and restrict the nutrients uptake (Kochian et al. 2005; Ali et al. 2016). Al toxicity significantly decreased the total soluble protein in citrus leaves (*Citrus grandis*) (Li et al. 2016b). Al toxicity cause hindrances in absorption as well as the transportation of major plants nutrients like P, K and Mg in maize and sorghum (Bhalerao and Prabhu 2013). In Arabidopsis a plant exposed to Al toxicity showed a remarkable decline in root elongation and increase ethylene evaluation and activity of ethylene reporters EBS:GUS in root apex (Sun et al. 2010). The effect of different toxic levels of HMs on plants is given in Table 6.1.

6.5 Defense Mechanisms Used by Plants against HMs Stress

Plants adopted a variety of defense mechanisms against HMs toxicity (Rasheed et al. 2020a, 2020b). In plants, the first defense line against HMs toxicity is the physical barrier; thick cuticles and trichomes are used by plants when they come across HMs stress (Wong et al. 2004). For instance, trichomes act as a storage site for heavy metals for their detoxification or they lead to the production of secondary metabolites. After crossing the biophysical barriers adopted by plants, once HMs enter into cells than plant employ various other defense mechanism to nullify the toxic effects of HMs (Emamverdian et al. 2015). Plants produce low molecular weight chelators, i.e., spermine, organic acids, mugineic acids, heat shock proteins (HSP), phytochelatins and some amino acids such as proline, histidine and ethylene, and salicylic acid (Emamverdian et al. 2015; Sharma and Dietz 2006). Heavy metals lead to the production of ROS which cause oxidation in plant cells. In the case of the failure of all of these mentioned strategies plants started developing several antioxidant defense response by producing ascorbate peroxidase (ABX), superoxide dismutase (SOD), carotenoids, catalase, and proline, which scavenge free radicals and reinforce the plant defense response (Rastgoo et al. 2011; Sharma et al. 2012).

6.6 Phytochelatins (PCs) Used by Plants as Defense Response to HMs Stress

Phytochelatins are a primary type of HMs chelators which belong to a family of Cys-rich peptides (Gupta et al. 2013). PCs are synthesized from GSH (non-transitionally) via trans-peptidation reaction carried out by enzymes. GSH is essential for PCs synthesis when they are exposed to HMs stress (Yadav 2010). PCs are not only involved in HMs stress tolerance but also against other stresses, such as salt and heat. In many plants, these PCs are used as biomarkers for the determination of HMs stress at an early stage (Saba et al. 2013). PCs are synthesized in cystol and from here they are shifted as heavy metals chelators to vacuole and their transport is facilitated by ABC-binding cassette (ABC). It is an ideal biochemical to safeguard plant against HMs stress (Rausch et al. 2007). Several reports demonstrated that plant produce PCs against a variety of metals such as Al, Ni, Cd, Cr and Cu (Manara 2012b).

In *Rubia tinctorum* plant PCs are produced in excessive amounts when they are subjected to different heavy metals stress like Cu and Cd (Cobbett and Goldsbrough 2002). The *Solanum nigrum* generation of PCs was increased in roots when plants were treated with 200 μmol^{-L} which inhibited the movement of excessive Cd from root to shoot (Fidalgo et al. 2013). In some rice cultivars which were subjected to various toxic levels of arsenic (AS), the production of PCs reduced the transport of As ions towards gains (Lemos Batista et al. 2014). The *Brassica juncea* exhibited three times higher accumulation of PCs when they were exposed to Cd for a long time (Heiss et al. 2003). The exposure of maize (*Zea mays*) plant to Cd stress for a longer time led to an increase in PCs in roots (Szalai et al. 2013). PCs with longer chains are known as solid binder with lead (Pb) in legumes (Piechalak et al. 2002). Two PCs genes were extracted in wheat (phytochelatins synthase) and *Arabidopsis thaliana* synthase (*AtPCS1*) (Cobbett 2000). The current studies have identified many PCs genes in crops like rice (*OsPCS1*) (Shen et al. 2010). The synthesis of these genes in transgenic plants leads to a very promising future for enhancing HMs stress tolerance. In Arabidopsis, the synthesis of PCs enhanced tolerance to many metals like Fe, Zn, Cu as compared to *L. cruciate* PCs synthesis (Degola et al. 2014). The *Bacopa monnieri* is a powerful HMs accumulator and it is suitable for phytoremediation (Rodrigo et al. 2016). Shukla et al. (2016) isolated the complexes of PCs from the root of *Rubia tinctorum* through the use of plasma atomic emission spectroscopy. This was further confirmed by the positive correlation of metals accumulation and the amount of PCs formed in response with this in *Brassica napus* and *Iglesia turino.*

Adrees et al. (2015a) identified that in *Saccharomyces cerevisiae, YCF1*, ABC protein transporters are involved in the detoxification of AS in vacuoles. In general, there are two ways of HMs detoxification in plants: symplastic as well as apoplastic. In most PCs HMs complexes used a symplastic path for their mobilization (Gupta et al. 2012). Mendoza-Cózatl et al. (2008) showed the symplastic movement of PC-Cd complexes in rape plant which indicated the four-times higher PC-Cd accumulation in xylem as compared to phloem. The expression of *AtPCS1* in tobacco, improved metals tolerance and the accumulation of Cd in rot by preventing its movement into shoots (Kotrba et al. 2009). In *T. caerulescens* an increase in the expression of PCs increased the rate of detoxification of Cd^{2+} in leaves. In some plants like *C. demersum*, PCs' formation showed no significant response towards HMs detoxification (Miyasaka et al. 2002). Clemens et al. (1999) reported that *TaPCS1*, *AtPCS1*, and *SpPCS* lead to tremendous increase in Cd tolerance in *S. cerevisiae*, and it was further confirmed that PCs complexes in yeast lead to detoxification of HMs stress.

6.7 Role of Proline in HMs Stress Tolerance in Plants

Plants accumulate a large amount of low molecular weight solutes in response to HMs stress (Siddique et al. 2018). Proline is considered to be the most vital osmoticum exists in cells when exposed to HMs stress (El-Beltagi et al. 2020). Plants accumulate proline as a result of HMs' stress (Bassi and Sharma 1993a); however, correlation among metals stress and proline accumulation is still unclear (Dar et al. 2016). The main role of proline is the scavenging of ROS, which are formed as an outcome of HMs' stress. Proline is produced by two consecutive reductions, which are catalyzed by pyroline-5-corboxylate synthase and reductase with glutamate (Dar et al. 2016). However, the synthesis of proline by the contribution of these two factors need more research. Proline plays a role in stabilizing the protein structure and maintains the cystolic pH which sustain the redox reaction status in cells (Aslam et al. 2017). Proline decreased the level of ROS in *Arabidopsis* when it was applied to roots, which showed its potential as a defense agent against HMs stress (Cuin and Shabala 2007). Sharma et al. (1998) observed that proline protect the glucose-6-phosphate dehydrogenase and nitrate reductase (NR) under Cd and Zn toxicity. In *Armaria*, it was observed that proline formed a complex with Cu and protected the plant from its toxicity. Likewise, in *Helianthus annus* proline protected the plants against Zn, Cd, Cu (Kastori et al. 1992) and in *Triticum aestivum* against Zn and Cu (Bassi and Sharma 1993b).

In roots of *Populus trichocarpa* the concentration of proline was higher when plant was treated with a high dose of Cd (Nikolić et al. 2008). Some studies have shown that some HMs-tolerant plants accumulated more proline in their shoot than in their roots (Zengin and Kirbag 2007). Proline accumulation and its role against HMs' stress is influenced by many factors, such as plant parts, experimental conditions, dose of metals stress, and duration of exposure (Emamverdian et al. 2015). However, the speedy induction of proline in roots which makes pro-metals complexes offers a most effective strategy to detoxify the HM effects rather than to allow them to reach above-ground parts. Theriappan et al. (2011) treated cauliflower (*Brassica oleracea* var. botrytis) seedling with three HMs, Cd, Zn and Hg. They noted that at level of 100 μM the accumulation of proline in roots was increased two times, which strongly evidenced its role against HMs' stress. Likewise, in another study conducted by Sagardoy et al. (2009), it was observed that when sunflower was treated with HMs stress, proline accumulation was enhanced. All of these findings indicated that the rate of proline accumulation depends on the concentration of HMs. The foliar spray of proline on plant parts is a useful technique to decrease the toxic effects of HMs and to enhance protective mechanisms in plants (Emamverdian et al. 2015). The exogenous application of Proline on chickpea under Cd toxicity boosted the action of certain enzymes, such as carbonic anhydrase, which plays a key role in Cd tolerance in plants (Hayat et al. 2013). In pea (*Pisum sativum*), the application of proline safeguards the plant against Ni toxicity by decreasing electrolyte leakage and enhancing the amount of compatible solutes (Shahid et al. 2014). All of these findings indicated that the rate of proline accumulation depends on the concentration of HMs. Hence, from the above-mentioned studies it may be concluded the proline can be used as an effective agent to enhance plant defense response, but further research would signify a promising future of HMs tolerance induced by proline.

6.8 Role of Metallothioneins (MTs) in HMs Stress Tolerance in Plants

MTs belong to group of Phyto-chelations which tend to bind the HMs via the thiol group of cysteine amino acid and play a key role in HMs detoxification in plants (Nikita et al. 2020). They protect plants against HMs by adopting a variety of mechanism, through ROS scavenging and sequestration (Huang and Wang 2010) but the exact way through which they mediate ROS homeostasis is still unclear (Hasan et al. 2017). Their proposed role is to keep the balance in homeostasis of transition metals ion and the sequestration of HMs and they safeguard plants from oxidative injury cause by HMs stress (Hossain et al. 2012). MTs are not an essential component for the completion of the plant life cycle, but they play a key role in primary ions homeostasis and their distribution in plants (Hasan et al. 2017). Generally, the sequestration of intracellular HMs in eukaryotic organisms also includes tie of HMs with cytosolic cysteine-rich MTs peptides and separation (Sácký et al. 2014). The grouping of low kinetic solidity and larger thermo-dynamics are key characteristics of metal and MTs complexes, which tie HMs, while a portion of metal ions is simply replaced for additional proteins (Maret 2000). Many transgenic plants which expressed MTs genes have been notched for superior HMs tolerance and they exhibit altered metal accretion or dissemination approaches (Tomas et al. 2015). The chelation of metal ions by MTs is well studied, but little is known about the ways of transportation of metals and MT complex from cytoplasm towards vacuole (Yang et al. 2011).

Irvine et al. (2017) presented a brilliant technique to develop a low-cost MT-biosensor that can promotes a significant rise the signal related with a metal. Such an unpretentious sensor technology could be possibly used in environmental checking, particularly in the zones with the metal uncleanness difficulties (Irvine et al. 2017). Banday et al. (2020) showed that MTs genes alterations are a trustworthy sign of the cellular and humoral immune reaction in *Cyprinus carpio*. Mekawy et al. (2020) worked with the metallothionein-like gene *OsMT-3a* and showed that this gene is a potential target for the genetic improvement of HMs' tolerance in rice. MTs have been divided into four types: type 1 of MTs mostly expressed in root, while type 2 is expressed in shoots, type 3 in leaves and type 4 in seeds (Lalelou et al. 2013; Sharma et al. 2010). Grennan (2011) stated that, in the *Arabidopsis* plant, there is every possibility that MT isoforms from types 1 and 2 (1a, 2a, and 2b) and 3 are responsible for chelation of copper, while MTs isoforms from type 4 (4a and 4b) acted as a zinc chelator.

In barley (*Hordeum valgare*) during the grain-filling and mature seed stage, it was proven that the main role of MT type 3 is to sustain the homeostasis of Zn as well as Cu, whereas MT type 4 was responsible for the storage of Zn (Hegelund et al. 2012). In the soybean plant, it was presented that MT1, MT2, and MT3 were involved in the reclamation of toxic concentration of Cd, while MT4 showed Zn-binding features (Pagani et al. 2012). It has been proposed that fluctuating kinds of MTs and types have different and overlying jobs in HMs purification (Balzano et al. 2020). More research is needed to find out the causes for these disparities and special activities of plant MTs to HM stress; however, it seems that dissimilarities in the hereditary construction of plants is due to multifarious varieties in the metal-binding areas of plant HMs. Some mechanisms are indicative of MTs stimulating the ability of transgenic plants in relations to reducing the generation of ROS and invigorating cellular antioxidant defense structure when it comes to depolluting unnecessary levels of HMs. Xia et al. (2012) indicated that the expression of *Elsholtzia haichowensis* metallothionein type 1 (*EhMT1*) in tobacco improved the tolerance of transgenic tobacco against Cu toxicity, but also lessened the production of hydrogen peroxide and enhanced peroxidase action (POD) in roots, leading to an increased capability of plants to deal with oxidative stress.

Zhou et al. (2014) verified that *TaMT3*, a metallothionein type 3 from *Tamarix androssowii*, transferred into tobacco improved tolerance to Cd stress via a noteworthy increase in SOD function, which increased the capacity of ROS cleaning in transgenic plants. It appears that the influence of appeared metallothioneins on different constituents of antioxidant structure of transgenic plants is diverse, which needs additional research. A gene (*OSMT1e-p*), a type1 MT found from a salt-tolerant cultivar of rice, showed tolerance against Cu and Zn stress when expressed in the transgenic plant of tobacco. They detected that tobacco plants that received the gene reduced the amount of Cu and Zn ions in leaves and roots and also hindered their movement (Kumar et al. 2012). Zhigang et al. (2006) determined that the expression of *BjMT2*, a metallothioneins type 2 from *Brassica juncea*, in *Arabidopsis thaliana* improved Cu and Cd tolerance at their seedlings stage but led to a dramatic decrease in roots growth when there was no HM contact. Further studies are necessary to thoroughly investigate the function of MTs in transgenic plants that act in a nonspecific way in host plants (Liu et al. 2020c).

6.9 Role of Arbuscular Mycorrhizal (AM) in HMs Stress Tolerance in Plants

AMF are soil-borne fungi that can pointedly increase the uptake of nutrients in plants as well as improve tolerance to many abiotic stresses (Sun et al. 2018). A total of four orders of AMF (Glomerales, Archaeosporales, Paraglomerales, and Diversisporales) have been recognized in this sub-phylum, which also comprises 25 genera (Redecker et al. 2013). Micorryzhal fungi (MF) form a symbiotic relationship with plants and the symbiotic association of AMF is a typical case of mutualistic association, which can control the growth and enlargement of plants. AMF are generally thought to back plant growth in soil polluted with HMs, owing to their ability to fortify the defense mechanism of AMF-interceded plants to stimulate growth as well as development (Begum et al. 2019).

Aguilera et al. (2014) observed that AMF increased the uptake of nutrients in wheat under Al stress. HMs can be removed by fungal hyphae of inner and outer origins (Ouziad et al. 2005) that has the capability to fix HMs in cell wall and stock them in vacuole (Punamiya et al. 2010) and therefore decreases the level of metal poisonousness in plants. The durable influences of AMF on plant growth under stressful circumstances are mainly due to the capacity of these fungi in boosting morphological as well as physiological processes that raise the biomass of plant and subsequently the uptake of significant immobile nutrients like Cu and Zn and thus condensed metal harmfulness in host plants (Miransari 2017). It has been reported that AMF fix Cd and Zn in the cell wall of mantle hyphae and cortical cells, thereby limiting their uptake and causing an increase in growth, yield and nutrient uptake (Garg and Chandel 2012).

It has been reported that that little Cd movement and poisonousness can also be removed with AMF by raising soil pH (Shen et al. 2006), refurbishing Cd in the extra-radical mycelium as studied by Janoušková and Pavlíková (2010) and binding Cd to glomalin, a glycoprotein. These studies reported that AMF were found to be effective in lowering the toxic levels of Cd in rice vacuole and cell wall which resulted in Cd reclamation. Likewise, improved Cd tolerance in alfalfa (*Medicago sativa* L.) was improved due to AMF (Wang et al. 2012). AMF reduces the HM toxicity through a variety of different methods, including the immobilization of HM ions, the precipitation of polyphosphate granules in soil, the adsorption to fungal cell wall chitin, and the chelation of HM inside fungus. Further studies are needed to shed more light on the role of AMF in HM tolerance in plants. In ryegrass (*Lolium perenne* L.), it was shown that symbiotic association of AM resulted in immobilization of Cd, Zn and Ni (Takács et al. 2001), and the same results were obtained for green gram (*Vigna radiata*) exposed to Zn toxicity (Shivakumar and Hemavani 2011). Some studies reported that variations in pH due to AM fungi are a main causal aspect to the restriction of metals in mycorrhizosphere area (Bano and Ashfaq 2013). Abad and Khara (2007) exhibited that wheat plants inhabited by diverse AM fungi classes exposed to lethal levels of Cd had noticeably more functional levels of protective antioxidants such as APX and GPX in their roots and shoots as compared to non-AM wheat.

In another study, mycorrhizal pigeon pea (*Cajanus cajan* L.) plants were grown in soil contaminated with Cd and Pb and it was noticed that there was a clear reduction in lipid peroxidation, electrolyte leakage and the enhanced activity of catalase and SOD (Garg and Aggarwal 2012). Rahmaty and Khara (2011) witnessed that Cr-stressed plants of maize blended with AM fungus (*Glomus intraradices*) showed better chlorophyll contents when compared with plants not treated with AM. Amanifar et al. (2014) detected that the concentration of proline in shoots in AM tomato exposed to Pb and the inoculation of AM was not considerably affected, but there was a prominent surge in root proline content. Main crops such as wheat, maize, and rice are shown to be host of AM fungi (Bhalerao 2013), which require the documentation and breeding of popular fungus species which are capable of enhancing tolerance to HM stress in plants. The pictorial display of mycorrhizal functions to control numerous processes in ecosystem and promotion of growth of plants under HMs stress condition is shown in Figure 6.3.

6.10 Some Other Type of Plant Defense Responses in HMs Stress Tolerance in Plants

6.10.1 Organic Acids (OA) and Amino Acids

There are many other factors, including OA and amino acids, which play a key role in plant defense responses towards HMs' stress (Kato et al. 2020). The uptake of Pb into the roots is also limited by a connection of Pb to carboxyl a group of mucilage uronic acid, and after probing root cell membranes, a small amount of Pb interferes with cellular gear and results in the improved thickness of cell walls (Pourrut et al. 2011). In plant cells, pectin tend to bind with Pb and reduce its toxicity (Fahr et al. 2013). After carrying HMs via xylem and confiscating ions in vacuoles, OA such as malate, citrate, and oxalate enhance HM tolerance in plants (Nunes-Nesi et al. 2014). Citrate tend to chelate Fe, and also to detoxify

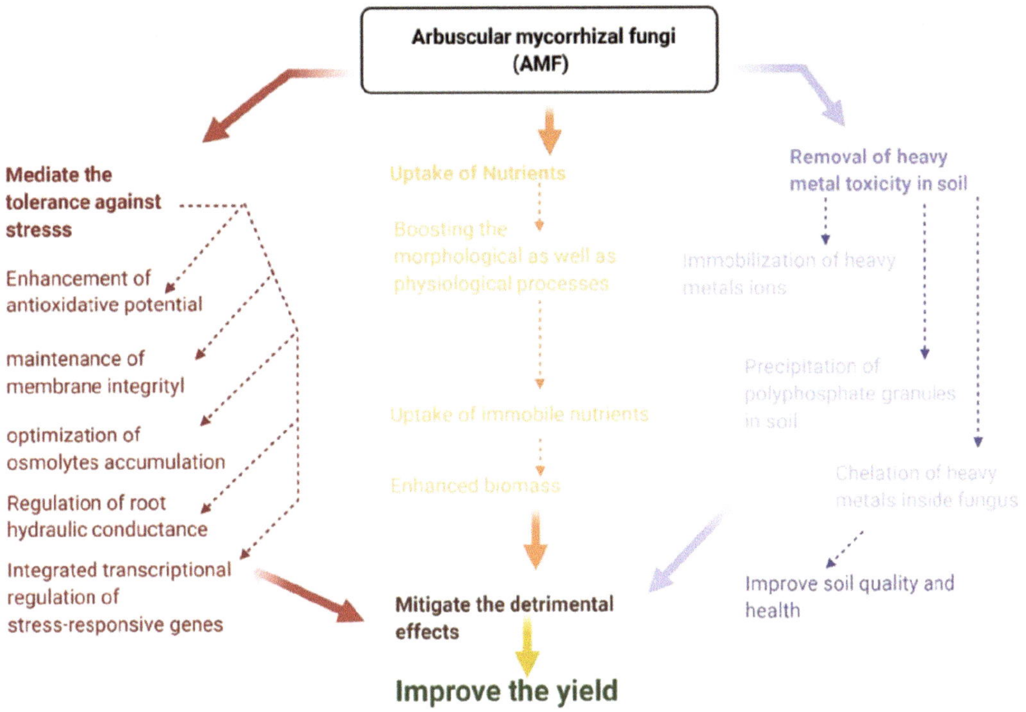

FIGURE 6.3 Pictorial display of mycorrhizal role to control numerous processes in ecosystem and promotion of growth of plants under HMs stress environment.

other HM such as Ni and Cd, while malate is involved in Zn chelation in Zn-tolerant plant species (Singh et al. 2015a). Histidine is reflected as a chelator of Ni in roots exudate (Richau et al. 2009). Several studies have shown that phylate plays a key role in the chelation of Mn, Fe and Zn (Dubey et al. 2018). There is still a need to thoroughly investigate the role of OA and amino acids in HMs chelation in order to improve abiotic stress in plants.

6.10.2 Antioxidant Defense System and Signaling Pathway

In plants the antioxidant defense system comprise superoxide dismutase, catalase, ascorbate peroxidase and glutathione and carotenoids (Dubey et al. 2018). In all aerobic organisms, SOD has a vital role in plant stress tolerance and delivers a defense against oxidative stress (Khan et al. 2017). There are three isozymes of SOD, Zn/Cu SOD, Fe SOD and Mn SOD, which exist in subcellular compartments (Fukai and Ushio-Fukai 2011). Catalase, vacuoles, cell wall, as well as cytosol, guaiacol peroxidase causes oxidation of aromatic electron donor such as guaiacol and thus assists in defense responses (Kumar et al. 2009). The signaling pathways play a key role in HMs detoxification. Thapa et al. (2012) espoused rice to Cd stress and observed a significant accumulation of ROS, abscisic acid and Ca. Likewise, barley and rice plants which were treated with Cd showed an increase in nitric production (Alemayehu et al. 2015). These findings suggest that nitric oxide is one of the most reliable factors to counteract HM toxicity and it might also function as a signaling molecule that is convoluted in the alleviation of the HM stress (Singh et al. 2015a).

6.10.3 Role of MicroRNA (miRNA) in HMs Stress Tolerance in Plants

MicroRNAs are a class of small noncoding RNAs that adversely control definite target mRNAs at the post-transcriptional level (Ding et al. 2020). It has been shown that numerous miRNAs are accountable for governing the expression level of proteins as well as transcription factors that are related with stress

TABLE 6.2

Candidate genes involved in various HMs stress tolerance in plants

Genes/Protein	Plants	HMs	References
AtfC1	Arabidopsis	Cd	(Song et al. 2004)
MuS1	*Nicotiana tobacum*	Cd	(Kim et al. 2011)
Multidrug resistance-associated protein	*Betula pendula*	Cu	(Keinänen et al. 2007)
Metallothionein (*MT2B*)	*Silene paradoxa*	Cu	(Mengoni et al. 2003)
PpMT2	*Physcomitrella patens*	Cu	(Liu et al. 2020c)
TonB-dependent receptor/protein	*Betula papyrifera*	Ni	(Theriault and Nkongolo 2017)
Metal transporter NRAMP (1–4)	*Noccaea*	Ni	(Visioli et al. 2012)
OsLCT1	Rice	Zn	(Adil et al. 2020)
HtNHX1	Rice	Al	(Li et al. 2020)
GsGIS3	Arabidopsis	Al	(Liu et al. 2020b)
CYP71A12	Arabidopsis	Cr	(Liu et al. 2020a)
OsPTR7	Rice	As	(Abedi and Mojiri 2020)
FRO2	Cucumber	Fe	(Ahammed et al. 2020)
AtMTP11	Arabidopsis	Mn	(Delhaize et al. 2007)
HvFeSOD	Barley	Co	(Lwalaba et al. 2020)

responses for HMs stress (Macovei et al. 2012). Sunkar et al. (2006) revealed that transgenic *Arabidopsis*, overexpressing Cu–Zn superoxide dismutase 2 (CSD2) which is miR398 resistant, added more Cu–Zn superoxide dismutase 2 (CSD2) mRNA, thus resulting in tolerance during HMs' stress. In a study, 19 miRNAs in rice after exposure to Cd stress in rice seedlings were identified by Ding et al. (2011); likewise Zhou et al. (2008) identified responsive miRNAs in *Medicago truncatula*. Under vital nutrient discrepancy and Hg stress, miR398 was found to be involved in tolerance in plants (Yang and Chen 2013). This describes the important role of miRNA398 in ROS stability during stress situations (Zhu et al. 2011). MiRNAs and their 16 targeted genes were identified in radish after exposure to Cr stress (Liu et al. 2015). All such studies showed that miRNA could play an essential role in HMs' stress tolerance in plant (Yadav et al. 2020). Several candidate genes responsible for HMs stress tolerance in plants are given in Table 6.2.

6.10.4 Role of Salicylic Acid (SA) in HMs Stress Tolerance in Plants

SA is monohydroxybenzoic lipophilic acid, a form of phenolic acid and a beta hydroxy acid (BHA). It is known as a controller of plant defense and a response to biotic as well as abiotic stress (Sharma et al. 2020b). SA plays a role in HMs' defense response in several ways. The addition of SA in a mixture with plant growth-promoting bacteria decreases Cr-induced oxidative damage in maize plants by improving activities of antioxidant as well as non-antioxidant enzymes (Islam et al. 2016). Song et al. (2014) described the SA-facilitated improvement in the activities of CAT and SOD enzymes in barley leaves under Zn and Cu toxicity stress. Further, the carbohydrate metabolism in Cr-treated maize plants is enhanced with SA (Islam et al. 2016). The use of SA-alleviated Cd toxicity stress in *Brassica juncea* plant and improved growth and photosynthesis. Moreover, the supplementation of SA-reduced ROS level by the consolidation of the antioxidant defense system in plants and offers firmness to plant membrane (Faraz et al. 2019). Hasanuzzaman et al. (2019) applied SA and observed an increase in growth and yield in *Brassica campestris* when they were exposed to Pb stress.

Application of SA-improved pigment contents and photosynthetic performance in tomato plants when exposed to Cd stress (Guo et al. 2018). SA also enhanced the chlorophyll content and carbon fixation activities in wheat under Cd stress (Moussa and El-Gamal 2010). Similarly, carbonic anhydrase was enhanced in *Brassica juncea* under Ni toxicity stress (Ventrella et al. 2009) and Mn (Parashar et al. 2014) stress when plants were exposed to SA at concentration of 10 uM SA. The alleviation of Cd toxicity mediated by SA application was observed in tomato. Rare investigations recommend that SA can indorse free radical scavenging of ROS by regulating antioxidant enzymes and the expression of some proteins and molecules such asOsWRKY45 as stated in rice by Chao et al. (2010). The scientific literature cited described the role of SA in the alleviation of HMs' toxicity via exogenous application. However, a little research work has been done about its endogenous application because of the complexity of cascade effects produced. Therefore, more genomic, transcriptomic, proteomic, and metabolomics investigations are essential (Sharma et al. 2020b) to notice SA responsive genes, proteins as well as metabolites changed by HMs' stress.

6.11 Conclusion and Future Research Directions

HMs' stress is a huge threat to food security and, therefore, putting a great hazard to all kinds of life on Earth anthropogenic activities are a major cause of increase in HMs in the environment. Plants come into contact with a variety of HMs' stress. HMs limit the seed germination, growth, development, yield, nutrient uptake, process of photosynthesis and pose the oxidative stress which has devastating impacts on overall plant performance.

HMs' stress result in the production of toxic ions which tend to replace important cations, which is the worst phenomena of HMs' stress. This deals with the fact that there is a significant similarity among the plants that showed perceptible toxic symptoms induced by HMs stress and those suffering from a lack of vital nutrients. The uptake of HMs depends on plant type, species and metals concentration. Plant use large number of defense mechanisms when they came into contact with HMs' stress. Noteworthy breakthrough and research done in molecular along with biological areas throw some light on to understand the some multifarious approaches of plants at both the cellular as well as the molecular level to battle HMs' toxicity. Functional multiplicity and molecular capability of PCs as well as MTs and are becoming fascinating when it reaches to HMs reclamation and retaining cellular ion equilibrium. PCs and MTs are possibly to contact straight or indirectly with the plant antioxidant defense system, and there are clear hints that the evaluation of transgenic plants with overexpression of PCs and MTs can show considerable HMs stress tolerance. For this reason, the use of transgenic plants and candidate genes can be successfully used for the purposes of phytoremediation. Proline has the capability of doing both chelating as well as antioxidant-related work, but its roles and efficiency are enormously different which are based on HMs type, concentration as well as plant variety, organ and tissue type.

The chapter discussed the fact that the role of AM symbiotic association with the defense system of plants to counter HM toxicity is crucial so that it can fringe or control several HMs defense events, such as the signaling of HM stress, chelation, control of ion homeostasis as well as companionable solutes expansion. AM as well as some other antioxidant molecules (like APX, SOD as well as CAT) possibly act in a joint way at toxic levels of HMs and increase HMs tolerance. Although promising interrelatedness among AM and proline is described, a good association of AM along with proline is not still clarified. We discussed the role of miRNA in HM tolerance in plants. It is shown that numerous miRNAs are accountable for governing expression levels of proteins as well as transcription factors that are related with stress responses for HMs stress. Therefore, identification of candidate genes encoded by miRNA need to be isolated and transfer into susceptible plant species to enhance HMs stress tolerance in plants. Several OA like, citrate and malate are important component of plant defense mechanism. Some amino acids like histidine are important chelator of HMs ions like, Fe, Zn and Mn and aid to plant defense system. There is still a need to thoroughly investigate the role of OA and amino acids in HMs chelation to improve abiotic stress in plants.

The scientific literature cited is about the role of SA in alleviation of HMs toxicity through its exogenous application, and little research has been presented about its endogenous application because of the complexity of the cascade effects produced. Studies indicated that there are some areas which will need further examination. It is therefore critical to study the influences of HMs' toxicity on plants through the immediate application of numerous stresses like drought, heat and salinity. Secondly, the raised level of trace gases present in atmospheric and also their likely contacts with HM toxicity should be studied. This study will offer an inclusive examination of reactions as well as an assessment of the efficiency of transgenic plant species to HM stress under the circumstances of the changing climate. We discussed literature that deals with the plants' defense response, mostly at the adult stage, so it is essential to carry out studies to evaluate plants at an early stage to screen for HMs' stress tolerance. At the end, it is essential to undertake a widespread study of the inspection and collection of appropriate AM species for an effective symbiotic association with plants to combat HM stress.

ACKNOWLEDGEMENTS

The author is thankful to the supervisor Prof Huijie Li for his support and guidance. Author is especially thankful to the all coauthors for their assistance.

FUNDING

The research was supported by the National Natural Science Foundation of China (71963020 and 31760350), the National Key Research and Development Program of China (2018YFD0301102), the Jiangxi natural Science foundation (20181BAA208055 and 20202BABL205020), the Key Research and Development Program of Jiangxi Province (20171ACF60018 and 20192ACB60003), the Jiangxi Agriculture Research System (JXARS-18) and Projects of Water Science and Technology of Jiangxi Province (KT201628).

CONFLICT OF INTEREST

Authors declared that that they have no conflict of interest.

REFERENCES

Abad AKJ, Khara J (2007) Effect of cadmium toxicity on the level of lipid peroxidation and antioxidative enzymes activity in wheat plants colonized by arbuscular mycorrhizal fungi. *Pakistan J Biol Sci 10*:2413–2417.

Abedi T, Mojiri A (2020) Arsenic uptake and accumulation mechanisms in rice species. *Plants 9*:129.

Adil MF, Sehar S, Chen G, Chen Z-H, Jilani G, Chaudhry AN, Shamsi IH (2020) Cadmium-zinc cross-talk delineates toxicity tolerance in rice via differential genes expression and physiological/ultrastructural adjustments. *Ecotoxicol Environ Saf 190*:110076.

Adrees M et al. (2015a) The effect of excess copper on growth and physiology of important food crops: a review. *Environ Sci Pollut Res 22*:8148–8162.

Adrees M et al. (2015b) Mechanisms of silicon-mediated alleviation of heavy metal toxicity in plants: a review. *Ecotoxicol Environ Saf 119*:186–197.

Aguilera P, Cornejo P, Borie F, Barea JM, von Baer E, Oehl F (2014) Diversity of arbuscular mycorrhizal fungi associated with *Triticum aestivum* L. plants growing in an Andosol with high aluminum level. *Agric Ecosyst Environ 186*:178–184.

Ahammed GJ et al. (2020) Melatonin alleviates iron stress by improving iron homeostasis, antioxidant defense and secondary metabolism in cucumber. *Scientia Horticulturae 265*:109205.

Ahn S-J, Matsumoto H (2006) The role of the plasma membrane in the response of plant roots to aluminum toxicity. *Plant Signal Behav 1*:37–45.

Al Khateeb W, Al-Qwasemeh H (2014) Cadmium, copper and zinc toxicity effects on growth, proline content and genetic stability of *Solanum nigrum* L., a crop wild relative for tomato; comparative study. *Physiol Mol Biol Plants 20*:31–39.

Alam S (1985) Effects of iron and manganese on the growth of rice and on the contents of these elements in rice plants. *Agronomie, Earth Sciences 5*(6):487–490.

Alaoui-Sossé B, Genet P, Vinit-Dunand F, Toussaint M-L, Epron D, Badot P-M (2004) Effect of copper on growth in cucumber plants (*Cucumis sativus*) and its relationships with carbohydrate accumulation and changes in ion contents. *Plant Sci 166*:1213–1218.

Alemayehu A, Zelinová V, Bočová B, Huttová J, Mistrík I, Tamás L (2015) Enhanced nitric oxide generation in root transition zone during the early stage of cadmium stress is required for maintaining root growth in barley. *Plant and Soil 390*:213–222.

Ali H, Khan E, Sajad MA (2013) Phytoremediation of heavy metals—concepts and applications. *Chemosphere 91*:869–881.

Ali M, Ullah Z, Mian IA, Khan N, Khan N, Adnan M, Saeed, M. (2016). Response of maize to nitrogen levels and seed priming. *Pure and Applied Biology 5*(3):578–587.

Amanifar S, Aliasgharzad N, Toorchi M, Zarei M (2014) Lead phytotoxicity on some plant growth parameters and proline accumulation in mycorrhizal tomato (*Lycopersicon esculentum* L.). *International Journal of Biosciences (IJB) 4*:80–88.

Ameen N et al. (2019) Biogeochemical behavior of nickel under different abiotic stresses: toxicity and detoxification mechanisms in plants. *Environmental Science and Pollution Research 26*:10496–10514.

Amjad M et al. (2020) Nickel toxicity induced changes in nutrient dynamics and antioxidant profiling in two maize (*Zea mays* L.) Hybrids. *Plants 9*:5.

Anjum S, Wang L, Farooq M, Hussain M, Xue L, Zou C (2011) Brassinolide application improves the drought tolerance in maize through modulation of enzymatic antioxidants and leaf gas exchange. *Journal of Agronomy and Crop Science 197*:177–185.

Anjum SA, Ashraf U, Imran K, Tanveer M, Shahid M, Shakoor A, Longchang W (2017) Phyto-toxicity of chromium in maize: oxidative damage, osmolyte accumulation, anti-oxidative defense and chromium uptake. *Pedosphere 27*:262–273.

Arif N et al. (2016) Influence of high and low levels of plant-beneficial heavy metal ions on plant growth and development. *Frontiers in Environmental Science 4*:69.

Asadi M, Saadatmand S, Khavari-Nejad RA, Ghasem-Nejad M, Fotokian MH (2015) Effect of Zinc (Zn) on some physiological characteristics of rice seedling. *Biomedical and Pharmacology Journal 5*:203–210.

Aslam M et al. (2017) Specific role of proline against heavy metals toxicity in plants. *Int J Pure Appl Biosci 5*:27–34.

Awasthi JP et al. (2017) Morpho-physiological analysis of tolerance to aluminum toxicity in rice varieties of North East India. *PloS one 12*: e0176357. https://doi.org/10.1371/journal.pone.0176357.

Aziz H, Sabir M, Ahmad HR, Aziz T, Zia-ur-Rehman M, Hakeem KR, Ozturk M (2015) Alleviating effect of calcium on nickel toxicity in rice. *CLEAN–Soil, Air, Water 43*:901–909.

Baccouch S, Chaoui A, El Ferjani E (2001) Nickel toxicity induces oxidative damage in Zea mays roots. *Journal of Plant Nutrition 24*:1085–1097.

Balzano S, Sardo A, Blasio M, Chahine TB, Dell'anno F, Sansone C, Brunet C (2020) Microalgae metallothioneins and phytochelatins and their potential use in bioremediation. *Frontiers in Microbiology 11*:517.

Banday UZ, Swaleh SB, Usmani N (2020) Heavy metal toxicity has an immunomodulatory effect on metallothionein and glutathione peroxidase gene expression in Cyprinus carpio inhabiting a wetland lake and a culture pond. *Chemosphere 126311.*

Bano SA, Ashfaq D (2013) Role of mycorrhiza to reduce heavy metal stress. *Natural Science 5*(12):16–20.

Bassi R, Sharma SS (1993a) Changes in proline content accompanying the uptake of zinc and copper by Lemna minor. *Annals of Botany 72*:151–154.

Bassi R, Sharma SS (1993b) Proline accumulation in wheat seedlings exposed to zinc and copper. *Phytochemistry 33*:1339–1342.

Begum N et al. (2019) Role of arbuscular mycorrhizal fungi in plant growth regulation: implications in abiotic stress tolerance. *Frontiers in Plant Science 10*:1–15.

Bhalerao SA (2013) Arbuscular mycorrhizal fungi: a potential biotechnological tool for phytoremediation of heavy metal contaminated soils. *Int J Sci Nat 4*:1–15.

Bhalerao SA, Prabhu DV (2013) Aluminium toxicity in plants: a review. *Journal of Applicable Chemistry 2*:447–474.

Bidhan R, Bhadra S (2014) Effects of toxic levels of aluminium on seedling parameters of rice under hydroponic culture. *Rice Science 21*:217–223.

Bojórquez-Quintal E, Escalante-Magaña C, Echevarría-Machado I, Martínez-Estévez M (2017) Aluminum, a friend or foe of higher plants in acid soils. *Frontiers in Plant Science 8*:1767.

Bouain N, Mushtak K, Aida R, Myriam D, Benoit L, Nibras B, Tahar G, Jean-Claude D, Pierre B, Chedly A, Hatem R (2014) Phosphate/zinc interaction analysis in two lettuce varieties reveals contrasting effects on biomass, photosynthesis, and dynamics of Pi transport. *BioMed Research International 2014*:1–9.

Bravin MN, Garnier C, Lenoble V, Gérard F, Dudal Y, Hinsinger P (2012) Root-induced changes in pH and dissolved organic matter binding capacity affect copper dynamic speciation in the rhizosphere. *Geochimica et Cosmochimica Acta 84*:256–268.

Brown PH, Welch RM, Madison JT (1990) Effect of nickel deficiency on soluble anion, amino acid, and nitrogen levels in barley. *Plant and Soil 125*:19–27.

Cabot C, Martos S, Llugany M, Gallego B, Tolrà R, Poschenrieder C (2019) A role for zinc in plant defense against pathogens and herbivores. *Frontiers in Plant Science 10*:1171.
Campanharo M, Monnerat PH, Espindula MC, Rabello WS, Ribeiro G (2010) Toxicity symptoms of nickel in common bean. *Revista Ciência Agronômica 41*:490–494.
Chao Y-Y, Chen C-Y, Huang W-D, Kao CH (2010) Salicylic acid-mediated hydrogen peroxide accumulation and protection against Cd toxicity in rice leaves. *Plant and Soil 329*:327–337.
Chen C, Huang D, Liu J (2009) Functions and toxicity of nickel in plants: recent advances and future prospects. *CLEAN–Soil, Air, Water 37*:304–313.
Chen J, Zou W, Meng L, Fan X, Xu G, Ye G (2019) Advances in the uptake and transport mechanisms and QTLs mapping of cadmium in rice. *International Journal of Molecular Sciences 20*:3417.
Chen L-S, Qi Y-P, Liu X-H (2005) Effects of aluminum on light energy utilization and photoprotective systems in citrus leaves. *Annals of Botany 96*:35–41.
Ciriaková A (2009) Heavy metals in the vascular plants of Tatra mountains. *Oecologia Montana 18*:23–26.
Clemens S, Kim EJ, Neumann D, Schroeder JI (1999) Tolerance to toxic metals by a gene family of phytochelatin synthases from plants and yeast. *The EMBO Journal 18*:3325–3333.
Cobbett C, Goldsbrough P (2002) Phytochelatins and metallothioneins: roles in heavy metal detoxification and homeostasis. *Annual Review of Plant Biology 53*:159–182.
Cobbett CS (2000) Phytochelatins and their roles in heavy metal detoxification. *Plant Physiology 123*:825–832.
Cui Y, Zhao N (2011) Oxidative stress and change in plant metabolism of maize (*Zea mays* L.) growing in contaminated soil with elemental sulfur and toxic effect of zinc. Plant, *Soil and Environment 57*:34–39.
Cuin TA, Shabala S (2007) Compatible solutes reduce ROS-induced potassium efflux in Arabidopsis roots. *Plant, Cell & Environment 30*:875–885.
Dalvi AA, Bhalerao SA (2013) Response of plants towards heavy metal toxicity: an overview of avoidance, tolerance and uptake mechanism. *Ann Plant Sci 2*:362–368.
Dar MI, Naikoo MI, Rehman F, Naushin F, Khan FA (2016) Proline accumulation in plants: roles in stress tolerance and plant development. In: Iqbal N, Nazar R, Khan A, Nafees (Eds.) *Osmolytes and plants acclimation to changing environment: emerging omics technologies*. Springer, pp 155–166.
Degola F et al. (2014) A Cd/Fe/Zn-responsive phytochelatin synthase is constitutively present in the ancient liverwort *Lunularia cruciata* (L.) Dumort. *Plant and Cell Physiology 55*:1884–1891.
Delhaize E et al. (2007) A role for the AtMTP11 gene of Arabidopsis in manganese transport and tolerance. *The Plant Journal 51*:198–210.
Dimkpa CO, Singh U, Adisa IO, Bindraban PS, Elmer WH, Gardea-Torresdey JL, White JC (2018) Effects of manganese nanoparticle exposure on nutrient acquisition in wheat (*Triticum aestivum* L.). *Agronomy 8*:158.
Ding Y, Chen Z, Zhu C (2011) Microarray-based analysis of cadmium-responsive microRNAs in rice (*Oryza sativa*). *Journal of Experimental Botany 62*:3563–3573.
Ding Y, Ding L, Xia Y, Wang F, Zhu C (2020) Emerging roles of microRNAs in plant heavy metal tolerance and homeostasis. *Journal of Agricultural and Food Chemistry 68*:1958–1965.
Dresler S, Hanaka A, Bednarek W, Maksymiec W (2014) Accumulation of low-molecular-weight organic acids in roots and leaf segments of *Zea mays* plants treated with cadmium and copper. *Acta Physiologiae Plantarum 36*:1565–1575.
Dubey S, Shri M, Gupta A, Rani V, Chakrabarty D (2018) Toxicity and detoxification of heavy metals during plant growth and metabolism. *Environmental Chemistry Letters 16*:1169–1192.
Ducic T, Polle A (2005) Transport and detoxification of manganese and copper in plants. *Brazilian Journal of Plant Physiology 17*:103–112.
El-Beltagi HS, Mohamed HI, Sofy MR (2020) Role of ascorbic acid, glutathione and proline applied as singly or in sequence combination in improving chickpea plant through physiological change and antioxidant defense under different levels of irrigation intervals. *Molecules 25*:1702.
El-Moneim DA, Contreras R, Silva-Navas J, Gallego FJ, Figueiras AM, Benito C (2014) Pectin methylesterase gene and aluminum tolerance in Secale cereale. *Environmental and Experimental Botany 107*:125–133.
Emamverdian A, Ding Y, Mokhberdoran F, Xie Y (2015) Heavy metal stress and some mechanisms of plant defense response. *The Scientific World Journal 2015*, 1–18.
Engwa GA, Ferdinand PU, Nwalo FN, Unachukwu MN (2019) Mechanism and health effects of heavy metal toxicity in humans. In: *Poisoning in the modern world-new tricks for an old dog?* IntechOpen.

Fahr M, Laplaze L, Bendaou N, Hocher V, El Mzibri M, Bogusz D, Smouni A (2013) Effect of lead on root growth. *Frontiers in Plant Science 4*:175.

Faraz A, Faizan M, Sami F, Siddiqui H, Hayat S (2019) Supplementation of salicylic acid and citric acid for alleviation of cadmium toxicity to *Brassica juncea. Journal of Plant Growth Regulation* 1–15.

Farid M, Shakoor MB, Ehsan S, Ali S, Zubair M, Hanif M (2013) Morphological, physiological and biochemical responses of different plant species to Cd stress. *International Journal of Chemical and Biochemical Sciences 3*:53–60.

Fecht-Christoffers MM, Braun H-P, Lemaitre-Guillier C, VanDorsselaer A, Horst WJ (2003) Effect of manganese toxicity on the proteome of the leaf apoplast in cowpea. *Plant Physiology 133*:1935–1946.

Fernando DR, Lynch JP (2015) Manganese phytotoxicity: new light on an old problem. *Annals of Botany 116*:313–319.

Fidalgo F, Azenha M, Silva AF, de Sousa A, Santiago A, Ferraz P, Teixeira J (2013) Copper-induced stress in *Solanum nigrum* L. and antioxidant defense system responses. *Food and Energy Security 2*:70–80.

Flora S, Mittal M, Mehta A (2008) Heavy metal induced oxidative stress & its possible reversal by chelation therapy. *Indian Journal of Medical Research 128*:501.

Fozia A, Muhammad AZ, Muhammad A, Zafar MK (2008) Effect of chromium on growth attributes in sunflower (*Helianthus annuus* L.). *Journal of Environmental Sciences 20*:1475–1480.

Führs H, Behrens C, Gallien S, Heintz D, Van Dorsselaer A, Braun H-P, Horst WJ (2010) Physiological and proteomic characterization of manganese sensitivity and tolerance in rice (*Oryza sativa*) in comparison with barley (*Hordeum vulgare*). *Annals of Botany 105*:1129–1140.

Fukai T, Ushio-Fukai M (2011) Superoxide dismutases: role in redox signaling, vascular function, and diseases. *Antioxidants & Redox Signaling 15*:1583–1606.

Gajewska E, Bernat P, Długoński J, Skłodowska M (2012) Effect of nickel on membrane integrity, lipid peroxidation and fatty acid composition in wheat seedlings. *Journal of Agronomy and Crop Science 198*:286–294.

Gajewska E, Skłodowska M (2009) Nickel-induced changes in nitrogen metabolism in wheat shoots. *Journal of Plant Physiology 166*:1034–1044.

Garg N, Aggarwal N (2012) Effect of mycorrhizal inoculations on heavy metal uptake and stress alleviation of *Cajanus cajan* (L.) Millsp. genotypes grown in cadmium and lead contaminated soils. *Plant Growth Regulation 66*:9–26.

Garg N, Chandel S (2012) Role of arbuscular mycorrhizal (AM) fungi on growth, cadmium uptake, osmolyte, and phytochelatin synthesis in *Cajanus cajan* (L.) Millsp. under NaCl and Cd stresses. *Journal of Plant Growth Regulation 31*:292–308.

Geszvain K et al. (2012) The molecular biogeochemistry of manganese (II) oxidation. *Biochemical Society Transactions 40*(6):1244–1248.

Gill SS, Tuteja N (2010) Reactive oxygen species and antioxidant machinery in abiotic stress tolerance in crop plants. *Plant Physiology and Biochemistry 48*:909–930.

Gopal R, Nautiyal N (2012) Growth, antioxidant enzymes activities, and proline accumulation in mustard due to nickel. *International Journal of Vegetable Science 18*:223–234.

Grennan AK (2011) Metallothioneins, a diverse protein family. *Plant Physiology 155*:1750–1751.

Gu H-H et al. (2012) Silicon-mediated amelioration of zinc toxicity in rice (*Oryza sativa* L.) seedlings. *Plant and Soil 350*:193–204.

Guo J et al. (2018) Effects of salicylic acid, Epi-brassinolide and calcium on stress alleviation and Cd accumulation in tomato plants. *Ecotoxicology and Environmental Safety 157*:491–496.

Gupta DK, Vandenhove H, Inouhe M (2013) Role of phytochelatins in heavy metal stress and detoxification mechanisms in plants. In: Dharmendra KG, Francisco JC, José MP (eds) *Heavy metal stress in plants.* Springer, pp 73–94.

Gupta SK, Rai AK, Kanwar SS, Sharma TR (2012) Comparative analysis of zinc finger proteins involved in plant disease resistance. *PLoS One 7*(8): e42578. https://doi.org/10.1371/journal.pone.0042578.

Hall J (2002) Cellular mechanisms for heavy metal detoxification and tolerance. *Journal of Experimental Botany 53*:1–11.

Hasan M, Cheng Y, Kanwar MK, Chu X-Y, Ahammed GJ, Qi Z-Y (2017) Responses of plant proteins to heavy metal stress—a review. *Frontiers in Plant Science 8*:1492.

Hasanuzzaman M, Matin MA, Fardus J, Hasanuzzaman M, Hossain MS, Parvin K (2019) Foliar application of salicylic acid improves growth and yield attributes by upregulating the antioxidant defense system in Brassica campestris plants grown in lead-amended soils. *Acta Agrobotanica 72*(2):1765.

Hassan MU et al. (2019) Nickel toxicity in plants: reasons, toxic effects, tolerance mechanisms, and remediation possibilities—a review. *Environ Sci Poll Res 26*:12673–12688.

Hayat S, Hayat Q, Alyemeni MN, Ahmad A (2013) Proline enhances antioxidative enzyme activity, photosynthesis and yield of *Cicer arietinum* L. exposed to cadmium stress. *Acta Botanica Croatica 72*:323–335.

Hegelund JN, Schiller M, Kichey T, Hansen TH, Pedas P, Husted S, Schjoerring JK (2012) Barley metallothioneins: MT3 and MT4 are localized in the grain aleurone layer and show differential zinc binding. *Plant Physiol 159*:1125–1137.

Heiss S, Wachter A, Bogs J, Cobbett C, Rausch T (2003) Phytochelatin synthase (PCS) protein is induced in *Brassica juncea* leaves after prolonged Cd exposure. *J Exp Bot 54*:1833–1839.

Hossain MA, Piyatida P, da Silva JAT, Fujita M (2012) Molecular mechanism of heavy metal toxicity and tolerance in plants: central role of glutathione in detoxification of reactive oxygen species and methylglyoxal and in heavy metal chelation. *J Bot 2012*:1–37.

Huang G-Y, Wang Y-S (2010) Expression and characterization analysis of type 2 metallothionein from grey mangrove species (*Avicennia marina*) in response to metal stress. *Aquatic Toxicol 99*:86–92.

Inostroza-Blancheteau C, Reyes-Díaz M, Aquea F, Nunes-Nesi A, Alberdi M, Arce-Johnson P (2011) Biochemical and molecular changes in response to aluminium-stress in highbush blueberry (*Vaccinium corymbosum* L.). *Plant Physiol Biochem 49*:1005–1012.

Irvine GW, Tan SN, Stillman MJ (2017) A simple metallothionein-based biosensor for enhanced detection of arsenic and mercury. *Biosensors 7*:14.

Islam F, Yasmeen T, Arif MS, Riaz M, Shahzad SM, Imran Q, Ali I (2016) Combined ability of chromium (Cr) tolerant plant growth promoting bacteria (PGPB) and salicylic acid (SA) in attenuation of chromium stress in maize plants. *Plant Physiol Biochem 108*:456–467.

Janoušková M, Pavlíková D (2010) Cadmium immobilization in the rhizosphere of arbuscular mycorrhizal plants by the fungal extraradical mycelium. *Plant Soil 332*:511–520.

Jianguo L, Zhongquan Z, Dinglong L, Wong MW (2010) *Toxicity of cadmium on different rice genotypes*. IEEE.

John R, Ahmad P, Gadgil K, Sharma S (2009) Heavy metal toxicity: Effect on plant growth, biochemical parameters and metal accumulation by *Brassica juncea* L. *International Journal of Plant Production 3*:65–76.

Karimi P, Khavari-Nejad R, Niknam V, Ghahremaninejad F, Najafi F (2012) The effects of excess copper on antioxidative enzymes, lipid peroxidation, proline, chlorophyll, and concentration of Mn, Fe, and Cu in Astragalus neo-mobayenii. *Sci World J 2012*:615670.

Kastori R, Petrović M, Petrović N (1992) Effect of excess lead, cadmium, copper, and zinc on water relations in sunflower. *J Plant Nutr 15*:2427–2439.

Kato FH, Carvalho MEA, Gaziola SA, Piotto FA, Azevedo RA (2020) Lysine metabolism and amino acid profile in maize grains from plants subjected to cadmium exposure. *Scientia Agricola* 77.

Keinänen SI, Hassinen VH, Kärenlampi SO, Tervahauta AI (2007) Isolation of genes up-regulated by copper in a copper-tolerant birch (*Betula pendula*) clone. *Tree Physiol 27*:1243–1252.

Khan MN, Mobin M, Abbas ZK, AlMutairi KA, Siddiqui ZH (2017) Role of nanomaterials in plants under challenging environments. *Plant Physiol Biochem 110*:194–209.

Kim Y-N et al. (2011) Cadmium resistance in tobacco plants expressing the MuSI gene. *Plant Biotechnol Rep 5*:323–329.

Kinraide TB (1997) Reconsidering the rhizotoxicity of hydroxyl, sulphate, and fluoride complexes of aluminium. *J Exp Bot 48*:1115–1124.

Kochian LV, Hoekenga OA, Pineros MA (2004) How do crop plants tolerate acid soils? Mechanisms of aluminum tolerance and phosphorous efficiency. *Annu Rev Plant Biol 55*:459–493.

Kochian LV, Pineros MA, Hoekenga OA (2005) The physiology, genetics and molecular biology of plant aluminum resistance and toxicity. In: *Root physiology: from gene to function*. Springer, pp 175–195.

Kotrba P, Najmanova J, Macek T, Ruml T, Mackova M (2009) Genetically modified plants in phytoremediation of heavy metal and metalloid soil and sediment pollution. *Biotechnol Adv 27*:799–810.

Kumar G et al. (2012) Clustered metallothionein genes are co-regulated in rice and ectopic expression of OsMT1e-Pconfers multiple abiotic stress tolerance in tobacco via ROS scavenging. *BMC Plant Biol 12*:107.

Kumar KA, Varaprasad P, Rao AVB (2009) Effect of fluoride on catalase, guiacol peroxidase and ascorbate oxidase activities in two verities of mulberry leaves (*Morus alba* L.). *Res J Earth Sci 1*:69–73.

Lalelou F, Shafagh-Kolvanagh J, Fateh M (2013) Effect of various concentrations of zinc on chlorophyll, starch, soluble sugars and proline in naked pumpkin (*Cucurbita pepo*). *Int J Farming Allied Sci 2*:1198–1202.

Lemos Batista B et al. (2014) Identification and quantification of phytochelatins in roots of rice to long-term exposure: evidence of individual role on arsenic accumulation and translocation. *J Exp Bot 65*:1467–1479.

Lešková A, Zvarík M, Araya T, Giehl RF (2020) Nickel toxicity targets cell wall-related processes and PIN2-mediated auxin transport to inhibit root elongation and gravitropic responses in Arabidopsis. *Plant Cell Physiol 61*:519–535.

Li H et al. (2016a) Do arbuscular mycorrhizal fungi affect cadmium uptake kinetics, subcellular distribution and chemical forms in rice? *Sci Total Environ 571*:1183–1190.

Li H, Yang L-T, Qi Y-P, Guo P, Lu Y-B, Chen L-S (2016b) Aluminum toxicity-induced alterations of leaf proteome in two citrus species differing in aluminum tolerance. *Int J Mol Sci 17*:1180.

Li W, Du J, Feng H, Wu Q, Xu G, Shabala S, Yu L (2020) Function of NHX-type transporters in improving rice tolerance to aluminum stress and soil acidity. *Planta 251*:1–13.

Lidon FC, Barreiro MG, Ramalho JC (2004) Manganese accumulation in rice: implications for photosynthetic functioning. *Journal of Plant Physiology 161*:1235–1244.

Liu J, Ding G, Gai Z, Zhang W, Han Y, Li W (2020a) Changes in the gene expression profile of Arabidopsis thaliana under chromium stress. *Ecotoxicol Environ Saf 193*:110302.

Liu W et al. (2015) Transcriptome-wide analysis of chromium-stress responsive microRNAs to explore miRNA-mediated regulatory networks in radish (*Raphanus sativus* L.). *Sci Rep 5*:14024.

Liu Y-T, Shi Q-H, Cao H-J, Ma Q-B, Nian H, Zhang X-X (2020b) Heterologous expression of a glycine soja C2H2 zinc finger gene improves aluminum tolerance in arabidopsis. *Int J Mol Sci 21*:2754.

Liu Y et al. (2020c) Heterologous expression of the metallothionein PpMT2 gene from Physcomitrella patens confers enhanced tolerance to heavy metal stress on transgenic Arabidopsis plants. *Plant Growth Reg 90*:63–72.

Lwalaba JLW et al. (2020) Physiological and molecular mechanisms of cobalt and copper interaction in causing phyto-toxicity to two barley genotypes differing in Co tolerance. *Ecotoxicol Environ Saf 187*:109866.

Aamer M, Muhammad U, Abid A, Su Q, Liu Y, Adnan R, Muhammad AUK, Tahir AK, Huang G (2018) Foliar application of glycinebetaine (GB) alleviates the cadmium (Cd) toxicity in spinach through reducing Cd uptake and improving the activity of antioxidant systems. *Appl Ecol Environ Res 16*:7575–7583.

Macovei A, Gill SS, Tuteja N (2012) microRNAs as promising tools for improving stress tolerance in rice. *Plant Signal Behav 7*:1296–1301.

Maksymiec W (1998) Effect of copper on cellular processes in higher plants. *Photosynthetica 34*:321–342.

Manara A (2012a) Plant responses to heavy metal toxicity. In: *Plants and heavy metals*. Springer, pp 27–53.

Manara A (2012b) Plants and heavy metals. *Signal Transduct 2754.*

Mantovi P, Bonazzi G, Maestri E, Marmiroli N (2003) Accumulation of copper and zinc from liquid manure in agricultural soils and crop plants. *Plant Soil 250*:249–257.

Maret W (2000) The function of zinc metallothionein: a link between cellular zinc and redox state. *J Nutr 130*:1455S–1458S.

Mathur S, Kalaji H, Jajoo A (2016) Investigation of deleterious effects of chromium phytotoxicity and photosynthesis in wheat plant. *Photosynthetica 54*:185–192.

McIlveen W, Negusanti J (1994) Nickel in the terrestrial environment. *Sci Total Environ 148*:109–138.

Mekawy AMM, Assaha DV, Ueda A (2020) Constitutive overexpression of rice metallothionein-like gene OsMT-3a enhances growth and tolerance of Arabidopsis plants to a combination of various abiotic stresses. *J Plant Res 133*:429–440.

Mendoza-Cózatl DG, Butko E, Springer F, Torpey JW, Komives EA, Kehr J, Schroeder JI (2008) Identification of high levels of phytochelatins, glutathione and cadmium in the phloem sap of *Brassica napus*. A role for thiol-peptides in the long-distance transport of cadmium and the effect of cadmium on iron translocation. *Plant J 54*:249–259.

Mengoni A, Gonnelli C, Hakvoort H, Galardi F, Bazzicalupo M, Gabbrielli R, Schat H (2003) Evolution of copper-tolerance and increased expression of a 2b-type metallothionein gene in *Silene paradoxa* L. populations. *Plant Soil 257*:451–457.

Michaud AM, Bravin M, Galleguillos M, Hinsinger P (2007) Copper uptake and phytotoxicity as assessed in situ for durum wheat (*Triticum turgidum durum* L.) cultivated in Cu-contaminated, former vineyard soils. *Plant Soil 298*:99–111.

Miransari M (2017) Arbuscular mycorrhizal fungi and heavy metal tolerance in plants. In: *Arbuscular mycorrhizas and stress tolerance of plants*. Springer, pp 147–161.

Miyasaka SC, Hamasaki RT, de la Pena RS (2002) *Nutrient deficiences and excesses in taro*. Honolulu, Hawaii: University of Hawaii.

Moussa H, El-Gamal SM (2010) Effect of salicylic acid pretreatment on cadmium toxicity in wheat. *Biol Plantarum 54*:315–320.

Mukhopadhyay MJ, Sharma A (1991) Manganese in cell metabolism of higher plants. *Botanical Rev 57*:117–149.

Mukta RH, Khatun MR, Nazmul Huda A (2019) Calcium induces phytochelatin accumulation to cope with chromium toxicity in rice (*Oryza sativa* L.). *J Plant Interac 14*:295–302.

Myśliwa-Kurdziel B, Prasad M, Strzałtka K (2004) *Photosynthesis in heavy metal stressed plants*. In: *Heavy metal stress in plants*. Springer, pp 146–181.

Nath K, Singh D, Shyam S, Sharma YK (2008) Effect of chromium and tannery effluent toxicity on metabolism and growth in cowpea (*Vigna sinensis* L. Saviex Hassk) seedling. *Res Environ Life Sci 1*:91–94.

Nikita G, Vinit S, Wang Y, Martina R, Weisman GA, Petris MJ (2020) Metallothioneins regulate ATP7A trafficking and control cell viability during copper deficiency and excess. *Sci Rep (Nature Publisher Group) 10*.

Nikolić N, Kojić D, Pilipović A, Pajević S, Krstić B, Borišev M, Orlović S (2008) Responses of hybrid poplar to cadmium stress: photosynthetic characteristics, cadmium and proline accumulation, and antioxidant enzyme activity. *Acta Biologica Cracoviensia Series Botanica 50*:95–103.

Nunes-Nesi A, Brito DS, Inostroza-Blancheteau C, Fernie AR, Araújo WL (2014) The complex role of mitochondrial metabolism in plant aluminum resistance. *Trends in Plant Science 19*:399–407.

Ouziad F, Hildebrandt U, Schmelzer E, Bothe H (2005) Differential gene expressions in arbuscular mycorrhizal-colonized tomato grown under heavy metal stress. *J Plant Physiol 162*:634–649.

Ozkay F, Kiran S, İsmail T, Kusvuran Ş (2014) Effects of copper, zinc, lead and cadmium applied with irrigation water on some eggplant plant growth parameters and soil properties.

Pagani M, Tomas M, Carrillo J, Bofill R, Capdevila M, Atrian S, Andreo C (2012) The response of the different soybean metallothionein isoforms to cadmium intoxication. *J Inorganic Biochem 117*:306–315.

Parashar A, Yusuf M, Fariduddin Q, Ahmad A (2014) Salicylic acid enhances antioxidant system in *Brassica juncea* grown under different levels of manganese. *Int J Biol Macromol 70*:551–558.

Parlak KU (2016) Effect of nickel on growth and biochemical characteristics of wheat (*Triticum aestivum* L.) seedlings. *NJAS-Wageningen J Life Sci 76*:1–5.

Patra M, Bhowmik N, Bandopadhyay B, Sharma A (2004) Comparison of mercury, lead and arsenic with respect to genotoxic effects on plant systems and the development of genetic tolerance. *Environ Exp Bot 52*:199–223.

Pätsikkä E, Aro E-M, Tyystjärvi E (1998) Increase in the quantum yield of photoinhibition contributes to copper toxicity in vivo. *Plant Physiol 117*:619–627.

Peško M, Kráľová K (2013) Physiological response of *Brassica napus* L. plants to Cu (II) treatment. Proceedings of ECOpole 7.

Piechalak A, Tomaszewska B, Baralkiewicz D, Malecka A (2002) Accumulation and detoxification of lead ions in legumes. *Phytochemistry 60*:153–162.

Pillay SV, Rao V, Rao K (1996) Effect of nickel toxicity in *Hyptis suareeolens* (L.) Poit. and *Helianthus annuus* L. *Ind J Plant Physiol 1*:153–156.

Pintro J, Taylor G (2004) Effects of aluminum toxicity on wheat plants cultivated under conditions of varying ionic strength. *J Plant Nutr 27*:907–919.

Pourrut B, Shahid M, Dumat C, Winterton P, Pinelli E (2011) Lead uptake, toxicity, and detoxification in plants. In: *Reviews of Environmental Contamination and Toxicology* Volume *213*. Springer, pp 113–136.

Prasad M (2004) Phytoremediation of metals in the environment for sustainable development. *Proc Indian Nat Sci Acad Part B 70*:71–98.

Punamiya P, Datta R, Sarkar D, Barber S, Patel M, Das P (2010) Symbiotic role of Glomus mosseae in phytoextraction of lead in vetiver grass [*Chrysopogon zizanioides* (L.)]. *J Hazard Mater 177*:465–474.

Rahmaty R, Khara J (2011) Effects of vesicular arbuscular mycorrhiza Glomus intraradices on photosynthetic pigments, antioxidant enzymes, lipid peroxidation, and chromium accumulation in maize plants treated with chromium. *Turkish J Biol 35*:51–58.

Ranieri E, Fratino U, Petruzzelli D, Borges AC (2013) A comparison between Phragmites australis and Helianthus annuus in chromium phytoextraction. *Water Air Soil Poll 224*:1465.

Ranieri E, Gikas P (2014) Effects of plants for reduction and removal of hexavalent chromium from a contaminated soil. *Water Air Soil Poll 225*:1981.

Rascio N, Navari-Izzo F (2011) Heavy metal hyperaccumulating plants: how and why do they do it? And what makes them so interesting? *Plant Sci 180*:169–181.

Rasheed A, Fahad S, Hassan MU, Tahir MM, Aamer M, Wu Z (2020a) A review on aluminum toxicity and quantitative trait loci maping in rice (*Oryza sative* L). *Appl Ecol Environ Res 18*:3951–3961.

Rasheed A, Fahad S, Aamer M, Hassan MU, Tahir MM, Wu Z (2020b) Role of genetic factors in regulating cadmium uptake, transport and accumulation mechanisms and quantitative trait loci mapping in rice. a review. *Appl Ecol Environ Res 18*:4005–4023.

Rastgoo L, Alemzadeh A, Afsharifar A (2011) Isolation of two novel isoforms encoding zinc-and copper-transporting P1B-ATPase from Gouan (*Aeluropus littoralis*). *Plant Omics J 4*:377–383.

Rausch T, Gromes R, Liedschulte V, Müller I, Bogs J, Galovic V, Wachter A (2007) Novel insight into the regulation of GSH biosynthesis in higher plants. *Plant Biology 9*:565–572.

Redecker D, Schüßler A, Stockinger H, Stürmer SL, Morton JB, Walker C (2013) An evidence-based consensus for the classification of arbuscular mycorrhizal fungi (Glomeromycota). *Mycorrhiza 23*:515–531.

Ren L, Liu P (2007) Review of manganese toxicity & the mechanisms of plant tolerance. *Acta Ecol Sinica 27*:357–367.

Richau KH et al. (2009) Chelation by histidine inhibits the vacuolar sequestration of nickel in roots of the hyperaccumulator Thlaspi caerulescens. *New Phytol 183*:106–116.

Rizwan M, Ali S, Rehman MZ, Maqbool A (2019) A critical review on the effects of zinc at toxic levels of cadmium in plants. *Environ Sci Poll Res 26*:6279–6289.

Rizwan M, Imtiaz M, Dai Z, Mehmood S, Adeel M, Liu J, Tu S (2017) Nickel stressed responses of rice in Ni subcellular distribution, antioxidant production, and osmolyte accumulation. *Environ Sci Poll Res 24*:20587–20598.

Rodrigo MAM, Anjum NA, Heger Z, Zitka O, Vojtech A, Pereira E, Kizek R (2016) Role of phytochelatins in redox caused stress in plants and animals. *Abiotic and Biotic Stress in Plants-Recent Advances and Future Perspectives*, 395.

Rosas A, Rengel Z, de la Luz Mora M (2007) Manganese supply and pH influence growth, carboxylate exudation and peroxidase activity of ryegrass and white clover. *J Plant Nutr 30*:253–270.

Rout GR, Samantaray S, Das P (2000) Effects of chromium and nickel on germination and growth in tolerant and non-tolerant populations of *Echinochloa colona* (L.) Link. *Chemosphere 40*:855–859.

Saba H, Jyoti P, Neha S (2013) Mycorrhizae and phytochelators as remedy in heavy metal contaminated land remediation. *Int Res J Environ Sci 2*:74–78.

Sácký J, Leonhardt T, Borovička J, Gryndler M, Briksí A, Kotrba P (2014) Intracellular sequestration of zinc, cadmium and silver in Hebeloma mesophaeum and characterization of its metallothionein genes. *Fungal Genet Biol 67*:3–14.

Sagardoy R, Morales F, López-Millán AF, Abadía A, Abadía J (2009) Effects of zinc toxicity on sugar beet (*Beta vulgaris* L.) plants grown in hydroponics. *Plant Biol 11*:339–350.

Schmidt SB, Husted S (2019) The biochemical properties of manganese in plants. *Plants 8*:381.

Schutzendubel A, Polle A (2002) Plant responses to abiotic stresses: heavy metal-induced oxidative stress and protection by mycorrhization. *J Exp Bot 53*:1351–1365.

Shahid M et al. (2015) Heavy metal stress and crop productivity. In: *Crop production and global environmental issues*. Springer, pp 1–25.

Shahid MA, Balal RM, Pervez MA, Abbas T, Aqeel MA, Javaid MM, Garcia-Sanchez F (2014) Exogenous proline and proline-enriched Lolium perenne leaf extract protects against phytotoxic effects of nickel and salinity in *Pisum sativum* by altering polyamine metabolism in leaves. *Turkish J Bot 38*:914–926.

Shahzad Z, Rouached H, Rakha A (2014) Combating mineral malnutrition through iron and zinc biofortification of cereals. *Comprehensive Rev Food Sci Food Saf 13*:329–346.

Shanmugam V et al. (2011) Differential expression and regulation of iron-regulated metal transporters in *Arabidopsis halleri* and *Arabidopsis thaliana*–the role in zinc tolerance. *New Phytologist 190*:125–137.

Sharma A et al. (2020a) Chromium bioaccumulation and its impacts on plants: an overview. *Plants 9*:100.
Sharma A, Patni B, Shankhdhar D, Shankhdhar S (2013) Zinc–an indispensable micronutrient. *Physiol Mol Biol Plants 19*:11–20.
Sharma A et al. (2020b) The role of salicylic acid in plants exposed to heavy metals. *Molecules 25*:540.
Sharma P, Jha AB, Dubey RS, Pessarakli M (2012) Reactive oxygen species, oxidative damage, and antioxidative defense mechanism in plants under stressful conditions. *J Bot 2012*.
Sharma R, Devi S, dan Dhyani P (2010) Comparative assessment of the toxic effects of copper and cypermethrin using seeds of *Spinacia Oleracea* L. plants. *Trop Ecol 51*:375–387.
Sharma SS, Dietz K-J (2006) The significance of amino acids and amino acid-derived molecules in plant responses and adaptation to heavy metal stress. *J Exp Bot 57*:711–726.
Sharma SS, Schat H, Vooijs R (1998) In vitro alleviation of heavy metal-induced enzyme inhibition by proline. *Phytochemistry 49*:1531–1535.
Sheldon A, Menzies N (2004) *The effect of copper toxicity on the growth and morphology of Rhodes grass (Chloris gayana) in solution culture.* In: *3rd Australian New Zealand Soils Conference*, 5–9 December, University of Sydney, Australia.
Shen G, Zhu C, Du Q (2010) Genome-wide identification of PHYTOCHELATIN and PHYTOCH_SYNTH domain-containing phytochelatin family from rice. *Electronic J Biol 6*:73–79.
Shen H, Christie P, Li X (2006) Uptake of zinc, cadmium and phosphorus by arbuscular mycorrhizal maize (*Zea mays* L.) from a low available phosphorus calcareous soil spiked with zinc and cadmium. *Environ Geochem Health 28*:111.
Shivakumar C, Hemavani C (2011) Effect of inoculation with arbuscular mycorrhizal fungi on green gram grown in soil containing heavy metal zinc. *J Exp Sci.*
Shukla D, Trivedi PK, Nath P, Tuteja N (2016) Metallothioneins and phytochelatins: role and perspectives in heavy metal (loid) s stress tolerance in crop plants. *Abiotic stress response in plants.* Wiley-VCH Verlag GmbH and Co. KGaA, Weinheim, Germany.
Siddique A, Kandpal G, Kumar P (2018) Proline accumulation and its defensive role under diverse stress condition in plants: an overview. *J Pure Appl Microbiol 12*:1655–1659.
Sihag S, Joshi U (2018) Toxic affects of chromium on some quality parameters of *Sorghum bicolor* (L.). *J Appl Nat Sci 10*:122–127.
Silva DM, Fonte NDS, Souza KRD, Rodrigues-Brandão I, Alves ITL, Donizeti J (2017) Relationship between manganese toxicity and waterlogging tolerance in *Zea mays* L. cv. Saracura. Acta Scientiarum. *Agronomy 39*:75–82.
Simões C, Melo J, Magalhaes J, Guimarães C (2012) Genetic and molecular mechanisms of aluminum tolerance in plants. *Embrapa Milho e Sorgo-Artigo em periódico indexado (ALICE).*
Singh AP et al. (2015a) Salicylic acid modulates arsenic toxicity by reducing its root to shoot translocation in rice (*Oryza sativa* L.). *Front Plant Sci 6*:340.
Singh D, Nath K, Sharma YK (2007) Response of wheat seed germination and seedling growth under copper stress. *J Environ Biol 28*:409.
Singh M, Kumar J, Singh S, Singh V, Prasad S, Singh M (2015b) Adaptation strategies of plants against heavy metal toxicity: a short review. *Biochem Pharmacol (Los Angel) 4*:2167-0501.1000161.
Singh S, Takkar P, Nayyar V (1989) Effect of cadmium on wheat as influenced by lime and manure and its toxic level in plant and soil. *Int J Environ Stud 33*:59–66.
Song W-Y et al. (2004) A novel family of cys-rich membrane proteins mediates cadmium resistance in Arabidopsis. *Plant Physiol 135*:1027–1039.
Song WY, Yang HC, Shao HB, Zheng AZ, Brestic M (2014) The alleviative effects of salicylic acid on the activities of catalase and superoxide dismutase in malting barley (*Hordeum uhulgare* L.) seedling leaves stressed by heavy metals. *Clean Soil Air Water 42*:88–97.
Sparrow L, Uren N (2014) Manganese oxidation and reduction in soils: effects of temperature, water potential, pH and their interactions. *Soil Res 52*:483–494.
Sreekanth T, Nagajyothi P, Lee K, Prasad T (2013) Occurrence, physiological responses and toxicity of nickel in plants. *Int J Environ Sci Technol 10*:1129–1140.
Stambulska UY, Bayliak MM, Lushchak VI (2018) Chromium (VI) toxicity in legume plants: modulation effects of rhizobial symbiosis. *BioMed Res Int 2018.*
Stass A, Wang Y, Eticha D, Horst W (2006) Aluminium rhizotoxicity in maize grown in solutions with Al3+ or Al (OH) 4– as predominant solution Al species. *J Exp Bot 57*:4033–4042.

Sun P, Tian Q-Y, Chen J, Zhang W-H (2010) Aluminium-induced inhibition of root elongation in Arabidopsis is mediated by ethylene and auxin. *J Exp Bot 61*:347–356.
Sun Z, Song J, Xa X, Xie X, Zhao B (2018) Arbuscular mycorrhizal fungal 14-3-3 proteins are involved in arbuscule formation and responses to abiotic stresses during AM symbiosis. *Front Microbiol 9*:91.
Sunkar R, Kapoor A, Zhu J-K (2006) Posttranscriptional induction of two Cu/Zn superoxide dismutase genes in Arabidopsis is mediated by downregulation of miR398 and important for oxidative stress tolerance. *Plant Cell 18*:2051–2065.
Sytar O, Kumar A, Latowski D, Kuczynska P, Strzałka K, Prasad M (2013) Heavy metal-induced oxidative damage, defense reactions, and detoxification mechanisms in plants. *Acta Physiol Plantarum 35*:985–999.
Szab-Nagy A (2015) Aluminium toxicity in winter wheat. *Acta Univ. Sapientiae, Alimentaria 8*:95–103.
Szalai G, Krantev A, Yordanova R, Popova LP, Janda T (2013) Influence of salicylic acid on phytochelatin synthesis in *Zea mays* during Cd stress. *Turkish J Bot 37*:708–714.
Takács T, Biró B, Vörös I (2001) Arbuscular mycorrhizal effect on heavy metal uptake of ryegrass (*Lolium perenne* L.) in pot culture with polluted soils. In: *Plant Nutrition*. Springer, pp 480–481.
Takkar P, Mann M (1978) Toxic levels of soil and plant zinc for maize and wheat. *Plant Soil 49*:667–669.
Tangahu BV, Abdullah S, Rozaimah S, Basri H, Idris M, Anuar N, Mukhlisin M (2011) A review on heavy metals (As, Pb, and Hg) uptake by plants through phytoremediation. *Int J Chem Eng 2011*:1–31.
Tao Y et al. (2018) Genome-wide association mapping of aluminum toxicity tolerance and fine mapping of a candidate gene for Nrat1 in rice. *PloS one 13*:e0198589.
Thapa G, Sadhukhan A, Panda SK, Sahoo L (2012) Molecular mechanistic model of plant heavy metal tolerance. *Biometals 25*:489–505.
Theriappan P, Gupta AK, Dhasarrathan P (2011) Accumulation of proline under salinity and heavy metal stress in cauliflower seedlings. *J Appl Sci Environ Manage 15*(2):251–255.
Theriault G, Nkongolo K (2017) Evidence of prokaryote like protein associated with nickel resistance in higher plants: horizontal transfer of TonB-dependent receptor/protein in Betula genus or de novo mechanisms? *Heredity 118*:358–365.
Tiwari S, Lata C (2018) Heavy metal stress, signaling, and tolerance due to plant-associated microbes: an overview. *Front Plant Sci 9*:452.
Todeschini V, Lingua G, D'agostino G, Carniato F, Roccotiello E, Berta G (2011) Effects of high zinc concentration on poplar leaves: a morphological and biochemical study. *Environ Exp Bot 71*:50–56.
Tomas M, Pagani M, Andreo C, Capdevila M, Atrian S, Bofill R (2015) Sunflower metallothionein family characterisation. Study of the Zn (II)-and Cd (II)-binding abilities of the HaMT1 and HaMT2 isoforms. *J Inorganic Biochem 148*:35–48.
Valko M, Morris H, Cronin M (2005) Metals, toxicity and oxidative stress. *Curr Med Chem 12*:1161–1208.
Vassilev A, Nikolova A, Koleva L, Lidon F (2011) Effects of excess Zn on growth and photosynthetic performance of young bean plants. *Journal of Phytology 3*(6):58–62.
Ventrella A, Catucci L, Piletska E, Piletsky S, Agostiano A (2009) Interactions between heavy metals and photosynthetic materials studied by optical techniques. *Bioelectrochemistry 77*:19–25.
Verma JP, Singh V, Yadav J (2011) Effect of copper sulphate on seed germination, plant growth and peroxidase activity of Mung bean (*Vigna radiata*). *Int J Bot 7*:200–204.
Viehweger K (2014) How plants cope with heavy metals. *Bot Stud 55*:35.
Visioli G, Vincenzi S, Marmiroli M, Marmiroli N (2012) Correlation between phenotype and proteome in the Ni hyperaccumulator Noccaea caerulescens subsp. caerulescens. *Environ Exp Bot 77*:156–164.
Wang H, Zhong G, Shi G, Pan F (2010) *Toxicity of Cu, Pb, and Zn on seed germination and young seedlings of wheat (Triticum aestivum L.)*. In: *International Conference on Computer and Computing Technologies in Agriculture. Springer*, pp 231–240.
Wang Y, Huang J, Gao Y (2012) Arbuscular mycorrhizal colonization alters subcellular distribution and chemical forms of cadmium in *Medicago sativa* L. and resists cadmium toxicity. *PLoS One 7*(11): e48669. https://doi.org/10.1371/journal.pone.0048669.
Watmough SA, Eimers MC, Dillon PJ (2007) Manganese cycling in central Ontario forests: response to soil acidification. *Appl Geochem 22*:1241–1247.
Wong HL, Sakamoto T, Kawasaki T, Umemura K, Shimamoto K (2004) Down-regulation of metallothionein, a reactive oxygen scavenger, by the small GTPase OsRac1 in rice. *Plant Physiol 135*:1447–1456.

Xia Y et al. (2012) Overexpression of Elsholtzia haichowensis metallothionein 1 (EhMT1) in tobacco plants enhances copper tolerance and accumulation in root cytoplasm and decreases hydrogen peroxide production. *J Hazard Mater 233*:65–71.

Xianghua X, Xincai C, Jiyan S, Yingxu C (2007) Effects of manganese on uptake and translocation of nutrients in a hyperaccumulator. *J Plant Nutr 30*:1737–1751.

Xu J-K, Yang L-X, Wang Z-Q, Dong G-C, Huang J-Y, Wang Y-L (2005) Effects of soil copper concentration on growth, development and yield formation of rice (*Oryza sativa*). *Rice Sci 12*:125–135.

Xu L-M et al. (2018) Transcriptomic responses to aluminum (Al) stress in maize. *J Integr Agric 17*:1946–1958.

Yadav S (2010) Heavy metals toxicity in plants: an overview on the role of glutathione and phytochelatins in heavy metal stress tolerance of plants. *South Afr J Bot 76*:167–179.

Yadav V et al. (2020) Roles of microRNAs in plant development and stress tolerance. In: *Plant Life Under Changing Environment*. Elsevier, pp 719–733.

Yang J, Wang Y, Liu G, Yang C, Li C (2011) Tamarix hispida metallothionein-like ThMT3, a reactive oxygen species scavenger, increases tolerance against Cd 2+, Zn 2+, Cu 2+, and NaCl in transgenic yeast. *Mol Biol Rep 38*:1567–1574.

Yang ZM, Chen J (2013) A potential role of microRNAs in plant response to metal toxicity. *Metallomics 5*:1184–1190.

Yusuf M, Fariduddin Q, Hayat S, Ahmad A (2011) Nickel: an overview of uptake, essentiality and toxicity in plants. *Bulletin of Environ Contamin Toxicol 86*:1–17.

Zahoor M, Afzal M, Ali M, Mohammad W, Khan N, Adnan M, Saeed M (2016a). Effect of organic waste and NPK fertilizer on potato yield and soil fertility. *Pure Appl Biol 5*(3):439–445.

Zahoor M, Khan N, Ali M, Saeed M, Ullah Z, Adnan M, Ahmad B (2016b). Integrated effect of organic waste and NPK fertilizers on nutrients uptake in potato crop and soil fertility. *Pure Appl Biol 5*(3):601–607.

Zeng F, Qiu B, Wu X, Niu S, Wu F, Zhang G (2012) Glutathione-mediated alleviation of chromium toxicity in rice plants. *Biol Trace Element Research 148*:255–263.

Zengin FK, Kirbag S (2007) Effects of copper on chlorophyll, proline, protein and abscisic acid level of sunflower (*Helianthus annuus* L.) seedlings. *J Environ Biol 28*:561.

Zengin FK, Munzuroglu O (2005) Effects of some heavy metals on content of chlorophyll, proline and some antioxidant chemicals in bean (*Phaseolus vulgaris* L.) seedlings. *Acta Biologica Cracoviensia Series Botanica 47*:157–164.

Zhang H, Jiang Z, Qin R, Zhang H, Zou J, Jiang W, Liu D (2014) Accumulation and cellular toxicity of aluminum in seedling of *Pinus massoniana*. *BMC Plant Biology 14*:264.

Zhigang A, Cuijie L, Yuangang Z, Yejie D, Wachter A, Gromes R, Rausch T (2006) Expression of BjMT2, a metallothionein 2 from Brassica juncea, increases copper and cadmium tolerance in *Escherichia coli* and *Arabidopsis thaliana*, but inhibits root elongation in Arabidopsis thaliana seedlings. *J Exp Bot 57*:3575–3582.

Zhou B, Yao W, Wang S, Wang X, Jiang T (2014) The metallothionein gene, TaMT3, from *Tamarix androssowii* confers Cd2+ tolerance in tobacco. *Int J Mol Sci 15*:10398–10409.

Zhou J et al. (2008) OsPHR2 is involved in phosphate-starvation signaling and excessive phosphate accumulation in shoots of plants. *Plant Physiol 146*:1673–1686.

Zhu C, Ding Y, Liu H (2011) MiR398 and plant stress responses. *Physiol Plantarum 143*:1–9.

Živković J, Ražić S, Arsenijević J, Maksimović Z (2012) Heavy metal contents in Veronica species and soil from mountain areas in Serbia. *J Serbian Chem Soc 77*:959–970.

Zou J, Wang M, Jiang W, Liu D (2006) Chromium accumulation and its effects on other mineral elements in Amaranthus *viridis* L. *Acta Biol Cracoviensia Series Bot 48*:7–12.

7

Soil Salinity and Climate Change

Abdel Rahman Mohammad Al-Tawaha
Al Hussein Bin Talal University, Jordan

Nezar Samarah
Jordan University of Science and Technology, Jordan

Aman Deep Ranga and Mayur S. Darvhankar
Lovely Professional University Phagwara, India

P. Saranraj
Sacred Heart College (Autonomous), India.

Alireza Pour-Aboughadareh
Agricultural Research, Education and Extension Organization (AREEO), Iran

Kadambot H.M. Siddique
The University of Western Australia, Australia

Amanullah and Imran
The University of Agriculture Peshawar, Pakistan

Ali M. Qaisi
University of Jordan, Jordan

Abdel Razzaq Al-Tawaha
University Putra Malaysia, Malaysia

Shah Khalid
The University of Agriculture Peshawar, Pakistan

Abdur Rauf
University of Swabi, Pakistan

Devarajan Thangadurai
Karnatak University, India

Jeyabalan Sangeetha
Central University of Kerala, India

Shah Fahad
The University of Haripur, Pakistan

Wafa'a A. Al-Taisan
Imam Abdulrahman Bin Faisal University, Saudi Arabia

Duraid K. A. Al-Taey
Al-Qasim Green University, Iraq

CONTENTS

7.1 Introduction

There are many varieties of salt-tolerant plants, which have in recent years become a source of optimism for the future cultivation of deserts and land that is not arable, either by benefiting directly from them and by cultivating them as crops that are capable of human consumption (as in the case of salicornia) or as fodder for livestock, or by benefiting indirectly from them (Turk and Tawaha 2002a, 2002b, 2002c). It is thought that it will be possible to transfer salt-resistant genes from these plants to traditional agricultural crops. Most of the salt, sodium, and boron accumulated in irrigated soils comes from irrigation water, reflecting the importance of good-quality irrigation (Turk and Tawaha 2001; Tawaha and Turk 2002). If the sodium is absorbed on clay pellets in greater quantities than usual, this will harm the soil consistency. If the absorbed sodium exceeds 10–20%, the natural properties of the soil will be compromised. In this case, the pH measurement may not be high. If the soil contains a significant percentage of boron, for example, this will prove highly damaging to many plants. Individual plants differ in terms of their tolerance to different soil salinity ratios. There have been studies of the extent to which plants are affected by the salinity of the soil, where the production of plants decreases, and plants are classified on the basis of reducing production by half, not the lack thereof. The smaller size of the plants cultivated in salty soil is a result of the high soil concentration of salts and, consequently, the plants do not benefit from the entire amount of water available, as the dissolved salts in the soil add an osmotic pressure that increases the moisture tension—usually—when there is a decrease in the percentage of water in the soil. There are ions that are more at risk than other dissolved ions in the soil. These ions include boron and chloride, and this risk depends on the type of plant and the presence of other ions counteracting the effect. Most plants are less tolerant of salinity, either during their germination phase or if they are very small (Tawaha et al. 2003; Turk et al. 2003a, 2003b, 2003c). This problem can be overcome through the use of appropriate irrigation and by choosing the appropriate method for planting, such as planting seedlings or seeds in the first third of the oblique distance between the top of the line and the bottom of the irrigation canal, where the salts accumulate at the top of the line away from the plant. Many researchers working in the field agree that the level of the groundwater, its source and salinity should be known if possible, especially if there is insufficient soil drainage (in which case a plan to drain the water must be developed). Salinity treatment is considered to be a straightforward process and work must be taken to control the existing salts in the soil and coexist with them so that they do not exceed the permissible limits through the integration of agricultural operations such as plowing, fertilizing, irrigation, drainage and salinity treatment.

7.2 Origin of Soil Salinity

The start of the 21st century was a period of profound climate change, followed by environmental pollution, the shortage of water resources, and the enhanced salinization of both soil and water (Shahbaz and Ashraf 2013; Ahmadi et al. 2018). According to the United Nation's Food and Agriculture Organization (FAO), the United States Salinity Laboratory (USSL), and others monitoring salinity, about one billion of the 13 billion hectares worldwide are now salt-affected. The loss of arable land due to salinization is estimated to be 10 million hectares per year (Anonymous 2019), a figure which is expected to increase through excessive use of groundwater for irrigation-related intensive farming and also the use of low-quality water in irrigation (Machado and Serralheiro 2017). Furthermore, it is estimated that around 33% of irrigated agricultural land and 20% of total cropland worldwide have become salinized due to inappropriate agricultural practices and it is predicted that, by 2050, as much as half of total cropland will be salinized (Nachshon 2018).

Soil salinity relates to the level of salt in the soil; the process of enhancing the salt substance is known as either salinization or the salt accumulation process. The imbalance in water and nutrient uptake due to osmotic effects and the increased toxicity resulting from the high ion concentrations are the major direct effects of salinity stress on plant growth. Hence, plans to resolve issues linked to soil salinization should focus on strategies to prevent salt accumulation, to remove accumulated salts, and to adapt plants to cope in saline environments (Rengasamy 2016). With this in mind, the profitable operation of salinized soils requires proper soil and irrigation organization and crops that can tolerate different levels of salinity stress. In the following section, we identify the sources of salt and how salinization occurs.

7.2.1 Salt Sources and Salinization Processes

The nature of soil salinization is related to the transportation of solutes within soil layers through several natural and anthropogenic processes. Although salts originate from a range of sources, the main source is the rocks and minerals present in the Earth's crust that weather with time. Furthermore, the ocean is a source for the redistribution of salts. To better understand the origin of soil salinity, below we discuss each salt source and the role it plays in soil salinization.

7.2.2 Natural Processes

Salts accumulate in the soil through several natural processes, including transport from parent material, weathering processes, floodwaters and waste effluent, and geological deposits or groundwater (Figure 7.1). Geological events and climatic changes (i.e., temperature and moisture) can enhance salt concentrations in groundwater and, consequently, soil layers when the dissolved salts transform at or near the surface through capillary pressure or evapotranspiration effects (Chari et al. 2012; van Beek and Toth 2012; Khan et al. 2016).

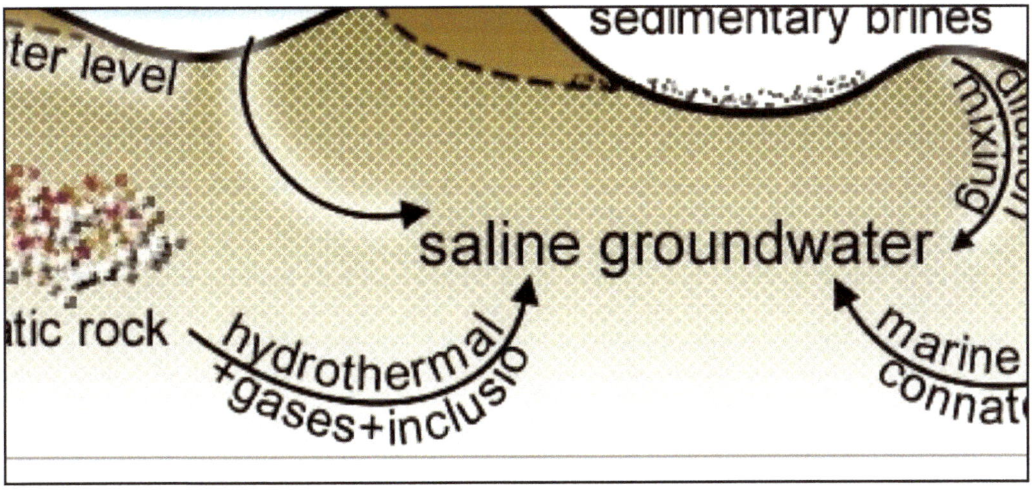

FIGURE 7.1 Soil salinization events from natural and human intervention processes (Daliakopoulos et al. 2016).

Seawater is another source of salinity which can contribute large amounts of salt to the soil. However, this process may be localized to small areas, particularly along the coast. The salt concentration in coastal soils can increase via seawater entry through estuaries and rivers, seawater high-tides in nearby surface soil, groundwater inflow, and the distribution of salt-enriched soils many kilometers inland from the coast and their deposition as dry fall-out or wash-out following showers (Zia-ur-Rehman et al. 2017). While these examples may only deposit small amounts of salt on the surface soil, over time this regular deposition leads to the salinization of the soil. In addition to parent material and seawater, several natural processes, such as brine deposits, shallow water tables, and restrictive drainage on some low-lying lands, may also contribute to salinization and the creation of salt flats. Saline flats occur when water on higher land permeates down to a sloped layer with relatively low levels of permeability in the soil. Water then flows laterally to nearby low-lying regions, where it creates a perched table. The lack of suitable drainage causes saline water to rise to the surface and evaporate, leaving behind salt deposits. When the water table is near the soil surface, especially in arid regions, water and salt will be transported upward by capillary action, and the upper soil profile and surface may become salinized as the water evaporates (Wallender and Tanji 2012).

7.2.3 Human Activities

Human intervention or activities are seen as a secondary process for salt accumulation in the soil (Figure 7.1). Human interventions include irrigation with salt-rich water and inappropriate irrigation practices, which are often coupled with weak drainage circumstances (Trnka et al. 2013). Arid climates are characterized by adverse evapotranspiration rates and inadequate and variable precipitation, as well as excessive heat. The overuse of irrigation water increases salt accumulation since it is not washed away by rainfall. This increasing trend of salinization is occurring in cultivated areas, with clay-textured soils and high rates of evaporation. These regions are known as prominent salinization hotspots. In addition, several other interventions—including the uneven distribution of irrigation water, an increase in the time of water ponding, a decrease sufficient drainage and an enhancement of the water table lever due to filtration from unlined canals—can mobilize salts that have accumulated in the deep layers to upper soil layers and increase the level of salinization (Barros et al. 2012; Daliakopoulos et al. 2016).

Soil pollution is another human intervention inducing salt accumulation in soil layers. The use of large amounts of chemical fertilizers and other inputs linked to irrigation and insufficient drainage has dramatically increased the amounts of various salts at the soil surface. This process is generally evident in intensive agriculture systems with limited leaching soils. Among the other sources of soil salinization are the mismanagement of wastewater treatments and industrial residues and the sewage from mining operations. Furthermore, the use of salt-based de-icing materials also contributes to the accumulation of salts in water and soil (Mateo-Sagasta and Burke 2011).

7.2.4 Global Climate Change

Global climate change will have an impact on many natural processes, such as changes to the hydrological cycle and rising sea levels in coastal areas. These changes, in turn, will result in a considerable increase in worldwide soil salinity, particularly in cropland areas (Hinkel et al. 2014). Increasing sea levels, along with groundwater overexploitation, will increase saltwater intrusion in coastal and inland aquifers from neighboring saline aquifers. In both semi-arid and arid climates, an increase in evapotranspiration is likely to increase the salinization of low groundwater (Taylor et al. 2012). Globally, there is great potential for 'salinity migration' in susceptible regions as a result of climate change and this is likely to affect many coasts and islands, especially regions with low rainfall and good evapotranspiration—for example, the Mediterranean (Koutroulis et al. 2013). Indeed, an intensified hydrological cycle may increase flooding and flash floods, which will increase the transfer of salts to soil in regions with geological substrates which are prone to releasing huge quantities of dissolved salts (Trnka et al. 2013). An increasing global mean temperature is another consequence of climate change, which will directly affect irrigation water requirements, and increase the salt content in soil following evaporation (Haddeland et al. 2014).

7.3 Effects of Salinity and Sodicity on Soil Properties

Environmental stresses are major limiting factors in the worldwide production of legumes and cereals (Warrence et al. 2003; Al-Rifaee et al. 2004; Tawaha and Turk 2004; Turk et al. 2004; Supanjani et al. 2005b; Supanjani et al. 2006; Abu-Darwish et al. 2009; Al-Tawaha and Seguin 2006; Al-Tawaha et al. 2006, 2007, 2010a, 2018a, 2018b, 2018c, 2018d; Tawaha and Odat 2010a; Abu Obaid et al. 2018). These constraints decrease crop yields and also constitute obstacles to the production of plants cultivated in the region which is unsuitable for cultivation. The salinity of the soil water can influence the physical properties of the soil by causing fine particles to bond together in aggregates; this is also known as flocculation and is useful in terms of soil aeration, which cause the growth of roots to spread deeper into the soil. When there is an increase in salts in the soil sector, the osmotic pressure increases in the area of propagation of plant roots, so that plants can resist these conditions. Furthermore, increasing its concentration is sufficient to cause ionic toxicity in plants. For example, the effect of boron on plants is considered to be qualitative, as it affects the growth of many plants, if its concentration is more than one component of the monomer/mole (Figure 7.2).

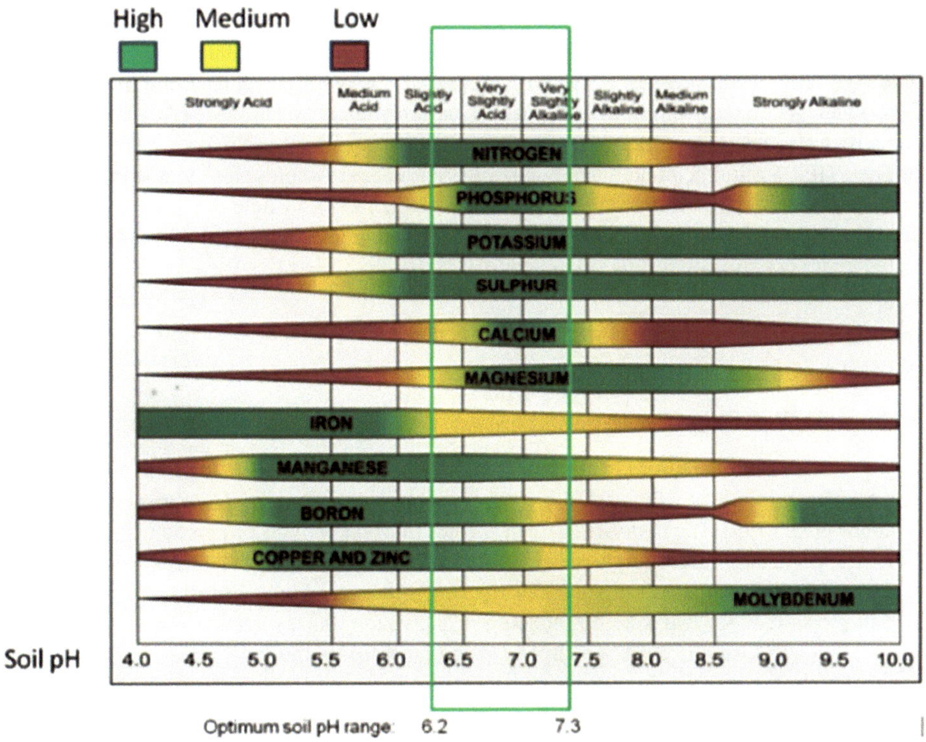

FIGURE 7.2 How pH affects soil fertility (Salinity Management, University of California).

7.4 Salt Stress Tolerance by Plant Growth-Promoting Rhizobacteria (PGPR)

Bacteria are the most numerous organisms, representing on average 6.108 cells per gram of soil and a weight of 10,000 kg/ha, equivalent to 5% of the dry weight of organic soil compounds. We then define the bacteria associated with plant roots such as rhizobacteria (Zhu 2002). These are generally very competitive strains capable of colonizing the root system rich in nutrients, throughout the plant's development cycle. If the plant releases organic compounds, conversely it takes up water and mineral elements essential to its metabolism (Janda et al. 1999). The exchanges between the plant and the soil are influenced by rhizobacteria, especially since their density and activity are high. Rhizobacteria are typical heterotrophs, so they require organic compounds as a source of energy (Strobel and Kuc 1995). Rhizobacteria indeed use many substrates from the plant: the cortical and epidermal cells of the roots which detach, the root mucilage polysaccharides, sugars and amino and organic acids of root exudates, etc. (Nemeth et al. 2002; Bakker et al. 1996). The abundance of bacteria in the soil is explained by their rapid multiplication and their ability to use a wide variety of substrates as sources of energy and nutrients. Rhizospheric microorganisms include symbionts (Rhizobia, actinobacteria and mycorrhizal fungi) and free saprophytes (Yalpani et al. 1994; Sharma et al. 1996). The microorganisms rhizosphere in general, and the diazotrophic bacteria in particular, exert various effects on plants. Furthermore, the association of bacteria with roots has important implications for plant health, productivity and soil quality (Glick 1995). These activities result from the synthesis of metabolites such as antibiotics, growth substances, acid hydrocyanic, lipopolysaccharides. Rhizobacteria known as PGPR directly stimulate growth plants by increasing the removal of nutrients from the soil, inducing and producing plant growth regulators and activating resistance mechanisms induced in plants (Shah 2003; Durrant and Dong 2004). Himmelbach et al. (2003) reported that the establishment of the association between PGPR and the plant is essential for the expression of beneficial effects. Many research groups are working on PGPRs to elucidate their modes of action. The use of plant growth-promoting rhizobacteria (PGPR) can prove beneficial. More specifically, the way in which the interaction of plants and PGPR affects crop production is not distinct. On the other hand, only a few strains of PGPR have been studied for their ability to improve the tolerance of plants to environmental stresses. Plants with reduced levels of ethylene, resulting from the inoculation of PGPR which generated 1-aminocyclopropane-1-carboxylic acid deaminase (ACC), showed tolerance to flood stress (Grichko and Glick 2001) and metallic contaminants (Nie et al. 2002; Belimov et al. 2005). The PGPR strain PaenibacilluspolymyxaB2 a protected Arabidopsis against Erwinia carotovora (biotic stress) and drought (abiotic stress) (Timmusk et al. 1999). In field experiments, sorghum plants inoculated with Azospirillumbrasilense Cd had a 15–18% increase in grain yield compared to plants not inoculated under dry soil conditions (Sarig et al. 1988).

7.5 Salinity Management Practices

7.5.1 Physiological Response to Salinity Stress

Plants respond to salinity stress by reducing plant growth, biomass, and crop yield (Maas 1993; Munns et al. 1995). The threshold at which salinity suppresses plant growth varies from species to species (Maas 1993; Munns et al. 1995). The majority of plant species (non-halophytes or glycophytes) have a threshold range of 1–10 dS/m as measured by the electrical conductivity of the saturated soil extracts (EC_e) (Maas 1993). Other species, called halophytes, can tolerate salinity at higher thresholds (>10 dS/m) (Maas 1993). Salt stress reduces leaf size by reducing cell division and expansion (Volkmar and Hu 1998). Leaf growth and expansion is more sensitive to salinity than root growth (Munns et al. 1988; Munns and Termaat 1986). In terms of the long-term exposure to salinity, the limiting factor for plant growth is the tolerance of fully expanded leaves to the maximum salt concentration in the growing medium (Munns and Termaat 1986). When salinity causes the leaf death rate to approach the new leaf expansion rate, continued growth is hindered by the reduction in the photosynthetic area (Munns and Termaat 1986; Volkmar and Hu 1998). The plant growth response to salinity can be divided into three phases: an osmotic response phase (osmotic) due to the accumulation of salt in the growing medium outside the roots causing a great reduction in growth rate; and a salt-specific response phase (ionic) due to the accumulation of salt to a toxic level within plants, causing a disruption in membrane integrity and function and a further decline in growth (Ashraf and Foolad 2013; Hanin et al. 2016; Munns et al. 1995; Munns and Tester 2008); nutrient ion imbalance due to the accumulation of sodium or chloride leading to a decline in the uptake of potassium, nitrate or phosphate, or impaired internal distribution of ions (Ashraf and Foolad 2013). During the first phase, wheat (*Triticum* spp.) and barley (*Hordeum vulgare*) genotypes that are different in salt tolerance showed similar reductions in leaf extension rates and dry matter production in response to salinity stress resulting from osmotic stress causing changes in metabolism or gene expressions (Munns et al. 1995). During the second phase (long-term exposure to salinity), the sensitive genotypes had higher reduction in growth, which is due to the accumulation of salts within plants causing leaf death (premature senescence of old leaves) and consequently reducing photosynthetic assimilates to growing parts (Munns 2002; Munns et al. 1995).

Salt solution reduces plant growth through a range of mechanisms (Volkmar and Hu 1998). Salts can reduce the rate of cell elongation by reducing cell wall extensibility and turgor pressure and increasing cell yield threshold (Volkmar and Hu 1998). Other researchers related the reduction in the leaf elongation rate for plants grown under a prolonged period of saline solution to factors other than reducing turgor pressure. The accumulation of salts in the root zone can also lower the osmotic and water potential of the soil solutions and consequently results in less water uptake by plants, alters the root cell turgor pressure, and induces a hormonal signal to the shoot to inhibit growth (Munns 2002; Volkmar and Hu 1998) and decreases the stomatal aperture (Brugnoli and Lauteri 1991; Munns and Tester 2008). The reduction in the photosynthetic rate under salt stress was also associated with the decrease in total chlorophyll content (Zhang et al. 2014).

7.5.2 Control of Salinity Stress on Plants

Plants tolerate salt stress through two main mechanisms: either by limiting the entrance salt by roots or by controlling the salt concentration and distribution with plant cells (Hanin et al. 2016). The mechanisms by which plants exclude salt solutes from plant leaf cytoplasm are the export of the solutes to the apoplast spaces between the cells and/or the isolation of solutes within the vacuole vessels (ion compartmentalization) (Hanin et al. 2016; Munns 2002; Volkmar and Hu 1998). Leaf cell growth is very sensitive to the accumulation of ions in the cell cytoplasm which can harm cell metabolism (ion toxicity) (Munns 2002; Volkmar and Hu 1998). Plant cells require energy for the two solute exclusion mechanisms, which can reduce the energy available for the biosynthetic processes (Volkmar and Hu 1998). Another mechanism by which salt tolerance genotypes differ from salt-sensitive genotypes is that the sensitive genotypes have a high rate of transport of Na and Cl from the roots to the shoots (Munns 2002).

Plant response to salinity in the growing medium also depends on other elements such as water, soil, crop developmental stage, and climate change (Maas 1993).

REFERENCES

Abu-Darwish MS, Abu Dieyeh ZH, Batarseh et al (2009) Trace element contents and essential oil yields from wild thyme plant (*Thymusserpyllum* L.) grown at different natural-variable environments, Jordan. *Journal of Food, Agriculture and Environment 7*(3&4):920–924.

Abu Obaid AM, Melnyk AV, Onichko VI et al (2018) Evaluation of six sunflower cultivars for forage productivity under salinity conditions. *Advances in Environmental Biology 12*(7):13–15.

Ahmadi J, Pour-Aboughadareh FO et al (2018) Screening wild progenitors of wheat for salinity stress at early stages of plant growth: insight into potential sources of variability for salinity adaptation in wheat. *Crop & Pasture Science 69*:649–658.

Al-Rifaee M, Turk MA, Tawaha AM (2004) Effect of seed size and plant population density on yield and yield components of local faba bean (*Viciafaba* L. major). *International Journal of Agriculture and Biology* *6*(2):294–299.

Al-Tawaha AM, Seguin P (2006). Effects of seeding date, row spacing, and weeds on soybean seed isoflavone concentration. *Canadian Journal of Plant Science* *86*:1079–1082.

Al-Tawaha AM, Seguin P, Smith DL et al (2006) Foliar application of elicitors alters isoflavone concentrations and other seed characteristics of field-grown soybean. *Canadian Journal of Plant Science* *86*:677–684.

Al-Tawaha AM, Seguin P, Smith DL et al (2007) Effects of irrigation on isoflavone concentrations of soybean grown in southwestern Québec. *Journal of Agronomy and Crop Science* *193*(4):238–246.

Al-Tawaha AM, Yadav SS, Turk M et al (2010) *Crop production and management technologies for drought prone environments*. In: *Climate change and drought management in cool season grain legume crops*. Cham: Springer.

Al-Tawaha, ARM, Odat N (2010a) Use of sorghum and maize allelopathic properties to inhibit germination and growth of wild barley (*Hordeum spontaneum*). *Notulae Botanicae, Horti Agrobotanici Cluj-Napoca* *38*(3):124–127.

Al-Tawaha AM, Odat N (2010b) Use of sorghum and maize allelopathic properties to inhibit germination and growth of wild barley (*Hordeumspontaneum*). *Notulae Botanicae Horti Agrobotanici Cluj-Napoca* *38*(3):124–127.

Al-Tawaha AR, Al-Karaki GN, Al-Tawaha AR et al (2018a) Effect of water flow rate on quantity and quality of lettuce (*Lactuca sativa* L.) in nutrient film technique (NFT) under hydroponics conditions. *Bulgarian Journal of Agricultural Science* *24*(5):791–798.

Al-Tawaha AR, Turk MA, Al-Tawaha AR et al (2018b) Using chitosan to improve growth of maize cultivars under salinity conditions. *Bulgarian Journal of Agricultural Science* *24*(3):437–442.

Al-Tawaha AM, Al-Tawaha AM, Alu'datt MH et al (2018c) Effects of soil type and rainwater harvesting treatments in the growth, productivity and morphological trains of barley plants cultivated in semi-arid environment. *Australian Journal of Crop Science* *12*(6):975–979.

Al-Tawaha A, Al-Karaki G, Al Tawaha A et al (2018d) Effects water flow rate on quantity and quality of lettuce (*Lactuca Sativa* L.) In nutrient film technique (Nft) under hydroponics conditions. *Bulgarian Journal of Agricultural Science* *24*(5):791–798.

Anonymous (2019) Data available from: https://biosalinity.org/global%20perspective.htm.

Ashraf M, Foolad MR (2013) Crop breeding for salt tolerance in the era of molecular markers and marker-assisted selection. *Plant Breeding* *132*:10–20.

Bakker PM, Ran LX, Pieterse CM et al (1996) Understanding the involvement of rhizobacteria mediated induction of systemic resistance in biocontrol of plant diseases. *Canadian Journal of Plant Pathology* *25*:5–9.

Barros R, Isidoro D, Aragues R (2012) Three study decades on irrigation performance and salt concentrations and loads in the irrigation return flows of La Violada irrigation district (Spain) Agriculture, *Ecosystems & Environment* *151*:44–52.

Belimov AA, Hontzeas H, Safronova VI et al (2005) Cadmium tolerant plant growth promoting bacteria associated with the roots of Indian mustard (*Brassica juncea* L. Czern.). *Soil Biology & Biochemistry* *37*:241–250.

Brugnoli E, Lauteri M (1991) Effects of salinity on stomatal conductance, photosynthetic capacity, and carbon isotope discrimination of salt-tolerant (*Gossypium hirsutum* L.) and salt-sensitive (*Phaseolus vulgaris* L.) C non-halophytes. *Plant Physiology* *95*:628–635.

Chari MM, Nemati F, Afrasiab P et al (2012) Prediction of evaporation from shallow water table using regression and artificial neural networks. *Journal of Agricultural Science* *5*:168.

Daliakopoulos IN, Tsanis IK, Koutroulis A et al (2016) The threat of soil salinity: A European scale review. *Science of the Total Environment* *573*:727–739.

Durrant WE, Dong X (2004) Systemic acquired resistance. *Annual Review of Phytopathology* *42*:185–209.

Grichko VP, Glick BR (2001) Amelioration of flooding stress by ACC deaminase containing plant growth promoting bacteria. *Plant Physiology and Biochemistry* *39*:11–17.

Haddeland I, Heinke J, Biemans H et al (2014) Global water resources affected by human interventions and climate change. *Proceedings of the National Academy of Sciences of the United States of America* *111*:3251–3256.

Hanin M, Ebel C, Ngom M et al (2016) New insights on plant salt tolerance mechanisms and their potential use for breeding. *Frontiers in Plant Science 7*:1787.

Himmelbach A, Yang Y, Grill E (2003) Relay and control of Abscisic acid signaling. *Current Opinion in Plant Biology 6*:470–479.

Hinkel J, Lincke D, Vafeidis AT et al (2014) Coastal flood damage and adaptation costs under 21st century sea-level rise. *Proceedings of the National Academy of Sciences of the United States of America 111*:3292–3297.

Janda T, Szalai G, Tari I et al (1999) Hydroponic treatment with salicylic acid decreases the effects of chilling injury in maize (*Zea mays* L.) plants. *Journal of Plant Sciences and Agricultural Research 208*:175–180.

Koutroulis AG, Tsanis IK, Daliakopoulos IN et al (2013) Impact of climate change on water resources status: a case study for Crete Island, Greece. *Journal of Hydrology 479*:146–158.

Khan A, Shah N, Muhammad M, Khan MS, Ahmad MS, Farooq M, Yousafzai AM (2016) Quantitative determination of lethal concentration Lc 50 of atrazine on biochemical parameters; total protein and serum albumin of freshwater fish grass carp (Ctenopharyngodon idella). *Polish J Environ Studies 25*(4):1555–1561.

Maas EV (1993) Plant growth response to salt stress. In: Lieth H, Al Masoom AA (eds) *Towards the rational use of high salinity tolerant plants*: Vol. *1* Deliberations about high salinity tolerant plants and ecosystems. Springer, Netherlands, Dordrecht, pp 279–291.

Machado RMA, Serralheiro RP (2017) Soil salinity: effect on vegetable crop growth. Management practices to prevent and mitigate soil salinization. *Horticulturae 3*:30.

Mateo-Sagasta J, Burke J (2011) Agriculture and water quality interactions: a global overview. SOLAW Background Thematic Report-TR08.

Munns R (2002) Comparative physiology of salt and water stress. *Plant, Cell & Environment 25*:239–250.

Munns R, Gardner P, Tonnet M et al (1988) Growth and development in NaCl-treated plants. II. Do NaCl concentrations in dividing or expanding tissues determine growth in barley? *Functional Plant Biology 15*:529–540.

Munns R, Schachtman D, Condon A (1995) The significance of a two-phase growth response to salinity in wheat and barley. *Functional Plant Biology 22*:561–569.

Munns R, Termaat A (1986) Whole-plant responses to salinity. *Functional Plant Biology 1*:143–160.

Munns R, Tester M (2008) Mechanisms of salinity tolerance. *Annual Review of Plant Biology 59*:651–681.

Nachshon U (2018) Cropland soil salinization and associated hydrology: Trends, processes and examples. *Water 10*:1030.

Nemeth M, Janda T, Horvath E et al (2002) Exogenous salicylic acid increases polyamine content but may decrease drought tolerance in maize. *Plant Science 162*:569–574.

Nie L, Shah S, Rashid A et al (2002) Phytoremediation of arsenate contaminated soil by transgenic canola and the plant growth promoting bacterium *Enterobacter cloacae* CAL2. *Plant Physiology and Biochemistry 40*:355–361.

Rengasamy P (2016) Soil salinization. In *Oxford Research Encyclopedia, Environmental Science*, Pichu Rengasamy, Oxford University Press, New York, NY.

Sarig S, Kapulnik Y, Nur I et al (1984) Response of non-irrigated *Sorghum bicolor* to *Azospirillum* inoculation. *Experimental Agriculture 20*:59–66.

Shah J (2003) The salicylic acid loop in plant defense. *Current Opinion in Plant Biology 6*:365–371.

Shahbaz M, Ashraf M (2013) Improving salinity tolerance in cereals. *Critical Reviews in Plant Sciences 32*:237–249.

Shakirova FM, Sakhabutdinova AR, Bezrukova MV et al (2003) Changes in the hormonal status of wheat seedlings induced by salicylic acid and salinity. *Plant Science 164*:317–322.

Sharma CP (2006) *Plant micronutrients*. Science Publisher, New Hampshire.

Sharma YK, Leon J, Raskin I et al (1996) Ozone induced responses in *Arabidopsis thaliana*: the role of salicylic acid in the accumulation of defense related transcripts and induced resistance. *Proceedings of the National Academy of Sciences of the United States 93*:5099–5104.

Shukla PS, Agarwal PK, Jha B (2012) Improved salinity tolerance of *Arachis hypogaea* (L.) by the Interaction of halotolerant plant growth promoting rhizobacteria. *J Plant Growth Regulation 31*(2):195–206.

Strobel NE, Kuc A (1995) Chemical and biological inducers of systemic acquired resistance to pathogens protect cucumber and tobacco from damage caused by paraquat and cupric chloride. *Phytopathology 85*:1306–1310.

Supanjani A, Yang MS, Tawaha ARM et al (2005a) Effect of Magnesium application on yield, mineral contents and active components of *Chrysanthemum coronarium* L. under hydroponics conditions. *Bioscience Research 2*(2):73–79.

Supanjani A, Tawaha AM, Min Yang MS et al (2005b) Role of calcium in yield and medicinal quality of *Chrysanthemum coronarium* L. *Journal of Agronomy 4*(3):188–192.

Tari I, Csiszar J, Szalai G et al (2002) Acclimation of tomato plants to salinity stress after a salicylic acid pretreatment. *Acta Biomaterialia 46*:55–56.

Tawaha AM, Turk MA (2002) Lentil (*Lens culinaris* Medic.) productivity as influenced by rate and method of phosphate placement in a Mediterranean environment. *Acta Agronomica Hungarica 50*(2): 197–201.

Tawaha AM, Singh VP, Turk MA et al (2003) A review on growth, yield components and yield of barley as influenced by genotypes, herbicides and fertilizer application. *Research on Crop 4*(1):1–9.

Tawaha AM, Turk MA (2004) Field pea seeding management for semi-arid Mediterranean conditions. *Journal of Agronomy and Crop Science 190*:86–92.

Taylor RG, Scanlon B, Doll P et al (2012) Ground water and climate change. *Nature Climate Change 3*:322–329.

Timmusk S, Nicander B, Granhall U et al (1999) Cytokinin production by *Paenibacilluspolymyxa*. *Soil Biology & Biochemistry 31*:1847–1852.

Trnka M, Kersebaum KC, Eitzinger J et al (2013) Consequences of climate change for the soil climate in Central Europe and the central plains of the United States. *Climatic Change 120*:405–418.

Turk MA, Tawaha, AM (2001) Common vetch (*Vicia sativa* L.) productivity as influenced by rate and method of phosphate fertilization in a Mediterranean environment. *Agricultura Mediterranea 131*:108–111.

Turk MA, Tawaha AM (2002a) Onion (*Allium cepa* L.) as influenced by rate and method of phosphorus placement. *Crop Research 23*(1):105–107.

Turk MA, Tawaha AM (2002b) Impact of seeding rate, seeding date and method of phosphorous application in faba bean (*Viciafaba* L. minor) in the absence of moisture stress *Biotechnology, Agronomy, Society and Environment 6*(3):171–178.

Turk MA, Tawaha AM (2002c). Inhibitory effects of aqueous extracts of black mustard on germination and growth of lentil. *Pakistan Journal of Agronomy 1*(1):28–30.

Turk MA, Tawaha AM, Shatnawi M (2003a) Lentil (*Lens culinaris* Medik) Response to plant density, sowing date, phosphorus fertilization and Ethephon application in the absence of moisture stress. *Journal of Agronomy and Crop Science 189*(1):1–6.

Turk MA, Tawaha, AM, Samara N (2003b) Effects of seeding rate and date, and phosphorus application on growth and yield of narbon vetch (*Vicia narbonensis*). *Agronomie 23*:1–4.

Turk MA, Shatnawi MK, Tawaha AM (2003c) Inhibitory effects of aqueous extracts of black mustard on germination and growth of alfalfa. *Weed Biology and Management 3*(1):37–40.

Turk MA, Tawaha AM, Lee KD (2004) Seed Germination and Seedling Growth of Three Lentil Cultivars under Moisture Stress. *Asian Journal of Plant Science 3*(3):394–397.

van Beek C, Toth G (2012) Risk assessment methodologies of soil threats in Europe. JRC Scientific and Policy Reports EUR 24097 EN. European Commission.

Volkmar KM, Hu SH (1998) Physiological responses of plants to salinity: A review. *Canadian Journal of Plant Science 78*:19–27.

Wallender WW, Tanji KK (2012) Agricultural salinity assessment and management. ASCE Manuals and Reports on Engineering Practice No. 71. ASCE, Reston, VA.

Warrence NJ, Pearson KE, and Bauder JW (2003) *Basics of salinity and sodicity effects on soil physical properties*. Montana State University.

Yalpani N, Enyedi AJ, Leon J (1994) Ultraviolet light and ozone stimulate accumulation of salicylic acid, pathogen related proteins and virus resistance in tobacco. *Planta 193*:372–376.

Zhang L, Ma H, Chen T, et al (2014). Morphological and physiological responses of cotton (*Gossypium hirsutum* L.) plants to salinity. *PLoS One* 9:e112807.

Zia-ur-Rehman M, Murtaza G, Qayyum M, Saqib M, Akhtar J (2017) Salt-affected soils: sources, genesis and management. In: Sabir M, Akhtar J, Hakeem KR (eds) *Soil science concept and applications*. University of Agriculture, Faisalabad, pp 191–216.

Zhu JK (2002) Salt and drought stress signal transduction in plants. *Ann Rev Plant Bio 53*:247–273.

8

Heavy Metal Toxicity and Plant Defense Responses

Nazish Huma Khan and Fazli Zuljalal
University of Swabi, Pakistan

Tooba Saeed
University of Peshawar, Pakistan

CONTENTS

8.1 Introduction

Rapid industrial and economic development has introduced widespread pollution into the natural global environment. Abiotic stresses, of which heat stress, drought, heavy metals and salinity are among the most important, have an adverse impact on globally sustainable crop production (Adnan et al. 2018a, 2018b; Adnan et al. 2019; Akram et al. 2018a, 2018b; Aziz et al. 2017a, 2017b; Habib ur et al. 2017; Hafiz et al. 2016; Hafiz et al. 2019; Kamarn et al. 2017; Muhammad et al. 2019; Sajjad et al. 2019; Saud et al. 2013; Saud et al. 2014; Saud et al. 2017; Saud et al. 2016; Shah et al. 2013; Saud et al. 2020; Qamar-uz et al. 2017; Wajid et al. 2017; Yang et al. 2017; Zahida et al. 2017; Depeng et al. 2018; Hussain et al. 2020; Hafiz et al. 2020a, 2020b; Shafi et al. 2020; Wahid et al. 2020; Subhan et al. 2020; Zafar-ul-Hye et al. 2020a, 2020b; Adnan et al. 2020; Ilyas et al. 2020; Saleem et al. 2020a, 2020b, 2020c; Rehman et al. 2020; Frahat et al. 2020; Wu et al. 2020; Mubeen et al. 2020; Farhana et al. 2020; Jan et al. 2019; Wu et al. 2019; Ahmad et al. 2019; Baseer et al. 2019; Hafiz et al. 2018; Tariq et al. 2018; Fahad and Bano 2012; Fahad et al. 2017; Fahad et al. 2013; Fahad et al. 2014a, 2014b; Fahad et al. 2016a, 2016b, 2016c, 2016d; Fahad et al. 2015a, 2015b; Fahad et al. 2018; Fahad et al. 2019a, 2019b; Hesham and Fahad 2020). The contamination of the environment with toxic metals and pollutants has become a serious issue in industrialized countries. These metals are carcinogenic in nature and pose a considerable threat to human health. The term "heavy metal" refers to the elements having high density and toxicity, even at low levels (Olaniran et al. 2013). The contamination of heavy metals refers to the excessive deposition of hazardous metals in the environment caused by human activities. Various anthropogenic activities of the biosphere showed a broad array of global phenomena such as intensive agriculture, extensive mining and a rapid increase in industrialization (Nagajyoti et al. 2010; Khan et al. 2016). Overpopulation and the accelerated rate of urbanization have not only put pressure on the natural available resources but also caused contamination of important components of life. One of the biggest implications of human-induced disturbance of natural cycles is the accumulation of heavy metals. The problem of heavy metals accumulation is a reason of nutritional, ecological and environmental pollution (Nagajyoti et al. 2010). These metals are highly toxic to both the wider environment and individual organisms. The organisms can be enriched through the food chain (Zojaji et al. 2014).

Currently, the problem of heavy metals stress is one of major abiotic stresses causing environmental pollution, since heavy metals are not degraded and converted into harmless compounds through biological processes (Gisbert et al. 2003). Hence, they persist longer in the environment and can also find their way into the food chain. They place a stress on the environment by producing toxic oxygen derivatives (Arora et al. 2002). Elevated levels of heavy metals pose serious effects on plant growth such as chlorosis, necrosis, stunning and low yield production (Emamverdian et al. 2015; Viehweger 2014).

Soil is an important component of the urban environment and has become a sink for various metals. Therefore, its management is the key to its quality. Soil contamination with heavy metals pose adverse impacts on its quality, productivity, crop quality and other biota dependent on it. Over recent years, soil quality and fertility have been badly damaged by heavy metals. Heavy metals from industrial, agricultural and municipal sources have also led to the increased contamination of the environment (Nagajyoti et al. 2010). Among metals, some are considered as essential and a few are non-essential. Metals such as iron (Fe), copper (Cu), zinc (Zn) and manganese (Mn) are regarded as being required by plants and animals in optimum quantities; under high concentrations, however, these metals are considered harmful to both plants and animals. By contrast, cadmium (Cd), chromium (Cr), nickel (Ni), arsenic (As) and lead (Pb) are not required and are therefore regarded as non-essential metals. Metal contamination in soil disrupts plants' absorption rates and retards plant growth and development processes. Heavy metals such as Cd, Cu, Zn, Pb and Ni can change the soil chemistry and pose harmful effects on the vegetation and organisms dependent on soil for its nutrition (Ali et al. 2011). Once the soil become contaminated with heavy metals, it is difficult to remediate it. Therefore, soil contamination has become a hot topic and a source of increased attention across the world.

8.2 Characteristics of Heavy Metals

The label heavy metals is given to all inorganic components belonging to the group of non-biodegradable elements with an atomic mass over 20 and a density higher than 5g/cm^3 (Rascio and Navari-Izzo 2011; Adnan et al. 2016). Heavy metals are classified into anionic metals (negatively charged particles) and cationic metals (positively charged particles). The most common and problem-causing cationic metals are mercury (Hg), cadmium (Cd), zinc (Zn), lead (Pb), nickel (Ni), chromium (Cr), copper (Cu), and the most common anionic metal is arsenic (As) (Olaniran et al. 2013; Khan et al. 2016). These metals have genotoxic, cytotoxic and mutagenic effects on plants, animals and humans through their contamination of food chains, water resources, soil, and surrounding atmosphere (Wuana and Okieimen 2011). Generally, heavy metals are represented through the term "trace elements". It is used for those metals that occur in small amounts in natural biological systems. This term is usually used a lot due to growing concern of environmental pollution. Other terms that are also used in reference to these elements are trace inorganics, trace metals, micronutrients and micro-elements. Heavy metals show a variety of properties, as are outlined below:

8.2.1 Widespread Distribution

The distribution pattern of heavy metals is extensive. Due to the current developmental process, these metals have increasingly occurred everywhere and have become a serious threat to almost every country. Of the world's ten most damaging environmental incidents, at least two events are related to heavy metals contamination.

8.2.2 Highly Reactive

Heavy metals contamination is difficult to notice because it possesses no specific color or odor. These elements do not damage the environment in a short time unless they exceed the environmental tolerance or its permissible fixed limit. In the case of higher availability, they can cause serious environmental damage. Therefore, their contamination is often referred as a 'Chemical Time Bomb'.

8.2.3 Complex Contamination

The contamination of soil, air, water or other natural resources are amplifying the complex contamination with a variety of metals. Currently, these cases are found to be common, posing harmful effects on the environment.

8.2.4 Remediation Potential

The behavior of heavy metals is different. In air and water, the metals contamination can be reversed by dilution and self-purification after switching off the pollution sources. While it is difficult to apply such techniques in soils contaminated with heavy metals. Such soils take a longer time period (estimated at between one hundred and two hundred years) to recover. Therefore, the remediation of heavy metals involves higher costs and much longer time periods.

8.3 Sources of Heavy Metals

Heavy metals are emitted into the environment through both natural and anthropogenic activities. They originate from various sources, including atmospheric deposition, sewage system, the improper handling of waste, and unscientific agricultural practices and mining activities. Some sources are discussed below:

8.3.1 Atmospheric Source

In the atmosphere, heavy metals originate from gas and dust generated by energy, transport, agriculture, metallurgy and construction materials. These metals enter the atmosphere in the form of aerosols and either remain suspended in the air or settle down into the soil through processes of natural sedimentation and precipitation.

8.3.2 Geological Process

The weathering of rock is a geological phenomenon involving the breaking of parent rock into smaller fragments. The rock is a big deposit of metals, such as Zn, Cd, Cr, Co, Mn, Ni, Hg and Pb. Due to unsuitable climatic conditions, rock breaks down and the weathered rock is converted into soil which contains these metals.

8.3.3 Volcanic Eruption

Volcanoes is the natural process through which metals can be emitted into the environment. They are reported to be the biggest emissions of toxic metals, along with a higher proportion of harmful gases. They emit high levels of Zn, Al, Hg, Ni, Cu and Pb in the form of wind dust and aerosols.

8.3.4 Agricultural Activities

Agricultural activities such as pesticides, liming, sewage, chemical sludge and irrigation water are the common sources for the introduction of metals into the soil. The long-term practices of agricultural inputs result in soil being contaminated with heavy metals. It can also cause contamination of the food chain, which causes health risks to living beings (Arao et al. 2010). Today, the intensive use of agrochemicals has become a common policy to get a high yield. Both organic and inorganic fertilizers are used to improve crop production. This practice brings a lot of heavy metals into the soil. Heavy metals are the most reported contaminants in agrochemicals and contain As, Hg, Cu, Zn and other metals in excess.

8.3.5 Industrial Activities

These sources include the mining, refining and manufacturing of products through the emission of several hazardous metals into the environment. The industrial emissions are in both air and liquid form, polluting the air and water quality. The industries do not treat the effluents and discharge them directly into water channels which lead to a deterioration in water quality and also effects the aquatic biota living within it. Cd, Cr, Ni, Pb, Zn and Cu are reported available in industrial emissions.

8.3.6 Vehicle Emissions

Vehicle emissions add heavy metals to the air in the form of dust, smoke and particulate particles. Automotive transport causes the serious contamination of the atmosphere and soils with heavy metals such as Zn, Pb, Cu, Cd, and Cr.

8.3.7 Sewage

Wastewater from municipal and industrial sources, such as chemical and sanitary works, deposit a huge quantity of heavy metals into water and soils. Their deposition in water and soil poses negative threats to the living biota and ecological systems.

8.3.8 Solid Waste

The improper disposal of solid waste from hospitals, commercial areas, households, construction and industries have become a source of major environmental concern. These wastes have complex composition, having a variety of hazardous metals and posing threats to the environment and natural resources. When these wastes are not handled and not disposed of properly, this can result in pollution problems (Su et al. 2014).

8.4 Heavy Metals Accumulation, Mobility and Uptake

The trend of industrialization and hostile human activities has increased the level of heavy metals in both terrestrial and aquatic environments. Such activities include smelting processes and the heating of metals at higher temperatures (Chen et al. 2016). Heavy metals exist in ionic, colloidal, dissolved and particulate forms. The higher accumulation of these toxic metals is available in soils ranging from small to as high as 100,000 mg/kg, depending upon the type of soil (Blaylock and Huang 2000). After entering the environment, heavy metals are transported by air and water, being deposited in soil and sediments, where they can become immobilized (Ozturk et al. 2008). Hence, the bonding may take significantly longer. The bioavailability of heavy metals by the plants is important. Heavy metals availability from soil to plant is influenced by the chemical characteristics of soil, such as pH, soil texture, organic content, and cation exchange capacity (Logan and Chaney 1983: Verloo and Eeckhout 1990). The pH balance is an important factor which affects the bioavailability of soil. Metals availability is high at lower pH and at higher pH the availability is low due to the formation of compounds having negligible solubility (Seregin and Ivanov 2001). Similarly, a higher level of organic matter and clay content in soil causes metals to be held for a longer time. Soil temperature is also an important property for metals accumulation in soil and uptakes by plants (Chang et al. 1987). After the mobilization of metals in soil, the root cells of plants captured these metals in the soil (Marschner 1995). The transportation of metals then starts by the cell wall, which behaves as an ion exchanger of comparatively low affinity. Then active transport of these metals begins across the plasma membrane (Hirsch et al. 1998). Once dissolved in soils, the metals are immobilized slowly. Although the availability of these metals is perhaps high in the beginning, but it decreases with the passage of time (Iqra et al. 2020; Akbar et al. 2020; Mahar et al. 2020; Noor et al. 2020; Bayram et al. 2020; Amanullah et al. 2020; Rashid et al. 2020; Arif et al. 2020; Amir et al. 2020; Saman et al. 2020; Muhammad et al. 2020; Md Jakir and Allah 2020; Farah et al. 2020; Sadam et al. 2020; Unsar et al. 2020; Fazli et al. 2020; Enamul et al. 2020; Gopakumar et al. 2020; Zia-ur-Rehman 2020; Ayman et al. 2020; Mohammad et al. 2020).

After the accumulation of metals in soils, their uptakes are accelerated by plant roots and leaves. The metals bound in tissue causing saturation that is directed by the rate of metals uptake. The efficiency of metals uptake by the plants is maximum at their low concentrations in the external medium (Greger et al. 1991; Greger 1997). Metals uptake through the root system is the common path. This occurs through plasmic interactions, while leaves take up metals in the form of gases and ions through stomata and cuticles, respectively (Lindberg et al. 1992; Marschner 1995). Some metals remain stuck as they are insoluble in the vascular system of plants; they form precipitates of sulphate, phosphate and carbonate, immobilizing them in intracellular and extracellular compartments (Raskin et al. 1997; Pourrut et al. 2011). Among the factors that affect plants' uptake of metals are plant morphology and anatomy, heavy metals speciation, and climatic conditions (such as temperature, humidity, and heat). The metals uptake can be enhanced by increased the availability of metals in the external environment (Abbruzzese et al. 2009).

8.5 Impacts of Heavy Metals

Heavy metals refer to some significant metals of biological toxicity, including Cd, Cr, Cu, Hg, Pb, Zn, As, and Ni. Several anthropogenic activities have increased their quality in the environment, causing environmental deterioration. Heavy meals show their negative impacts on soil, plants and humans as discussed below.

8.5.1 Impacts on Soil

The higher presence of heavy metals shows their toxic effects on agricultural soil. Heavy metals in the soil affect both microbial and enzymatic activities. Microbial biomass is an important indicator in determining the extent of soil contamination. The contamination of soil with heavy metals can have a significant impact on the microbial performance by stimulating the microbial biomass. Enzymes in the soil are responsible for the decomposition of organic matter and also regulate nutrient cycling. The availability of heavy metals in soil disrupt the enzymes and reduce them by 10–50 times with the increase of heavy metals availability (Chander et al. 1995).

8.5.2 Impacts on Plants

Agricultural soils contaminated with heavy metals pose serious threats to any agricultural crops/plants grown on that soil. As the heavy metals stay in soil extensively are called as soil pollutants which put chronic and acute problems on plants (Salt et al. 1995). Heavy metals in low quantity will not affect the plants. When the concentration of metals is too high, it will cause the poisoning or even the death of the plant. Such as higher availability of Cd affects the process of photosynthesis, water and nutrient uptakes in plants, chlorosis, browning tips and finally death in case of higher availability (Mohanpuria et al. 2007).

8.5.3 Impacts on Humans

Heavy metals find their way into human bodies through several pathways, such as the inhalation of dust, absorption through the skin, the drinking of water and the ingestion of food. All of these sources are involved in the contamination of the water and atmosphere in the food chain. The intake of heavy metals has negative impacts on human health: for example, the intake of Pb damages kidneys, the reproductive system, the urinary and nervous systems, immunity, the physiological process of cells and gene expression (Yabe et al. 2010).

8.6 Soil Contamination with Heavy Metals

The problem of contamination with heavy metals has become a widespread problem across the world. Human activities have introduced plenty of toxic metals into both urban and agricultural soils, although the behavior of these contaminating elements is different in urban and agricultural soils. Heavy metals from agricultural soils enter into humans/animals through food chain via the ingestion of crops. The intake of heavy metals through the soil-crop system is a major route for damaging human's health. By contrast, in urban soils, heavy metals find their way directly into human body through ingestion or absorption of skin (Aeliona et al. 2008). The contamination of agricultural soils with heavy metals is caused by the use of agrochemicals, sewage sludge, industrial effluents, smelting minerals, automobiles exhaust and waste treatment. By contrast, sources involved in the contamination of urban soils are transportation, industrial waste, household waste, weathered and construction material and atmospheric precipitation (Montagne et al. 2007). Literature revealed that the soil quality in both urban and agricultural settings has been quickly getting worse due to industrial development in China. About 2.9% of the soil in China has been contaminated either moderately or severely (Su et al. 2014). Similarly, the soils of Spain, Hong Kong, Wuxi, Changsha were found serious for heavy metals contamination. Heavy metals such as Cd, Cr, Zn, Pb, Hg, Cu, Ni have been detected commonly in soils which show potentially toxic effects on soil and thus on plants and animals. Cadmium is toxic and nonessential and may prove poisonous for soil, plants and animals. Its availability affects the process of photosynthesis, water and nutrient uptakes in plants, chlorosis, browning tips and finally death in case of higher availability (Mohanpuria et al. 2007). Zinc is a micronutrient required by the plants, but its higher availability inhibits plants' metabolic functions leading to retardation in plants' growth and development (Choi et al. 1996; Ebbs and Kochian 1997; Fontes and Cox 1998; Uddin et al. 2016). It may cause chlorosis in plant leaves. Excess availability of zinc replaces Mn and Cu, causing their deficiencies in plant shoots. Another effect of zinc toxicity in plants is the phosphorus deficiency, which is characterized by the appearance of a purplish-red coloring of the leaves (Lee et al. 1996). Copper, an important micronutrient, plays a vital role in assimilation of CO_2 and ATP synthesis. The enhanced industrial and mining activities increased the level of Cu availability in ecosystems. The smelting of copper ores also adds Cu to soil. An excess of Cu in results in cytotoxic responses in soils, which leads to stress and injuries in plants (Demirevska-kepova et al. 2004). The exposure of plants to copper causes oxidative stress, which damages the macromolecules and disturbs the metabolic pathways (Hegedus et al. 2001). The intensive inputs of Hg into agricultural lands can also have toxic effects on plants' health. Hg is phytotoxic to plant cells and can cause serious injuries to plants, such as the disruption of cellular metabolism (Han et al. 2006). Chromium is a toxic metal which is discharged by the tanning industry and can cause serious impacts on soil, water and sediments. Chromium in its Cr (IV) form is very toxic and carcinogenic to humans. In plants, it causes misbalancing of nutrients, slows down plant growth, and causes chlorosis in young leaves and root injury (Scoccianti et al. 2006). Lead is one of the most widely distributed metals in the Earth's crust. The mining, smelting and municipal sewage add Pb into soil stops the enzymatic activities, posing its adverse effects on plants growth and their morphology (Sharma and Dubey 2005). Ni is found in naturally in soils while human activities such as agricultural practices, the burning of fossil fuels, sewage and mining also put Ni into soils. Soils with Ni contamination is a major problem and sometimes it may exceed up to twenty or thirty times higher than the overall range. An excess of Ni in soil causes nutrient imbalancing in soils and the disorder of cell membranes and chlorosis in plants (Izosimova 2005).

8.7 Impacts of Heavy Metals on Various Parts of Plants

Plants are quite sensitive to both an excess or a deficiency of heavy metals that are essential to plants. While the higher availability of some metals such as Cd, Hg, Cr and As are highly poisonous for plants, increasing levels of heavy metals affect the entire mechanism of plants' growth. However, not all heavy metals are toxic to plants. Some, such as Cu and Zn, are important for plants' development and growth within certain limits. When these metals exceed these higher limits, they pose a serious danger to plants' developmental process (Afonne and Ifediba 2020). Some heavy metals are considered to be harmful at low concentrations and cause contamination of the food chain. These metals have the potential to accumulate inside human fatty tissues and bone matrix responsible for renal, physiological and cardiovascular problems (El-Kady and Abdel-Wahhab 2018).

8.7.1 Impact of Heavy Metals on Plant Cells

During metabolism, the production of reactive oxygen species (ROs) inside the plant cell is a natural process. But the higher availability of heavy metals in plants causes stress in plant cells, which tend to cause the overproduction of these oxygen species and lead to oxidative stress. It disrupts the important process of plant growth (growth of pollen tube), the growth of plant fiber, the loosening of the cell wall, and the fruit ripening process (Berni et al. 2019). Literature revealed that the plant species of Sorghum when applied with certain bacterial strains (*Alcaligenes faecalis, Bacillus cereus, Alcaligenes faecalis*) were found effective to ignore such toxic effects of heavy metals on plant cells (El-Meihy et al. 2019).

8.7.2 Impacts of Heavy Metals on Plant Shoots and Leaves

An excess of heavy metals has a serious impact on shoot growth and plant height. This is attributed to the slow growth of root systems and the reduced availability of essential nutrients and water to the higher parts of the plant. The chromium availability affects the cellular metabolism of plants, tending to reduce plant height and also growth (Shanker et al. 2005). Similarly, high levels of copper and lead disrupt the morphology of leaves (Rizvi and Khan 2018).

8.7.3 Heavy Metals Impact on Plant Roots

Heavy metals are in direct contact with plant roots due to the exposure of waste and polluted soil or water. Therefore, roots are considered to be the most affected parts. Heavy metals toxicity causes the growth of roots as they don't elongate, decrease in root vessels and diameter, damage the root hairs and root tips, effect the epidermis and hypodermis and formation of lateral and adventitious roots etc. (Punz and Sieghardt 1993). As roots are responsible for the heavy accumulation of toxic metals, it is here that the effects are greatest. Such accumulation also disturbs the water balance and the uptake of essential nutrients and their transport to other parts of the plants (Rizvi and Khan 2018).

8.8 Heavy Metals Impacts on the Germination of Plants

Germination of seeds is an important process in the plant life cycle which is affected by the presence of heavy metals. The heavy metals pose different effects in different plant species depending upon the composition and structure of the seed coating. In germination, heavy metals depend upon the capability to enter the seed coat and reach the embryonic tissues. An excess of Cu and Cd blocks the uptake of water, meaning that germination does not take place. The seed germination teste is usually used to know the metals toxicity in plants (Moosavi et al. 2012; Kranner and Colville 2011).

As heavy metals stress affects the overall system of plant and, more specifically, its growth. In the response, plants also mechanisms which they can use as a defense against heavy metals. These mechanisms include the following:

i) Plants retard heavy metals absorption by binding them to the cell wall or other ligands and thereby reducing their concentration in plants, thereby abating the harmful effects of toxic metals.
ii) To avoid the impacts of heavy metals on plants' mesophyll, plants absorbed these metals in the trichomes of the epidermis. Trichomes serve as storage sites for detoxification purposes and secrete different secondary metabolites to ameliorate the harmful effects of metals.
iii) To detoxify the toxic metals, plants precipitate them in a special site within plants. Plants initiate various cellular defense mechanisms to abolish the harmful effects of metals.
iv) Production of different kinds of proteins that are induced by toxic metals to resist the impact of environmental stress on plants.
v) Activate the anti-oxidization enzymes and remove free radicals to prevent the damages to plants. These enzymes are responsible to control the overproduction of ROS (Zhang et al. 1999).

8.9 Remediation of Heavy Metals

To reduce the harmful impacts of heavy metals on soil, plants and humans, it is important to remove them. Different techniques are used to minimize the toxic impacts of heavy metals, as discussed below.

8.9.1 Engineering Remediation

This method refers to control the heavy metals in soil through physical and chemical methods. Through this method, the contaminated soil may be replaced, removed or isolated. This includes the removal of existing soil with the addition of fresh clean soil or through blending with latter. The replacement of old soli is applied on a low scale. Soil isolation is the isolation of polluted soils with unpolluted ones. All these techniques need manpower, material resources, high cost and can be applied on small areas.

8.9.2 Adsorption

This is the chemical method in which all the heavy metals from soil can be fixed and adsorbed by clay minerals, furnace slag and so on.

8.9.3 Soil Leaching

This is the cleaning of polluted soil by specific reagents. It involves the removal of the heavy metals complex and then a recycling of the metals extracting solution.

8.9.4 Electro-Kinetic Remediation

This is a new technique of heavy metals remediation from soil. In this method, an electric field gradient is built by applying a DC-voltage on both sides of an electrolytic tank containing polluted soil. All of the pollutants are removed and taken up to the processing chamber from the soil. This method is economically effective and can be applied well in soils with low permeability.

8.9.5 Bioremediation

This is the effective removal technique of heavy metals from soil. It includes the phytoremediation, microbial remediation and animal remediation techniques. Phytoremediation is the removal of heavy metals from soil by specific plants grown on contaminated soils. More than 400 plant species have been identified as being effective for the remediation technique, including the majority from the Brassica, Cruciferea, Thlaspi and Alyssums families. In the microbial technique, contaminated soils are treated with microorganisms such as fungi and bacteria using biotechnology. One specie of fungus, *Gomus intraradices*, is considered to be effective in remediation techniques. In the area of animal remediation, organisms such as earthworms and maggots are used to remove heavy metals from soils.

8.9.6 Pyrolysis and Composting

To control the heavy metals stress in plants, pyrolysis has been considered to be an effective method which is helpful in the production of ash, tar, biochar and bio oil. Similarly, composting is an important remedy against heavy metals through the use of compost of any type (primary, secondary/aging compost) to treat the soil up to 50% (Liu et al. 2019, 2020).

8.10 Conclusion

Like all living organisms, plants are also sensitive to heavy metals. They show their responses to both the low and the high availability of metals. The contamination of soil by agricultural activities has become an increasing problem, having an adverse impact on the quality of plant health. Industrial emissions, domestic discharges, agricultural activities, vehicle emissions, atmospheric sources, waste-handling practices, mining activities, weathering and volcanic eruptions are the key sources for the emissions of heavy metals into the environment. Such metals are referred to as soil pollutants due to their widespread availability in soil and their toxic effects on plants. In response to plant toxicity, mechanisms are available to cope with metals toxicity through the activation of reduced forms of oxygen. Similarly, phytoremediation has been considered to be an effective way of heavy metals removal from contaminated soil. There is a clear need for further advances in technology to develop cost-effective ways of heavy metals removal.

REFERENCES

Abbruzzese G, Beritognolo I, Muleo R, Piazzai M, Sabatti M, Mugnozza GS, Kuzminsky, E. (2009). Leaf morphological plasticity and stomatal conductance in three Populus alba L. genotypes subjected to salt stress. *Environmental and Experimental Botany* *66*(3):381–388.

Adnan M, Shah Z, Arif M, Khan M J, Mian IA, Sharif M, Saleem, N (2016). Impact of rhizobial inoculum and inorganic fertilizers on nutrients (NPK) availability and uptake in wheat crop. *Canadian Journal of Soil Science* *96*(2):169–176.

Adnan M, Zahir S, Fahad S, Arif M, Mukhtar A, Imtiaz AK, Ishaq AM, Abdul B, Hidayat U, Muhammad A, Inayat-Ur R, Saud S, Muhammad ZI, Yousaf J, Amanullah, HMH, Wajid N (2018a) Phosphate-solubilizing bacteria nullify the antagonistic effect of soil calcification on bioavailability of phosphorus in alkaline soils. *Sci Rep 8*:4339. https://doi.org/10.1038/s41598-018-22653-7.

Adnan M, Shah Z, Sharif M, Rahman H. (2018b). Liming induces carbon dioxide (CO_2) emission in PSB inoculated alkaline soil supplemented with different phosphorus sources. *Environ Sci Poll Res 25*(10):9501–9509.

Adnan M, Fahad S, Khan IA, Saeed M, Ihsan MZ, Saud S, Riaz M, Wang D, Wu C. (2019). Integration of poultry manure and phosphate solubilizing bacteria improved availability of Ca bound P in calcareous soils. *Biotech 9*(10):368.

Adnan M, Fahad S, Muhammad Z, Shahen S, Ishaq AM, Subhan D, Zafar-ul-Hye M, Martin LB, Raja MMN, Beena S, Saud S, Imran A, Zhen Y, Martin B, Jiri H, Rahul D (2020) Coupling Phosphate-Solubilizing Bacteria with Phosphorus Supplements Improve Maize Phosphorus Acquisition and Growth under Lime Induced Salinity Stress. *Plants 9*(900). doi:10.3390/plants9070900.

Aeliona CM, Davisa HT, McDermott S. 2008. Metal concentrations in rural topsoil in South Carolina: potential for human health impact. *Science of the Total Environment, 402*:149–156.

Afonne OJ, Ifediba EC (2020) Heavy metals risks in plant foods–need to step up precautionary measures. *Current Opinion in Toxicology.*

Ahmad S, Kamran M, Ding R, Meng X, Wang H, Ahmad I, Fahad S, Han Q (2019) Exogenous melatonin confers drought stress by promoting plant growth, photosynthetic capacity and antioxidant defense system of maize seedlings. *Peer J 7*:e7793. http://doi.org/10.7717/peerj.7793.

Akbar H, Timothy JK, Jagadish T, Golam M, Apurbo KC, Muhammad F, Rajan B, Fahad S, Hasanuzzaman M (2020) Agricultural Land Degradation: Processes and Problems Undermining Future Food Security. in: Fahad S, Hasanuzzaman M, Alam M, Ullah H, Saeed M, Khan AK, Adnan M (Ed.), *Environment, Climate, Plant and Vegetation Growth.* Springer Publ Ltd, Switzerland, 17–62. https://doi.org/10.1007/978-3-030-49732-3.

Akram R, Turan V, Hammad HM, Ahmad S, Hussain S, Hasnain A, Maqbool MM, Rehmani MIA, Rasool A, Masood N, Mahmood F, Mubeen M, Sultana SR, Fahad S, Amanet K, Saleem M, Abbas Y, Akhtar HM, Waseem F, Murtaza R, Amin A, Zahoor SA, ul Din MS, Nasim W (2018a) Fate of Organic and Inorganic Pollutants in Paddy Soils. In: Hashmi MZ, Varma A (eds) *Environmental Pollution of Paddy Soils, Soil Biology*. Springer International Publishing AG, Switzerland, pp. 197–214.

Akram R, Turan V, Wahid A, Ijaz M, Shahid MA, Kaleem S, Hafeez A, Maqbool MM, Chaudhary HJ, Munis, MFH, Mubeen M, Sadiq N, Murtaza R, Kazmi DH, Ali S, Khan N, Sultana SR, Fahad S, Amin A, Nasim W (2018b) Paddy land pollutants and their role in climate change. In: Hashmi MZ, Varma A (eds) *Environmental Pollution of Paddy Soils, Soil Biology*. Springer International Publishing AG, Switzerland, pp 113–124.

Ali S, Bai P, Zeng F, Cai S, Shamsi IH, Qiu B, Wu F, Zhang G (2011) The ecotoxicological and interactive effects of chromium and aluminum on growth, oxidative damage and antioxidant enzymes on two barley genotypes differing in Al tolerance. *Environ Exp Bot 70*:185–191.

Amanullah, SK, Imran, HAK, Muhammad A, Abdel RA, Muhammad A, Fahad S, Azizullah S, Brajendra P (2020) Effects of Climate Change on Irrigation Water Quality. in: Fahad S, Hasanuzzaman M, Alam M, Ullah H, Saeed M, Khan AK, Adnan M (Ed.), *Environment, Climate, Plant and Vegetation Growth.* Springer Publ Ltd, Switzerland, 123–132. https://doi.org/10.1007/978-3-030-49732-3.

Amir M, Muhammad A, Allah B, Sevgi Ç, Haroon ZK, Muhammad A, Emre A (2020) Bio fortification Under Climate Change: The Fight Between Quality and Quantity. in: Fahad S, Hasanuzzaman M, Alam M, Ullah H, Saeed M, Khan AK, Adnan M (Ed.), *Environment, Climate, Plant and Vegetation Growth.* Springer Publ Ltd, Switzerland, 173–228. https://doi.org/10.1007/978-3-030-49732-3.

Arao T, Ishikawa S, Murakam IM, et al. 2010. Heavy metal contamination of agricultural soil and counter measures in Japan. *Paddy and Water Environment, 8*(3):247–257.

Arif M, Talha J, Muhammad R, Fahad S, Muhammad A, Amanullah, KA, Ishaq AM, Bushra K, Fahd R (2020). Biochar: a Remedy for Climate Change. in: Fahad S, Hasanuzzaman M, Alam M, Ullah H, Saeed M, Khan AK, Adnan M (Ed.), *Environment, Climate, Plant and Vegetation Growth.* Springer Publ Ltd, Switzerland, 151–172. https://doi.org/10.1007/978-3-030-49732-3.

Arora A, Sairam RK, Srivastava GC (2002) Oxidative stress and antioxidative system in plants. *Curr Sci 82*:1227–1238.

Ayman EL, Akbar H, Celaleddin B, Muhammad A, Sohidul Islam M, Shah F, Oksana S, Fatih C, Ram S, and Murat E (2020). Consequences of Salinity Stress on the Quality of Crops and Its Mitigation Strategies for Sustainable Crop Production: An Outlook of Arid and Semi- arid Regions. in: Fahad S, Hasanuzzaman M, Alam M, Ullah H, Saeed M, Khan AK, Adnan M (Ed.), *Environment, Climate, Plant and Vegetation Growth*. Springer Publ Ltd, Switzerland, 503–534. https://doi.org/10.1007/978-3-030-49732-3.

Aziz K, Daniel KYT, Fazal M, Muhammad ZA, Farooq S, Fan W, Fahad S, Ruiyang Z (2017a) Nitrogen nutrition in cotton and control strategies for greenhouse gas emissions: a review. *Environ Sci Pollut Res 24*:23471–23487. https://doi.org/10.1007/s11356-017-0131-y.

Aziz K, Daniel KYT, Muhammad ZA, Honghai L, Shahbaz AT, Mir A, Fahad S (2017b) Nitrogen fertility and abiotic stresses management in cotton crop: a review. *Environ Sci Pollut Res 24*:14551–14566. https://doi.org/10.1007/s11356-017-8920-x.

Baseer M, Adnan M, Fazal M, Fahad S, Muhammad S, Fazli W, Muhammad A, Amanullah, Depeng W, Saud S, Muhammad N, Muhammad Z, Fazli S, Beena S, Mian AR, Ishaq AM (2019) Substituting urea by organic wastes for improving maize yield in alkaline soil. *J Plant Nut*. doi.org/10.1080/01904167.2019.1659344.

Bayram AY, Seher Ö, Nazlican A (2020) Climate Change Forecasting and Modeling for the Year of 2050. in: Fahad S, Hasanuzzaman M, Alam M, Ullah H, Saeed M, Khan AK, Adnan M (Ed.), *Environment, Climate, Plant and Vegetation Growth*. Springer Publ Ltd, Switzerland, 109–122. https://doi.org/10.1007/978-3-030-49732-3.

Berni R, Luyckx M, Xu X, Legay S, Sergeant K, Hausman JF, Guerriero G (2019). Reactive oxygen species and heavy metal stress in plants: impact on the cell wall and secondary metabolism. *Environ Experiment Bot 161*:98–106.

Blaylock JM, Huang JW (2000) Phytoextraction of metals. In: Raskin I, Ensley BD (eds) *Phytoremediation of toxic metals: using plants to clean up the environment*. Wiley, New York.

Chander K, Brookes PC, Harding SA (1995) Microbial biomass dynamics following addition of metal-enriched sewage sludge to a sandy loam. *Soil Biol Biochem 27*(11):1409–1421.

Chang AC, Page AL, Warneke JE (1987) Long-term sludge application on cadmium and zinc accumulation in Swiss chard and radish. *J Environ Qual 16*:217–221.

Chen B, Stein AF, Castell N, Gonzalez-Castanedo Y, de la Campa AS, de la Rosa JD (2016) Modeling and evaluation of urban pollution events of atmospheric heavy metals from a large Cu-smelter. *Sci Tot Environ 539*, 17–25.

Choi JM, Pak CH, Lee CW (1996) Micronutrient toxicity in French marigold. *Journal of Plant Nutrition 19*:901–916.

Demirevska-kepova K, Simova-Stoilova L, Stoyanova Z, Holzer R, Feller U (2004) Biochemical changes in barely plants after excessive supply of copper and manganese. *Environ Experiment Bot 52*, 253–266.

Depeng W, Fahad S, Saud S, Muhammad K, Aziz K, Mohammad NK, Hafiz MH, Wajid N (2018) Morphological acclimation to agronomic manipulation in leaf dispersion and orientation to promote "Ideotype" breeding: evidence from 3D visual modeling of "super" rice (*Oryza sativa* L.). *Plant Physiol Biochem*. https://doi.org/10.1016/j.plaphy.2018.11.010.

Ebbs SD, Kochian LV (1997) Toxicity of zinc and copper to Brassica species: implications for phytoremediation. *J Environ Qual 26*:776–781.

El-Kady AA, Abdel-Wahhab MA (2018) Occurrence of trace metals in foodstuffs and their health impact. *Trend Food Sci Technol 75* 36–45.

El-Meihy RM, Abou-Aly HE, Youssef AM, Tewfike TA, El-Alkshar EA (2019) Efficiency of heavy metals-tolerant plant growth promoting bacteria for alleviating heavy metals toxicity on sorghum. *Environment Experiment Bot 162*:295–301.

Emamverdian A, Ding Y, Mokhberdoran F, Xie X (2015) Heavy Metal Stress and Some Mechanisms of Plant Defense Response. Hindawi Publishing Corporation. *Scientific World Journal 18*:Article ID 756120. http://dx.doi.org/10.1155/2015/756120.

Enamul H, Shoeb AZM, Mallik AH, Fahad S, Kamruzzaman MM, Akib J, Nayyer S, Mehedi A, Swati AS, Yeamin A, Most SS (2020) Measuring Vulnerability to Environmental Hazards: Qualitative to Quantitative. in: Fahad S, Hasanuzzaman M, Alam M, Ullah H, Saeed M, Khan AK, Adnan M (Ed.), *Environment,*

Climate, Plant and Vegetation Growth. Springer Publ Ltd, Switzerland, 421–452. https://doi.org/10.1007/978-3-030-49732-3.

Fahad S, Bano A (2012) Effect of salicylic acid on physiological and biochemical characterization of maize grown in saline area. *Pak J Bot 44*:1433–1438.

Fahad S, Chen Y, Saud S, Wang K, Xiong D, Chen C, Wu C, Shah F, Nie L, Huang J (2013) Ultraviolet radiation effect on photosynthetic pigments, biochemical attributes, antioxidant enzyme activity and hormonal contents of wheat. *J Food Agri Environ 11*(3&4):1635–1641.

Fahad S, Hussain S, Bano A, Saud S, Hassan S, Shan D, Khan FA, Khan F, Chen Y, Wu C, Tabassum MA, Chun MX, Afzal M, Jan A, Jan MT, Huang J (2014a) Potential role of phytohormones and plant growth-promoting Rhizobacteria in abiotic stresses: consequences for changing environment. *Environ Sci Pollut Res 22*(7):4907–4921. https://doi.org/10.1007/s11356-014-3754-2.

Fahad S, Hussain S, Matloob A, Khan FA, Khaliq A, Saud S, Hassan S, Shan D, Khan F, Ullah N, Faiq M, Khan MR, Tareen AK, Khan A, Ullah A, Ullah N, Huang J (2014b) Phytohormones and plant responses to salinity stress: a review. *Plant Growth Regul 75*(2):391–404. https://doi.org/10.1007/s10725-014-0013-y.

Fahad S, Hussain S, Saud S, Tanveer M, Bajwa AA, Hassan S, Shah AN, Ullah A, Wu C, Khan FA, Shah F, Ullah S, Chen Y, Huang J (2015a) A biochar application protects rice pollen from high-temperature stress. *Plant Physiol Biochem 96*:281–287.

Fahad S, Nie L, Chen Y, Wu C, Xiong D, Saud S, Hongyan L, Cui K, Huang J (2015b) Crop plant hormones and environmental stress. *Sustain Agric Rev 15*:371–400.

Fahad S, Hussain S, Saud S, Hassan S, Chauhan BS, Khan F et al. (2016a) Responses of rapid viscoanalyzer profile and other rice grain qualities to exogenously applied plant growth regulators under high day and high night temperatures. *PLoS One 11*(7):e0159590. https://doi.org/10.1371/journal.pone.0159590.

Fahad S, Hussain S, Saud S, Khan F, Hassan S, Nasim W, Arif M, Wang F, Huang J (2016b) Exogenously applied plant growth regulators affect heat-stressed rice pollens. *J Agron Crop Sci 202*:139–150.

Fahad S, Hussain S, Saud S, Hassan S, Ihsan Z, Shah AN, Wu C, Yousaf M, Nasim W, Alharby H, Alghabari F, Huang J (2016c) Exogenously applied plant growth regulators enhance the morphophysiological growth and yield of rice under high temperature. *Front Plant Sci 7*:1250. https://doi.org/10.3389/fpls.2016.01250.

Fahad S, Hussain S, Saud S, Hassan S, Tanveer M, Ihsan MZ, Shah AN, Ullah A, Nasrullah KF, Ullah S, Alharby H NW, Wu C, Huang J (2016d) A combined application of biochar and phosphorus alleviates heat-induced adversities on physiological, agronomical and quality attributes of rice. *Plant Physiol Biochem 103*:191–198.

Fahad S, Bajwa AA, Nazir U, Anjum SA, Farooq A, Zohaib A, Sadia S, Nasim W, Adkins S, Saud S, Ihsan MZ, Alharby H, Wu C, Wang D, Huang J (2017) Crop production under drought and heat stress: plant responses and management options. *Front Plant Sci 8*:1147. https://doi.org/10.3389/fpls.2017.01147.

Fahad S, Muhammad ZI, Abdul K, Ihsanullah D, Saud S, Saleh A, Wajid N, Muhammad A, Imtiaz AK, Chao W, Depeng W, Jianliang H (2018) Consequences of high temperature under changing climate optima for rice pollen characteristics-concepts and perspectives. *Archives Agron Soil Sci.* doi:10.1080/03650340.2018.1443213.

Fahad S, Rehman A, Shahzad B, Tanveer M, Saud S, Kamran M, Ihtisham M, Khan SU, Turan V, Rahman MHU (2019a) Rice responses and tolerance to metal/metalloid toxicity. In: Hasanuzzaman M, Fujita M, Nahar K, Biswas, JK (eds) *Advances in rice research for abiotic stress tolerance*. Woodhead Publ Ltd, Cambridge, pp 299–312.

Fahad S, Adnan M, Hassan S, Saud S, Hussain S, Wu C, Wang D, Hakeem KR, Alharby HF, Turan V, Khan MA, Huang J (2019b) Rice responses and tolerance to high temperature. In: Hasanuzzaman M, Fujita M, Nahar K, Biswas JK (eds) *Advances in rice research for abiotic stress tolerance*. Woodhead Publ Ltd, Cambridge, pp 201–224.

Farah R, Muhammad R, Muhammad SA, Tahira Y, Muhammad AA, Maryam A, Shafaqat A, Rashid M, Muhammad R, Qaiser H, Afia Z, Muhammad AA, Muhammad A, Fahad S (2020). Alternative and Non-conventional Soil and Crop Management Strategies for Increasing Water Use Efficiency. in: Fahad S, Hasanuzzaman M, Alam M, Ullah H, Saeed M, Khan AK, Adnan M (Ed.), *Environment, Climate, Plant and Vegetation Growth*. Springer Publ Ltd, Switzerland, 323–338. https://doi.org/10.1007/978-3-030-49732-3.

Farhana G, Ishfaq A, Muhammad A, Dawood J, Fahad S, Xiuling L, Depeng W, Muhammad F, Muhammad F, Syed AS (2020) Use of crop growth model to simulate the impact of climate change on yield of various wheat cultivars under different agro-environmental conditions in Khyber Pakhtunkhwa, Pakistan. *Arabian J Geosci 13*:112. https://doi.org/10.1007/s12517-020-5118-1.

Farhat A, Hafiz MH, Wajid I, Aitazaz AF, Hafiz FB, Zahida Z, Fahad S, Wajid F, Artemi C (2020) A review of soil carbon dynamics resulting from agricultural practices. *J Environ Manage 268*(2020):110319.

Fazli W, Muhmmad S, Amjad A, Fahad S, Muhammad A, Muhammad N, Ishaq AM, Imtiaz AK, Mukhtar A, Muhammad S, Muhammad I, Rafi U, Haroon I, Muhammad A (2020) Plant-Microbes Interactions and Functions in Changing Climate. in: Fahad S, Hasanuzzaman M, Alam M, Ullah H, Saeed M, Khan AK, Adnan M (Ed.), *Environment, Climate, Plant and Vegetation Growth*. Springer Publ Ltd, Switzerland, 397–420. https://doi.org/10.1007/978-3-030-49732-3.

Fontes RLS, Cox FR (1998) Zinc toxicity in soybean grown at high iron concentration in nutrient solution. *Journal of Plant Nutrition 21*:1723–1730.

Gisbert C, Ros R, Deharo A, Walker DJ, Pilarbernal M, Serrano R, Navarro-Avino J (2003) A plant genetically modified that accumulates Pb is especially promising for phytoremediation. *Biochem Biophys Res Commun 303*:440–445.

Gopakumar L, Bernard NO, Donato V (2020) Soil Microarthropods and Nutrient Cycling. in: Fahad S, Hasanuzzaman M, Alam M, Ullah H, Saeed M, Khan AK, Adnan M (Ed.), *Environment, Climate, Plant and Vegetation Growth*. Springer Publ Ltd, Switzerland, 453–472. https://doi.org/10.1007/978-3-030-49732-3.

Greger M (1997) *Willow as phytoremediator of heavy metal contaminated soil. Proceedings of the 2nd International Conference on Element Cycling in the Environment*, Warsaw, pp. 167–172.

Greger M, Brammer E, Lindberg S, Larson G, Ildestan-Almquist J (1991) Uptake and physiological effects of cadmium in sugar beet (*Beta vulgaris*) related to mineral provision. *J Exp Bot 42*:729–737.

Habib ur R, Ashfaq A, Aftab W, Manzoor H, Fahd R, Wajid I, Aminul I, Vakhtang S, Muhammad A, Asmat U, Abdul W, Syeda RS, Shah S, Shahbaz K, Fahad S, Manzoor H, Saddam H, Wajid N (2017) Application of CSM-CROPGRO-Cotton model for cultivars and optimum planting dates: Evaluation in changing semi-arid climate. *Field Crops Res*. http://dx.doi.org/10.1016/j.fcr.2017.07.007.

Hafiz MH, Wajid F, Farhat A, Fahad S, Shafqat S, Wajid N, Hafiz FB (2016) Maize plant nitrogen uptake dynamics at limited irrigation water and nitrogen. *Environ Sci Pollut Res 24*(3):2549–2557. https://doi.org/10.1007/s11356-016-8031-0.

Hafiz MH, Farhat A, Shafqat S, Fahad S, Artemi C, Wajid F, Chaves CB, Wajid N, Muhammad M, Hafiz FB (2018) Offsetting land degradation through nitrogen and water management during maize cultivation under arid conditions. *Land Degrad Dev* 1–10. doi:10.1002/ldr.2933.

Hafiz MH, Muhammad A, Farhat A, Hafiz FB, Saeed AQ, Muhammad M, Fahad S, Muhammad A (2019) Environmental factors affecting the frequency of road traffic accidents: a case study of sub-urban area of Pakistan. *Environ Sci Pollut Res*. https://doi.org/10.1007/s11356-019-04752-8.

Hafiz MH, Farhat A, Ashfaq A, Hafiz FB, Wajid F, Carol Jo W, Fahad S, Gerrit H (2020a) Predicting Kernel Growth of Maize under Controlled Water and Nitrogen Applications. *Int J Plant Prod*. https://doi.org/10.1007/s42106-020-00110-8.

Hafiz MH, Abdul K, Farhat A, Wajid F, Fahad S, Muhammad A, Ghulam MS, Wajid N, Muhammad M, Hafiz FB (2020b) Comparative effects of organic and inorganic fertilizers on soil organic carbon and wheat productivity under arid region. *Commun Soil Sci Plant Anal*. doi:10.1080/00103624.2020.1763385.

Han FX, Su Y, Monts DL, Waggoner AC, Plodinec JM (2006) Binding, distribution, and plant uptake of mercury in a soil from Oak Ridge, Tennessee, USA. *Sci Total Environ 368*:753–768.

Hegedus A, Erdei S, Horvath G (2001) Comparative studies of H_2O_2 detoxifying enzymes in green and greening barley seedlings under cadmium stress. *Plant Sci 160*:1085–1093.

Hesham FA, Fahad S (2020) Melatonin application enhances biochar efficiency for drought tolerance in maize varieties: Modifications in physio-biochemical machinery. *Agron J* 1–22.

Hirsch RE, Lewis BD, Spalding EP, Sussman MR (1998) A role for the AKT1 potassium channel in plant nutrition. *Science 280*:918–921.

Hussain MA, Fahad S, Rahat S, Muhammad FJ, Muhammad M, Qasid A, Ali A, Husain A, Nooral A, Babatope SA, Changbao S, Liya G, Ibrar A, Zhanmei J, Juncai H (2020) Multifunctional role of brassinosteroid and its analogues in plants. *Plant Growth Regul*. https://doi.org/10.1007/s10725-020-00647-8.

Ilyas M, Mohammad N, Nadeem K, Ali H, Aamir HK, Kashif H, Fahad S, Aziz K, Abid U (2020) Drought tolerance strategies in plants: a mechanistic approach. *J Plant Growth Regul*. https://doi.org/10.1007/s00344-020-10174-5.

Iqra M, Amna B, Shakeel I, Fatima K, Sehrish L, Hamza A, Fahad S (2020) Carbon Cycle in Response to Global Warming. in: Fahad S, Hasanuzzaman M, Alam M, Ullah H, Saeed M, Khan AK, Adnan M (Ed.), *Environment, Climate, Plant and Vegetation Growth*. Springer Publ Ltd, Switzerland, pp. 1–16. https://doi.org/10.1007/978-3-030-49732-3.

Izosimova A (2005) Modelling the interaction between calcium and nickel in the soil–plant system. *FAL Agri Res 288*:99.

Jan M, Muhammad A, Adnan NS, Muhammad Y, Javaid I, Xiuling L, Depeng W, Fahad S (2019) Modulation in growth, gas exchange, and antioxidant activities of salt-stressed rice (*Oryza sativa* L.) genotypes by zinc fertilization. *Arab J Geosci 12*:775. https://doi.org/10.1007/s12517-019-4939-2.

Kamarn M, Wenwen C, Irshad A, Xiangping M, Xudong Z, Wennan S, Junzhi C, Shakeel A, Fahad S, Qingfang H, Tiening L (2017) Effect of paclobutrazol, a potential growth regulator on stalk mechanical strength, lignin accumulation and its relation with lodging resistance of maize. *Plant Growth Regul 84*:317–332. https://doi.org/10.1007/ s10725-017-0342-8.

Khan A, Yousafzai A M, Shah N, Ahmad MS, Farooq M, Aziz F, Jawad SM (2016). Enzymatic Profile Aactivity of Grass Carp (Ctenopharyngodon Idella) After Exposure to the Pollutant Named Atrazine (Herbicide). *Pol J Environ Stud 25*(5):2001–2006.

Kranner I, Colville L (2011) Metals and seeds: biochemical and molecular implications and their significance for seed germination. *Environ Exp Bot 72*:93–105.

Lee CW, Choi JM., Pak CH (1996) Micronutrient toxicity in seed geranium (Pelargonium×hortorum Baley). *J Am Soc Hortic Sci 121*:77–82.

Lindberg SE, Meyers TP, Taylor Jr GE, Turner RR, Schroeder WH (1992) Atmosphere-surface exchange of mercury in a forest: results of modeling and gradient approached. *J Geophys Res 97*:2519–2528.

Liu L, Wang S, Guo X, Wang H (2019) Comparison of the effects of different maturity composts on soil nutrient, plant growth and heavy metal mobility in the contaminated soil. *J Environ Manage 250*: 109525.

Liu Z, Wang LA, Xiao H, Guo X, Urbanovich O, Nagorskaya L, Li X (2020) A review on control factors of pyrolysis technology for plants containing heavy metals. *Ecotoxicol Environment Safety 191*:110181.

Logan TJ, Chaney RL (1983) Metals. In: Page AL (ed) *Utilization of municipal wastewater and sludge on land*. University of California, Riverside, pp. 235–326.

Mahar A, Amjad A, Altaf HL, Fazli W, Ronghua L, Muhammad A, Fahad S, Muhammad A, Rafiullah, IAK, Zengqiang Z (2020) Promising Technologies for Cd-Contaminated Soils: Drawbacks and Possibilities. in: Fahad S, Hasanuzzaman M, Alam M, Ullah H, Saeed M, Khan AK, Adnan M (Ed.), *Environment, Climate, Plant and Vegetation Growth*. Springer Publ Ltd, Switzerland, 63–92. https://doi.org/10.1007/978-3-030-49732-3.

Marschner H (1995) *Mineral nutrition of higher plants*. Academic Press, Cambridge.

Md Jakir H, Allah B (2020) Development and Applications of Transplastomic Plants; A Way Towards Eco-Friendly Agriculture. in: Fahad S, Hasanuzzaman M, Alam M, Ullah H, Saeed M, Khan AK, Adnan M (Ed.), *Environment, Climate, Plant and Vegetation Growth*. Springer Publ Ltd, Switzerland, 285–322. https://doi.org/10.1007/978-3-030-49732-3.

Mohammad I, Munir A, Adel RA, Mutair A, and Muhammad I (2020) Advances in Pyrolytic Technologies with Improved Carbon Capture and Storage to Combat Climate Change. in: Fahad S, Hasanuzzaman M, Alam M, Ullah H, Saeed M, Khan AK, Adnan M (Ed.), *Environment, Climate, Plant and Vegetation Growth*. Springer Publ Ltd, Switzerland, 535–576. https://doi.org/10.1007/978-3-030-49732-3.

Mohanpuria P, Rana NK, Yadav SK (2007) Cadmium induced oxidative stress influence on glutathione metabolic genes of *Camellia sinensis* (L.) O. Kuntze. *Environ Toxicol 22*:368–374.

Montagne D, Cornu S, Bourennane H (2007) Effect agricultural practices on trace-element distribution in soil. *Commun Soil Sci Plant Anal 38*:473–491.

Moosavi SE, Gharineh MH, Afshari RT, Ebrahimi A (2012) Effects of some heavy metals on seed germination characteristics of canola (*Barassica napus*), wheat (*Triticum aestivum*) and safflower (*Carthamus tinctorious*) to evaluate phytoremediation potential of these crops. *J Agric Sci 4*:1–19.

Mubeen M, Ashfaq A, Hafiz MH, Muhammad A, Hafiz UF, Mazhar S, Muhammad S, Asad A, Amjed A, Fahad S, Wajid N (2020) Evaluating the climate change impact on water use efficiency of cotton-wheat in semi-arid conditions using DSSAT model. *J Water Clim Change* doi:10.2166/wcc.2019.179/622035/jwc2019179.pdf.

Muhammad Z, Abdul MK, Abdul MS, Kenneth BM, Muhammad S, Shahen S, Ibadullah J, Fahad S (2019) Performance of Aeluropus lagopoides (mangrove grass) ecotypes, a potential turfgrass, under high saline conditions. *Environ Sci Pollut Res*. https://doi.org/10.1007/s11356-019-04838-3.

Muhammad T, Amna F, Amna B, Barira Z, Xitong Z, Ling-Ling C (2020) Effectiveness of Conventional Crop Improvement Strategies vs. Omics. in: Fahad S, Hasanuzzaman M, Alam M, Ullah H, Saeed M, Khan AK, Adnan M (Ed.), *Environment, Climate, Plant and Vegetation Growth*. Springer Publ Ltd, Switzerland, 253–284. https://doi.org/10.1007/978-3-030-49732-3.

Nagajyoti PC, Lee KD, Sreekanth TVM (2010) Heavy metals, occurrence and toxicity for plants: a review. *Environ Chem Lett 8*(3):199–216.

Noor M, Naveed ur R, Ajmal J, Fahad S, Muhammad A, Fazli W, Saud S, Hassan S (2020) Climate Change and Costal Plant Lives. in: Fahad S, Hasanuzzaman M, Alam M, Ullah H, Saeed M, Khan AK, Adnan M (Ed.), *Environment, Climate, Plant and Vegetation Growth*. Springer Publ Ltd, Switzerland, 93–108. https://doi.org/10.1007/978-3-030-49732-3.

Olaniran AO, Balgobind A, Pillay B (2013) Bioavailability of heavy metals in soil: impact on microbial biodegradation of organic compounds and possible improvement strategies. *Int J Mol Sci 14*:10197–10228. doi:10.3390/ijms140510197.

Ozturk M, Yucel E, Gucel S, Sakcali S, Aksoy A (2008) Plants as bio-monitors of trace elements pollution in soil. In: Prasad MNV (eds) *Trace elements: environmental contamination, nutritional benefits and health implications*, Wiley, New York, pp 723–744.

Pourrut B, Shahid M, Dumat C, Winterton P, Pinelli E (2011) Lead uptake, toxicity, and detoxification in plants. *Reviews of Environmental Contamination and Toxicology Volume 213*. Springer, New York, NY, pp 113–136.

Punz WF, Sieghardt H (1993) The response of roots of herbaceous plant species to heavy metals. *Environ Experiment Botany 33*(1):85–98.

Qamar-uz Z, Zubair A, Muhammad Y, Muhammad ZI, Abdul K, Fahad S, Safder B, Ramzani PMA, Muhammad N (2017) Zinc biofortification in rice: leveraging agriculture to moderate hidden hunger in developing countries. *Arch Agron Soil Sci 64*:147–161. https://doi.org/10.1080/03650340.2017.1338343.

Rascio N, Navari-Izzo F (2011) Heavy metal hyper accumulating plants: how and why do they do it? And what makes them so interesting? *Plant Sci 180*(2):169–181.

Rashid M, Qaiser H, Khalid SK, Mohammad I, Zhang A, Muhammad A, Shahzada SI, Rukhsanda A, Ghulam AS, Shahzada MM, Sarosh A, Muhammad FQ (2020). Prospects of Biochar in Alkaline Soils to Mitigate Climate Change. in: Fahad S, Hasanuzzaman M, Alam M, Ullah H, Saeed M, Khan AK, Adnan M (Ed.), *Environment, Climate, Plant and Vegetation Growth*. Springer Publ Ltd, Switzerland, 133–150. https://doi.org/10.1007/978-3-030-49732-3.

Raskin I, Smith RD, Salt DE (1997) Phytoremediation of metals: using plants to remove pollutants from the environment. *Curr Opin Biotechnol 8*:221–226.

Rehman M, Fahad S, Saleem MH, Hafeez M, Muhammad H, Liu F, Deng G (2020) Red light optimized physiological traits and enhanced the growth of ramie (*Boehmeria nivea* L.). *Photosynthetica 58*(4):922–931.

Rizvi A, Khan MS (2018). Heavy metal induced oxidative damage and root morphology alterations of maize (*Zea mays* L.) plants and stress mitigation by metal tolerant nitrogen fixing Azotobacter chroococcum. *Ecotoxicol Environment Safety 157*:9–20.

Sadam M, Muhammad T, Ghulam M, Muhammad SK, Faiz AJ (2020). Role of Biotechnology in Climate Resilient Agriculture. in: Fahad S, Hasanuzzaman M, Alam M, Ullah H, Saeed M, Khan AK, Adnan M (Ed.), *Environment, Climate, Plant and Vegetation Growth*. Springer Publ Ltd, Switzerland, 339–366. https://doi.org/10.1007/978-3-030-49732-3.

Sajjad H, Muhammad M, Ashfaq A, Waseem A, Hafiz MH, Mazhar A, Nasir M, Asad A, Hafiz UF, Syeda RS, Fahad S, Depeng W, Wajid N (2019) Using GIS tools to detect the land use/land cover changes during forty years in Lodhran district of Pakistan. *Environ Sci Pollut Res*. https://doi.org/10.1007/s11356-019-06072-3.

Saleem MH, Fahad S, Adnan M, Mohsin A, Muhammad SR, Muhammad K, Qurban A, Inas AH, Parashuram B, Mubassir A, Reem MH (2020a) Foliar application of gibberellic acid endorsed phytoextraction of copper and alleviates oxidative stress in jute (*Corchorus capsularis* L.) plant grown in highly copper-contaminated soil of China. *Environ Sci Pollution Res*. https://doi.org/10.1007/s11356-020-09764-3.

Saleem MH, Rehman M, Fahad S, Tung SA, Iqbal N, Hassan A, Ayub A, Wahid MA, Shaukat S, Liu L, Deng G (2020b) Leaf gas exchange, oxidative stress, and physiological attributes of rapeseed (*Brassica napus* L.) grown under different light-emitting diodes. *Photosynthetica 58*(3):836–845.

Saleem MH, Fahad S, Shahid UK, Mairaj D, Abid U, Ayman ELS, Akbar H, Analía L, Lijun L (2020c) Copper-induced oxidative stress, initiation of antioxidants and phytoremediation potential of flax (*Linum usitatissimum* L.) seedlings grown under the mixing of two different soils of China. *Environ Sci Poll Res*. https://doi.org/10.1007/s11356-019-07264-7.

Salt DE, Prince RC, Pickering IJ, Raskin I (1995) Mechanisms of cadmium mobility and accumulation in Indian mustard. *Plant Physiol 109*:1427–1433.

Saman S, Amna B, Bani A, Muhammad T, Rana MA, Muhammad SK (2020) QTL Mapping for Abiotic Stresses in Cereals. in: Fahad S, Hasanuzzaman M, Alam M, Ullah H, Saeed M, Khan AK, Adnan M (Ed.), *Environment, Climate, Plant and Vegetation Growth*. Springer Publ Ltd, Switzerland, 229–252. https://doi.org/10.1007/978-3-030-49732-3.

Saud S, Chen Y, Long B, Fahad S, Sadiq A (2013) The different impact on the growth of cool season turf grass under the various conditions on salinity and drought stress. *Int J Agric Sci Res 3*:77–84.

Saud S, Li X, Chen Y, Zhang L, Fahad S, Hussain S, Sadiq A, Chen Y (2014) Silicon application increases drought tolerance of Kentucky bluegrass by improving plant water relations and morph physiological functions. *Sci World J 2014*:1–10. https://doi.org/10.1155/2014/ 368694.

Saud S, Chen Y, Fahad S, Hussain S, Na L, Xin L, Alhussien SA (2016) Silicate application increases the photosynthesis and its associated metabolic activities in Kentucky bluegrass under drought stress and post-drought recovery. *Environ Sci Pollut Res 23*(17):17647–17655. https://doi.org/10.1007/s11356-016-6957-x.

Saud S, Fahad S, Yajun C, Ihsan MZ, Hammad HM, Nasim W, Amanullah, Arif M and Alharby H (2017) Effects of nitrogen supply on water stress and recovery mechanisms in Kentucky Bluegrass plants. *Front Plant Sci 8*:983. doi:10.3389/fpls.2017.00983.

Saud S, Fahad S, Cui G, Chen Y, Anwar S (2020) Determining nitrogen isotopes discrimination under drought stress on enzymatic activities, nitrogen isotope abundance and water contents of Kentucky bluegrass. *Sci Rep 10*:6415. https://doi.org/10.1038/s41598-020-63548-w.

Scoccianti V, Crinelli R, Tirillini B, Mancinelli V, Speranza A (2006) Uptake and toxicity of Cr (Cr3+) in celery seedlings. *Chemosphere 64*:1695–1703.

Seregin IV, Ivanov VB (2001) Physiological aspects of cadmium and lead toxic effects on higher plants. *Russian J Plant Physiol 48*(4):523–544.

Shafi MI, Adnan M, Fahad S, Fazli W, Ahsan K, Zhen Y, Subhan D, Zafar-ul-Hye M, Martin B, Rahul D (2020) Application of Single Superphosphate with Humic Acid Improves the Growth, Yield and Phosphorus Uptake of Wheat (*Triticum aestivum* L.) in Calcareous Soil. *Agron 10*:1224. doi:10.3390/agronomy10091224.

Shah F, Lixiao N, Kehui C, Tariq S, Wei W, Chang C, Liyang Z, Farhan A, Fahad S, Huang J (2013) Rice grain yield and component responses to near 2°C of warming. *Field Crop Res 157*:98–110.

Shanker AK, Cervantes C, Loza-Tavera H, Avudainayagam S (2005) Chromium toxicity in plants. *Environ Int 31*:739–753.

Sharma P, Dubey RS (2005) Lead toxicity in plants. *Brazil J Plant Physiol 17*:35–52.

Su C, Jiang L, Zhang W (2014) A review on heavy metal contamination in the soil worldwide: situation, impact and remediation techniques. *Environ Skept Critic 3*(2):24–38.

Subhan D, Zafar-ul-Hye M, Fahad S, Saud S, Martin B, Tereza H, Rahul D (2020) Drought stress alleviation by ACC Deaminase producing *Achromobacter xylosoxidans* and *Enterobacter cloacae*, with and without Timber Waste Biochar in Maize. *Sustain 12*(6286). doi:10.3390/su12156286.

Tariq M, Ahmad S, Fahad S, Abbas G, Hussain S, Fatima Z, Nasim W, Mubeen M, ur Rehman MH, Khan MA, Adnan M (2018). The impact of climate warming and crop management on phenology of sunflower-based cropping systems in Punjab, Pakistan. *Agri Forest Met 15*(256):270–282.

Uddin MN, Ali M, Muhammad MF, Ahmad N, Jamil J, Kalsoom MA, Khan A (2016) Characterizing microbial populations in petroleum-contaminated soils of Swat district, Pakistan. *Polish J Environ Studies 2*(4):11691–11697.

Unsar N, Muhammad R, Syed HMB, Asad S, Mirza AQ, Naeem I, Muhammad H, Fahad S, Shafqat S (2020). Insect Pests of Cotton Crop and Management Under Climate Change Scenarios. in: Fahad S, Hasanuzzaman M, Alam M, Ullah H, Saeed M, Khan AK, Adnan M (Ed.), *Environment, Climate, Plant and Vegetation Growth*. Springer Publ Ltd, Switzerland, 367–396. https://doi.org/10.1007/978-3-030-49732-3.

Verloo M, Eeckhout M (1990) Metal species transformations in soil: an analytical approach. *Int J Environ Anal Chem 39*:170–186.

Viehweger K (2014) How plants cope with heavy metals. *Botanical Studies 55*(35):1–12.

Wahid F, Fahad S, Subhan D, Adnan M, Zhen Y, Saud S, Manzer HS, Martin B, Tereza H, Rahul D (2020) Sustainable management with Mycorrhizae and phosphate solubilizing bacteria for enhanced phosphorus uptake in calcareous soils. *Agri 10*(334). doi:10.3390/agriculture10080334.

Wajid N, Ashfaq A, Asad A, Muhammad T, Muhammad A, Muhammad S, Khawar J, Ghulam MS, Syeda RS, Hafiz MH, Muhammad IAR, Muhammad ZH, Muhammad Habib ur R, Veysel T, Fahad S, Suad S, Aziz K, Shahzad A (2017) Radiation efficiency and nitrogen fertilizer impacts on sunflower crop in contrasting environments of Punjab. *Pakistan Environ Sci Pollut Res 25*:1822–1836. https://doi.org/10.1007/s11356-017-0592-z.

Wu C, Tang S, Li G, Wang S, Fahad S, Ding Y (2019) Roles of phytohormone changes in the grain yield of rice plants exposed to heat: a review. *Peer J 7*:e7792. doi:10.7717/peerj.7792.

Wu C, Kehui C, She T, Ganghua L, Shaohua W, Fahad S, Lixiao N,Jianliang H, Shaobing P, Yanfeng D (2020) Intensified pollination and fertilization ameliorate heat injury in rice (*Oryza sativa* L.) during the flowering stage. *Field Crops Res 252*:107795.

Wuana RA, Okieimen FE (2011) Heavy metals in contaminated soils: a review of sources, chemistry, risks and best available strategies for remediation. *ISRN Ecol 2011*:Article ID 402647, 20.

Yabe J, Ishizuka M, et al. (2010) Current levels of heavy metal contamination in Africa. *J Vet Med Sci 72*(10):1257–1263.

Yang Z, Zhang Z, Zhang T, Fahad S, Cui K, Nie L, Peng S, Huang J (2017) The effect of season-long temperature increases on rice cultivars grown in the central and southern regions of China. *Front Plant Sci 8*:1908. https://doi.org/10.3389/fpls.2017.01908.

Zafar-ul-Hye M, Muhammad N, Subhan D, Fahad S, Rahul D, Mazhar A, Ashfaq AR, Martin B, Jĭrí H, Zahid HT, Muhammad N (2020a) Alleviation of cadmium adverse effects by improving nutrients uptake in bitter gourd through cadmium tolerant Rhizobacteria. *Environment 7*(54). doi:10.3390/environments7080054.

Zafar-ul-Hye M, Muhammad T, Muhammad A, Fahad S, Martin B, Tereza D, Rahul D, Subhan D (2020b) Potential role of compost mixed biochar with rhizobacteria in mitigating lead toxicity in spinach. *Scientific Rep 10*:12159. https://doi.org/10.1038/s41598-020-69183-9.

Zahida Z, Hafiz FB, Zulfiqar AS, Ghulam MS, Fahad S, Muhammad RA, Hafiz MH, Wajid N, Muhammad S (2017) Effect of water management and silicon on germination, growth, phosphorus and arsenic uptake in rice. *Ecotoxicol Environ Saf 144*:11–18.

Zhang Y, Cai T, Burkard G (1999) Research advances on the mechanisms of heavy metal tolerance in plants. *Acta Botanica Sinica 41*(5):453–457.

Zia-ur-Rehman M (2020) Environment, Climate Change and Biodiversity. in: Fahad S, Hasanuzzaman M, Alam M, Ullah H, Saeed M, Khan AK, Adnan M (Ed.), *Environment, Climate, Plant and Vegetation Growth*. Springer Publ Ltd, Switzerland, 473–502. https://doi.org/10.1007/978-3-030-49732-3.

Zojaji F, Hassani AH, Sayadi MH. 2014. Bioaccumulation of chromium by *Zea mays* in wastewater-irrigated soil: an experimental study. *Proc Int Acad Ecol Environ Sci 4*(2):62–67.

9

Soil Biodiversity and Climate Change

Hafiz Muhammad Rashad Javeed and Mazhar Ali
COMSATS University Islamabad, Pakistan

Shahid Ibni Zamir
University of Agriculture Faisalabad, Pakistan

CONTENTS

9.1 What is Soil Biodiversity?

Soil biodiversity is the term given to the mixture of different living organisms in the soil system. These living organisms control the soil's biological activities after interacting with plants and other small animals. There is a wide variety of life in the soil, ranging from genes to large communities and sometime micro-aggregates develop a large landscape. Soil organisms can be easily divided into three main groups, i.e. macro, meso and micro fauna. The major soil biomass consists of microorganisms such as bacteria, algae and fungi.

Soil is a more diverse system than any other system on the Earth. Many organisms are involved in forming a soil food web, including bacteria, fungi, beetles, mites, worms, ants, springtails, nematodes and spiders. These organisms improve the levels of soil plant nutrition, its resistance to erosion, the entry and storage of water and the breaking down of complex organic residues. The diverse organisms in the soil system interact: in this way, they maintain the delicate check and balance in the soil food web through population survival, control and mobility from season to season.

9.2 What is Climate Change?

Climate change is the significant change in either its mean state of the climate and persistence for a longer period of time (typically for a decade or more). Furthermore, climate change may be to the result of either natural processes persistent anthropogenic activities in the atmosphere or in the land. Moreover, any abnormal variation in the climate and the long-term effects of such a variation on the Earth is also referred as climate change. Climate is one of the major components that helps in plant growth and the species distributions. Any change in the soil-root environment that influences the microbial processes and biotic soil conditions is known as climate-induced changes which are usually shown in soil-plant feedback. Moreover, the consequences for biodiversity and potential feedback and ecosystem functioning are the major stimuli of climate change.

9.3 The Effect of Climate Change on Soil Environment and Fertility

Global climate change is disturbing the soil organic biomass concentration, soil temperature and soil hydrology. These postulated changes result in a decline in the capacity of soil to hold water and also increase the levels of evapotranspiration, leading to physical degradation in the soil and also a loss of nutrients. Carbon dioxide emissions into the atmosphere have resulted in an increase in atmospheric and soil temperatures. The change in temperature will have provided more nutrients to the plant; on the other hand, it will also have increased the levels of night time respiration. The shortened growth cycle of crops led to the production of poor-quality food and may also increase rates of malnutrition. Thus, the agro-ecosystem required the introduction of long duration cultivars or frequent changes in the cropping pattern or fallow land due to unproductive periods. Long duration crops had more time than short duration crops to adjust in the changing climate. That's why agriculture is shifting toward a perennial cropping system rather than an annual cropping system. A perennial cropping system is better able to cope with the turbulence of environmental factors.

The increase in CO_2 pressure in the soil air would decrease the O_2 level in the soil air, resulting in an impairment of the root function. The increase in the stable gas exchange system between the atmosphere and soil through an increase in the numbers of stable biopores which tend to maintain O_2 and CO_2 at safe levels in the soil system. Such gas exchange phenomena do not affect some crops such as jute and rice. These two crops have their own specific gas exchange mechanisms in their stem which would be unaffected through elevated levels of moisture or CO_2 in the rooting zone. The moisture variation in the soil limited the activities of soil organisms and thereby disturbed the biogeocycles and soil ecological functioning. These changes will force the plant to either increase or decrease growth, root exudation and biomass, mycorrhizal colonization, symbiotic relationships and root nitrogen fixers. This would have a negative effect on N supply to the rooting zone and decrease the level of microbial and root activity in the soil.

Hence, lower down the release of plant nutrients release from weathering of minerals in the soil. Similarly, weak mycorrhizal interaction led to poor phosphate uptake. Sometime higher atmospheric CO_2 concentration (up to a certain limit) triggered the plant defense system to produce a more intensive root system. Some literature has suggested that higher soil temperatures can counteract this to increase the decomposition of stable organic matter contents while at the same time further stimulating the microbial activities. At higher temperatures, soil organic dynamics and microorganisms may undergo some temporary competition for plant nutrients. On occasion, this temporary competition would have been recorded as a plant's negative response to elevated CO_2 concentration. If CO_2 increased gradually (in range between 350 and 450 ppm) over decades (as two decades ago), then an increase in soil organic matter dynamics and microbial activities in soils would be positive for the plant-soil system. Over a longer period, higher CO_2 conditions tend to result in a slowing of the activities of both termites and earthworms. There is no prior reason in the literature why the higher CO_2 conditions subjected to more nutrients demand by the crops.

Greater levels of microbial activity tend to increase the nutrients cycling through the soil's biological system. This increases the root growth and biomass (at plant ambient temperature) and tends to increase the soil organic matter contents, which also forces the immobilization and cycling of plant nutrients in the soil on a temporary basis. Some studies have reported that the C/N ratio increases under high CO_2 concentration, which will result in a slowing down of the decomposition of organic matter and the remobilization of nutrients from the soil. Furthermore, it also decreases the nutrients by root mat. Increased CO_2 concentration increases soil stability through the production of polysaccharides as soil stabilizers through higher levels of activity of soil macro fauna, including beetles and earthworms. All these activities improved both bypass flow and infiltration rate through the increased production of biopores. The greater soil stability and the increased infiltration rate improved the resilience of the soil against climate change factors and hence conserves soil fertility.

9.4 The Effect of Climate Change on Soil Microbiota

In general, the increase in atmospheric temperature corresponds to an increase in microbial activity. Climate-based abiotic regulations are important determinants of microbial activities. The overall effect of climate on soil microorganisms can be perceived through the seasonal dynamics of microbial populations. These dynamics are due to the fact that the growth, activity and composition of microbial communities are susceptible to the two main factors regulated by climate: temperature and moisture. Growth and activity rates are individual characteristics of microbial communities and may vary independently. Changing soil temperature causes changes in microbial-mediated nitrification and denitrification dynamics in the soil environment, possibly due to changes in nitrification and denitrification populations. Occasional disturbances in the soil environment can lead to community changes and metabolic activity in microorganisms involved in soil nutrients and to either an increase or a decrease in the survival and differentiation of soil-mediated pathogenic microorganisms, such as *Salmonella typhimurium*. Therefore, the decrease in temperature will usually result in a decrease in the growth and activity levels of microorganisms in winter. High temperatures are usually harmful to most microorganisms. In fact, some species of chemical engineers are able to withstand such adverse conditions by entering into passive forms that can resist high temperatures than active individuals. It should be noted that there is much uncertainty as to how different microbial groups (and organisms) react to changes in temperature. The observed seasonal changes in soil microbial activity are often associated with changes in the microbial community structure.

Climate change across the whole globe is a fact and this has a substantial impact on every type of soil community and also on the other functions of soil. Climatic parameters such as temperature and humidity have experienced modifications due to the impacts of climate change. The communities of soil are more widespread than the other environment on ecosystem, when all life forms are taken into account. Soil biodiversity consists of all varieties of microorganisms, such as bacteria, viruses, fungi, algae and the other micro fauna like nematodes and protozoa. The present-day distribution of communities is changing due to the cumulative impacts and community invasions and human alterations to the environment.

Temperature and humidity are also essential among the determinants of population structure and the functioning of the biological regulators. The key effects on nematodes and micro arthropods have been identified, and are extremely important to predict the impact of average temperature rises due to climate change or other, more local impacts, such as fires. Nematode sensitivity to temperature and soil humidity depends on their metabolic state. Each class of species has a different survival strategy under intense environmental conditions and may either develop cysts or enter dormant stages, enabling them to survive the most extreme changes in soil temperature and humidity. The overall effect of climate on soil microorganisms can be interpreted through the microbial community seasonal dynamics. These dynamics are the result of microbial communities' growth, activity, and composition, which are sensitive to the two main climate-regulated factors: temperature and humidity. The rates of growth and development are individual features of microbial communities and may differ independently. This means that climatic conditions favoring high levels of microbial activity do not always facilitate high microbial growth and the associated increase in biomass.

Soil organisms contribute to a wide variety of services required for the sustainable functioning of all ecosystems, act as the primary driving agents of nutrient cycles, regulate soil organic matter, soil carbon sequestration and greenhouse gas emissions [1]. Modifying the soil physical properties and the water system, increasing the amount and efficiency of nutrient acquisition by vegetation and increasing plant health. Soil organisms participate in the soil food web, where each tropical level serves as food for the next tropical level [2]. Generally, a soil is dependent on the depletion of food web sources and dead organic matter (detritus). The sustainability of the performance of the environment is dependent on the stability of the soil food web. Soil microorganisms perform a wide range of functions: they decompose organic matter, release nutrients into plant-derived forms and weaken toxic residues; They form symbiotic associations with plant roots, act as antagonists for pathogens, affect the weather and mineral solubility, and contribute to soil structure and integrity.

The time scale of microbial metabolism is generally between 0.2 and 6 years for the microbial biomass in the soil, with a greater rate of turnover in the soil compared to the majority of the organic matter more than 40 years. The overall effect of climate on soil microorganisms can be accounted for by the seasonal dynamics of microbial populations. The growth, activity and composition of microbial communities are sensitive to two main factors that are regulated by the environment: temperature and humidity. Growth and activity rates may vary independently of individual characteristics of microbial communities. Climate conditions that favor high levels of microbial activity do not always facilitate high microbial growth and associated growth biomass.

In general, an increase in atmospheric temperature is equivalent to an increase in microbial activity. Therefore, due to the decreased temperature, microbial growth and development usually decline in winter time. Nonetheless, these predicted seasonal dynamics can change in specific soil environments, and in tundra soils, at low temperature, microbial biomass is at its highest in late winter time [3]. Therefore, even if there is a generally positive association between temperature and microbial growth and development, temperature responses may also depend on the species of chemical engineers present in the microbial community and the temperature range considered.

Interactions between plant and other soil communities may be unpredictable and their consequences of natural alterations in climate because of the temperature response of cycling of carbon procedure the minor alterations in temperature may result in an abundant release of soil carbon back into the atmosphere. The comparative significance of direct versus indirect effects of climate change on soil carbon still unresolved specifically in ecosystems which are opposite from one state to another. It is admitted from the above study that the indirect impacts of climate change on microorganisms mediated through plants are stronger than the direct impacts of climate on forming microbial and other communities' composition and activity.

9.5 Resilience Against Soil Physicochemical Degradation

Almost all soil types do not have intrinsic resilience against soil physicochemical degradation. In a proper soil environment such moisture, nutrients and cation exchange capacity increase the soil cover near the ground level led to soil-vegetation systems resilient against the physical degradation. In such soils, faunal activities are very high because of the adequate organic matter addition and greater macrospores that drain off the extra moisture through these macrospores without leaching most of nutrients from the soul mass. The climate change factors do not change the soil pH rapidly in almost all soils due to moderate or high cation exchange capacity. In most of the soils, the ongoing decomposition rate of the organic matter maintains CO_2 concentration in the soil air with respect to atmospheric CO_2. In these soils, $CaCO_3$ solubility and its activity in the soil water is determined by the pressure of CO_2 in the soil air. The leaching of $CaCO_3$ from soil macrospores is positively related to the rate of organic matter decomposition and the water movement through soil is negatively related to the gas diffusion rate. Moreover, some clay fractions have variable chargers on their surfaces that will decrease the acidification.

9.6 Impact of Climate Change on Soil Biodiversity

Most of the soil ecosystems in the developing countries have undergone degradation and the level of biological diversity has fallen as a result. So far, many projects have been launched to conserve and manage the biological diversity focusing on the loss of above ground species; the reduction in the soil biodiversity has gone unnoticed [4]. Therefore, it is very important to understand and dissect the mechanisms of the soil ecosystem and biodiversity functioning to maintain the above ground diversity [5]. Soil management practices that reorganize and promote soil biodiversity and its functions, and which also strengthen the link with above-ground communities leading to an improvement in the crop production factors, such as soil structure, soil fertility, nutrients and water retention [6,7]. The link between the above and below ground communities have some important consequences for the direction and rate of above ground vegetation change [7] that is key in restoration ecology.

Changes in the structure of the microbial community can change different soil functions. Ecosystem functions such as nitrogen fixation, nitrification, denitrification and methanogenesis are controlled by different microbial groups. Changes in the relative abundance of organisms which control specific processes have a direct effect on the rate of the process. However, certain processes, like nitrogen minerals, at coarse levels are more linked to abiotic factors, such as temperature and moisture, than the microbial community structure because of the diversity of organisms driving them.

Global climate change can have a significant impact on all the biodiversity and related services in the soil. These impacts may be directly or indirectly linked to climate parameter alterations (e.g., temperature, humidity). By considering all living types, soil biodiversity is more extensive than any other ecosystem on the globe. The soil biota comprises examples of all microorganism classes such as fungi, bacteria, algae, and viruses, as well as microfauna such as protozoa and nematodes. Today, under the combined effects of climate change, biological invasions and direct human environmental modifications, disturbance regimes are changing drastically. However, assessing and predicting how soil communities will respond to these disturbances remains extremely difficult. Environmental variability is an integral part of ecosystem dynamics, and some perturbations are inevitable.

Climate change can exacerbate these seasonal disturbances, extending the limits toward those of extreme events. Inside the global system, soil biodiversity plays a very important role, and ongoing research continues to highlight this. Under the combined effects of climate change and biological invasions, perturbation regimes are changing drastically today. Climate-based abiotic controls are important determinants of microbial activities. The overall effect of climate on soil microorganisms can be perceived through the microbial populations' seasonal dynamics. These dynamics result from the fact that growth, activity and composition of microbial communities are susceptible to the two main factors regulated by climate: temperature and moisture. Rates of growth and activity are individual features of microbial communities and may vary independently. Increasing soil temperatures can strongly influence the soil fauna by altering the occurrence and composition of soil bacteria and fungi and influencing plant physiology and the function of the population. Soil mesofauna is especially susceptible to environmental changes and therefore considers an outstanding bioindicator.

In general, an increase in atmospheric temperature is equivalent to an increase in microbial activity. Changing soil temperature will likely alter microbial-mediated nitrification and denitrification dynamics in the soil environment due to a shift in the population of nitrifiers and denitrifiers Sometimes perturbations in the soil environment could lead to community shifts and altered metabolic activity in microorganisms involved in soil nutrient cycling, and to increasing or decreasing survival and virulence of soil-mediated pathogenic microorganisms like *Salmonella typhimurium*. Therefore, due to the fall in temperature, microbial growth and development usually decrease in the winter. In general, exceedingly high temperatures are deleterious to many microorganisms. Indeed, some species of chemical engineers may survive such adverse conditions by entering survival inactive forms, which may resist high temperatures better than active individuals. It is worth emphasizing that there is indeed a great deal of uncertainty about how different microbial groups (and fauna) are reactive to temperature.

The seasonal changes observed in soil microbial activity are also often associated with modifications in microbe's community composition. An increase in atmospheric CO_2 may lead to changes in soil ecology, primarily by changing the distribution of above and below ground nutrients. For example, an increase in atmospheric CO_2 could lead to increased plant growth because carbon dioxide (CO_2) is the molecular photosynthetic component [8]. The litter production rate can thus be increased and the litter composition may be altered, which can in turn change its digestibility. Such changes can influence the composition of organic matter in soil. As a result, the total supply of carbon and nitrogen can be influenced by changed litter production between plants and micro-organisms. In addition, higher carbon dioxide emissions can cause an increased root growth that has significant consequences for soil structure and soil biota [9].

9.7 The Influence of Climate Change on Soil Functions

The climate-changing activities affect soil processes both indirectly, such as changes in soil moisture contents via change transpiration, and directly, such as the influence of temperature on the organic matter contents but also increase greenhouse gas emissions and thus contributing to an increase in the levels of climate change gases. An increase in atmospheric temperature tends to accelerate the loss of soil carbon, which ultimately disturbs the function of soil in areas such as top soil water holding capacity, stability, soil structure, erosion and nutrients availability. The integrated influence of climate change increased the crop yields of crops such as sunflowers, sugar beet and winter wheat as increased radiation use efficiency, a longer growing season and increased CO_2 fertilization. These things usually help the C3 species pathway but not the C4 species.

Moreover, elevated CO_2 enhanced the dry weight and size of most C3 plants and plant characteristics. In addition, the relatively higher acumination of assimilated materials are partitioned into vegetative components (stem and petioles) during plant development in order to support the light harvesting leaves. The harvest index is usually decreased with an increase of temperature and CO_2 concentration. Therefore, an increase in yield was seen in sunflowers, whereas a smaller increase or a decrease in yield was noted in the oil seeds and potatoes and there was even a deterioration in the quality of some horticultural crops. Moreover, an increase in the yield of grass specie plants can also be expected. Increases in both CO_2 levels and temperature in the atmosphere will increase plant growth in the short-term growth period species.

The nitrogen cycle, one of the important nutrient cycles, is heavily dependent on soil biota. Most of the organisms on earth, including plants, cannot use nitrogen in its gaseous form. It is fixed by free living microbes such as cyanobacteria and different genera of bacteria and actinomycetes, or by symbiotic microbes such as rhizobium which form root nodules in legumes. This cycle of nitrogen fixation transforms gaseous nitrogen into ammonia, which can be used by plants. Alternatively, a large fraction of this ammonia is converted into other plant types that are available first into nitrites (NO_2^-) and then into nitrates (NO_3^-). A process known as denitrification consists in the conversion of nitrogen products such as nitrates and nitrites back to nitrogen gas. This process takes place in anaerobic conditions where the bacteria use nitrogen for anaerobic respiration in the absence of oxygen. The nitrogen cycle has very important implications for agriculture and the environment because it impacts soil fertility.

9.8 Impact of Climate Change on Soil's Physical Properties

9.8.1 Soil Water

An increase in soil air temperature tends to increase the evapotranspiration loss of water from the soil biopores. Most of the time, precipitation can fluctuate soil moisture rapidly and also the response timescale within a few hours. The four plausible climate-vegetation-hydrology-land complex phenomenon are based on the soil moisture regime and the soil moisture balance. Generally, the rate of transpiration is directly correlated to the drought sensitivity of the plant with this physiological, ecological and the environmental consequences. Somewhat, the direct influence of climate change is mitigated with the vegetation characteristics such as density, type, dynamics, biomass production and species composition and root characteristics. Moreover, the soil forming process such as organic matter turnover, structure formation, runoff, infiltration, drainage and percolation, weathering, clay translocation, podsolization and gleying are significantly affected by the soil moisture contents.

9.8.2 Soil Temperature

Changes in the soil air temperature are very important, but they are rarely discussed and reported in the literature. It has been observed that there is a close relationship between air and soil temperature; generally, an increase rise in air temperature will inevitably tend to rise in the soil temperature. Many factors, such as the gain or loss of soil radiation, head conduction among the soil profiles through water and gas media, and the process of evaporation are also responsible for changes in soil temperature. However, the rise in soil temperature above normal levels will accelerate the soil processes, such the rate of decomposition of organic matter in the soil. At even higher temperatures, the soil may be scorched. There may also be quicker nutrient release, higher microbial activity, an accentuation in the chemical weathering of minerals and a higher rate of nitrification. In addition, soil air temperature will also decide about the adaptation and management of the vegetation cover above ground.

9.8.3 Soil Structure and Texture Differentiation

Soil structure is an important soil physical property that helps to explain the level of soil stability. Soil structure is responsible for the smoothing function of many processes, such as the exchange of gases and pollution or contaminants, water movement, soil fauna, nutrients, seepage, water quality, building foundations and, finally, the emergence of crops. Moreover, the soil structure is mainly influenced by the quality and amount of organic matter contents present in the soil profiles. A reduction in soil's organic matter resulted in poor soil aggregate stability, increases in soil compaction, higher infiltration rates, and increased runoff and erosion.

Both diffuse and local sources can cause soil contamination. The first group is contaminated by the atmosphere, by running water or by the soil. This may lead to acidification, eutrophication, and other serious damage. Different sources may also be used specifically for the application of chemicals (fertilizers, nutrients and waste sloes), and sometimes even heavy metals. Local sources of contamination may have various origins and are usually linked to industrial activity. Global factors, including global climate change, intensive land use and industrialization, as well as population growth, led to the degradation and depletion of fertile soils. Underdeveloped countries, in particular, are suffering severely from the process of land degradation. Such countries are also causing accelerated soil erosions through the exploitation of natural resources in environmentally sensitive areas. Climate change often raises the erosion rate of already degraded soils, leading to the worsening soil conditions.

9.8.4 Soil Biological Parameters

Soil biological parameters are mainly dependent on the soil organic matter (SOM) contents as it stabilized the soil structure, stability, nutrient storage, organic matter turnover, and also its capacity to hold oxygen. Organic matter provides the habitat and food to all soil fauna and microflora that improve the health and productivity of soil. In addition, SOM contents are highly susceptible to any change in soil moisture and soil air temperature. Moreover, soil physical parameters are also disturbed due to a decline in the SOC contents. SOC is capable as acting as both a sink and a source of carbon in the soil. The amount of organic carbon stored in the soils that is over three times higher that is found in the atmosphere [10]. A higher CO_2 level will increase crop productivity due to the soil moisture and soil nitrogen should not be limited factors [11,12].

Soil fauna and flora are vital components of all soil types, driving the breakdown, retention, incorporation of plant remains, soil porosity, soil structure and nutrients cycling. Overall, global warming may have no direct impact on the composition of the soil fauna and flora because of their broad temperature optimum. However, the migration of vegetative cover may affect soil fauna and flora through increased temperature. Generally, climate change enhanced the CO_2 concentration in the atmosphere which results in an increase in plant growth and nitrogen fixation, denitrification and nitrogen immobilization, an increase in soil aggregate and mycorrhizal associations and, finally, the weathering of soil minerals. The balance between CO_2 emission and plant growth will be very important under the changing climate. A rise in temperature leads to an increase in soil air temperature, which has both direct and indirect impacts on the soil microbial activities that emit greenhouse gases to the atmosphere resulted in global warming.

9.8.5 Soil Chemical Parameters

The most rapid response of soil against climate change is the loss of nutrients and salts where there are increases in leaching and salinization. Normally, total soil organic matter (SOM), soil carbon (C) and soil nitrogen (N) have inverse correlations with temperature, since biological activity and the decomposition of organic matter normally decreases as temperatures rise [13].

9.8.6 Acidification, Sodicity and Salinization Problem in Soil

Due to a rise in temperature in some parts of the globe there has been more evapotranspiration and less rainfall, which has led to sodicity and salinization problems in the soils.. Significant increases in rainfall will lead to increases in leaching, loss of nutrients and increasing acidification, depending on the buffering pools existing in soils. The direction of change towards increased leaching or increased evaporation will depend on the extent to which rainfall and temperature change and consequent changes to land use and its management. In either case, the situation could lead to important changes in soils.

Increased salinization and alkalization would occur in areas where evaporation increased or rainfall decreased. Transient salinity increases as capillary rise dominates, bringing salts into the root zone on sodic soils. Leaching during episodic rainfall events may be limited due to surface sealing. Increased subsoil drying increases the concentration of salts in the soil solution. Conversely, the severity of saline scalds due to secondary salinization may abate as groundwater levels fall in line with reduced rainfall; this development could have significant impacts on large areas of semi-arid zones. In areas where salinity is a result of recharge processes, salinization would increase if the upstream recharging rainfall increased. Increasing atmospheric CO_2 concentration can reduce the impact of salinity on plant growth.

Bush and his co-workers anticipated impacts of climate change in the coastal lowland acid sulfate soils. The anticipated impacts of climate change are warmer conditions, an increasing proportion of rainfall to occur from heavy falls, increasing occurrence of drought in many regions, increasing frequency of intense tropical cyclones, rising sea levels and the frequency of extreme high seas (e.g., storm surges). All of these predicted impacts have direct relevance to coastal acid sulfate soils landscapes, through either exacerbating sulfide oxidation by drought, re-instating reductive geochemical processes or changing the export and mobilization of contaminants. The interaction of specific land management factors such as man-made drainage will also play a significant role in how the predicted impacts of climate change affect these landscapes. Understanding the potential impacts of climate change for coastal lowland acid sulfate soils is particularly important, given the utility of these areas for agriculture and urban communities, their unique capacity to cause extreme environmental degradation and their sensitivity to climatic factors such as temperature and hydrology and susceptibility to sea-level inundation.

9.9 Mitigation/Adaptation Measures

The effects of climate change could be counteracted by increasing the nutrients release, which will result in changes to plant productivity. Soil management practices have an important and sometimes immediate impact on soil biodiversity and as a result affect ecosystem services. A major approach explaining the change in soil micro biota with the intensity of management practices is associated with the input of organic matter. Organic matter runs the soil food web and depends on the feeding cycle of bacteria (low C:N ratio) or fungi (high C: N ratio). Greater crop residues input to grasslands promotes fungal-dominant microbial communities and the greater diversity of nematodes and micro arthropods. Increased microbial activity also enhances biological regulators, reducing nematodes and soil pathogens. In contrast, in densely maintained (fertilized) grasslands or crops, microbial communities are apathetic and opportunistic bacterial dominant communities undergo changes. In turn, it is favorable for nourishing opportunistic bacteria. The method of soil farming is disruptive to fungal hype and large earthworm species, which travel to the soil surface to obtain plant materials for food, such as anemia earthworms. The abundance of biomass and anesthetic earthworm in conventional treated soils decreases by a factor of 1.3–3 compared to the biologically managed type. Conventional maintenance will result in ground gas and ground drainage.

9.9.1 The Introduction of Perennial Crops

As the population grows, the demand for food, land and energy also increases, leading to the threat of food security. Food security, along with climate change, becomes a hindrance for the maintenance of sustainable agricultural practices to overcome the challenges of growing populations. Therefore, many researchers agree that agriculture could be a great threat to the biodiversity and ecosystem functions of any human activities [14]. Currently, more than two-thirds of the world's cultivated areas are under monoculture cropping systems of annual crops. But the increasing production of non-food items have more negative effects on the potential annual crop yields [15]. This expansion of croplands for higher energy demands has increased the excessive use of chemical fertilizers, pesticides, and wastewater irrigation, which in turn is degrading the soil quality and soil functions.

Compared to annual crops, perennial crop species have the tendency to overcome many agricultural problems like soil erosion, nutrient losses, soil carbon sequestration and soil microbial activities. Furthermore, in terms of potential yields, the cultivation of perennial crops in mixed cropping systems improves the performance of different cropping systems [16]. As the cultivation of perennial crops plays a vital role in achieving sustainable agriculture, it is necessary in the present era to shift cropping systems from annual crops to perennial crops. This may also be helpful in mitigating the effects of, or adapting to, climate changes.

Perennial crops are supposed to have longer growing seasons and deeper rooting depths. They have an ability to draw water from deeper soil layers due to their deeper roots. They intercept more radiation, retain more precipitation, have high nutrient use efficiency and produce larger amounts of biomass; in turn, this reduces soil erosion and nutrient leaching problems. Moreover, perennial crops require less cultural operations for example, land preparation, heavy farm machinery, application of fertilizers and pesticides/insecticides. Regrowth of plants parts i.e. stems and leaves of perennial crops, after harvesting, can be used as feed for the livestock or as biofuels, or can be incorporated within the soil to minimize erosion and nutrient losses of soil. To increase the land coverage by perennial crops, plant more trees and grass/pastures along river banks. This will improve the quality of eroded lands as well as the accumulation of contaminants released in the rivers, streams and seas from agricultural wastes. Perennial cropping practices greatly improve cropping systems in terms of agricultural and environmental sustainability, and in the future significant agricultural and environmental benefits of perennial crops will be realized [17].

Perennial crops are a source of feed for livestock, biofuels and a habitat to support regional biodiversity [18]. In addition, because perennials do not need to be reseeded annually and have extensive root systems, they can provide a source of system resilience in changing climatic conditions. Among the perennial crops, kernza (*Thinopyrum intermedium*) is the novel perennial grain and fodder crop that has been used as a cool-season forage throughout the USA and Canada and also across its native range in Eurasia [19]. Kernza is an intermediate wheatgrass species with the potential to provide multi-ecosystem services and have become more recent commercially available to the farmers mostly of the developed countries.

9.9.2 Mulching/Light Soil Sealing

Mulching can protect the soil surface from erosion and increase its fertility. Mulch is planted at the beginning of the growing season, and can be re-planted as needed. Mulches play an important role in achieving the sustainability goal in modern agricultural production systems. These can help in water and soil conservation, and in achieving the non-chemical control of insect pests, diseases, and weeds, manage soil salinity and add nutrients to the soil. Many types of materials, including crop residues (such as crop residues, straw, bark, etc.), can be used as mulch, but also compost, sewage sludge, compost, rubber or plastic films. Mulching, which consists of covering the soil surface with organic and sometimes inorganic material, is an age-old practice and was traditionally used to control soil moisture, soil temperature, nutrient losses, soil salinity and erosion. With the advancement of modern agriculture, this practice dwindled significantly; however, it is now gaining importance once again in the context of sustainable agriculture [20].

Among the mulches, black plastic mulch has widespread use in orchards or greenhouse conditions, and also has a role to play in arable farming. For almost 70 years, both black and transparent plastics have been used as mulches, particularly in vegetable production. The use of degradable or biodegradable mulches has been suggested in the wake of environmental demerits of black or other colored plastic mulches. The degradable mulches are made of starch or cellulose obtained from crops such as potato, wheat, sugarcane, maize or other crops. During or towards the end of the crop-growing season, these mulches are decomposable to water, carbon dioxide, residual polymers or other minerals within the soil layers. Organic mulch decomposes over time and as it breaks down it slowly adds nutrients to the soil. Thus, contributes to the long-term nutrients' availability in the soil. Moreover, as the organic mulching decays overtime in the soil, it improves the soil organic matter content which, in turn, enhances the water-holding capacity of the soil [21]. Compost made from farm waste or wastes from other sources (such as municipal solid waste) has also been used as mulch. Compost mulch not only adds nutrients to the soil, but also covers the soil to reduce evaporation, improves soil moisture status, and enhances productivity and yields.

9.9.3 Slow-Release Fertilizers

The excessive use of certain inorganic fertilizers makes it easier for microorganisms to use nitrogen and other nutrients that promote microbial activity. This increases the rate of decomposition of low-quality organic inputs and organic soil matter, resulting in a constant deterioration of soil organic matter resulting in soil structure loss and a decline in soil-holding capacity. The problem of greenhouse gas concentrations raises the need for a high level of management strategies towards the effective use/application of chemical fertilizers in various crop systems. When and where necessary, the use of fertilizer can increase soil fertility by increasing soil organic matter or by delaying its degradation process. Insufficient and/or inappropriate use of synthetic fertilizers may reduce crop yields, resulting in lower organic carbon in the soil, lower organic matter, and significantly lower soil productivity. The slow release of nitrogen fertilizers plays an important role in decreasing residual soil nitrates (NO_3^-), which reduce the risk of increased emissions of nitrous oxides.

Most of the fertilizers, especially nitrogen used in crops, are unavailable to crop plants or may be lost through leaching and are processed and released as nitrates and/or nitrites. Enhancing the use of fertilizer crop efficiencies leads to both environmental and economic benefits. Different management strategies, including slow release fertilizers, coated fertilizers, and nitrogen inhibitors, significantly reduce the loss of nitrogen through leaching and volatilization. Also, the application methods of fertilizers should be improved by alternative use of fertilizer application methods, e.g. band placement, deep placement and precision application of fertilizers, can minimize the overuse and/or application losses of fertilizers.

9.9.4 Choice of Crop Species

Crop selection is an important factor for cultivation because it defines the type of habitat available to soil plant biota. For example, legumes act as natural fertilizers by improving soil nitrogen concentrations, thanks to the symbiotic relationship they established with Rhizobia. Various cultivars of the same genus may be grown as a single crop in mixtures (varietal mixtures). For example, the cultivation of a mixture of crop species with the same growing duration which can concurrently be grown and harvested, but which react differently to different water regimes, is a technique that will tackle the varying rainfall patterns and improve the yield stability of the crop. The shifting of sowing dates of various crop cultivars is an important management technique in improving grain yields. Sowing dates can lead to crop production being compatible with precipitation patterns and improves water use efficiencies, thus providing a break from frost risk.

9.10 Agroforestry Systems

Reduction in CO_2 emissions due to the problem of global warming is a major issue on global level. Agroforestry programs have gained a particular attention in terms of climate change mitigation and food security. The substitution of renewable energy sources, the most important of which is currently fuel-wood, for fossil fuels is a key strategy in this regard. The prospect of using additional trees as carbon sinks is closely connected to this strategy. Agroforestry plays a significant role in mitigation of the atmospheric accumulation of greenhouse gases. In sub-humid tropical regions agroforestry play an important role in carbon sequestration as it increases the resilience against mid-season drought stresses resulting from changing climatic conditions. The soil's organic carbon can be improved by planting trees and grass species in degraded soils. Agroforestry systems contribute to the mitigation of climate change by storing carbon over the ground in the form of plant biomass and under the ground in the form of soil organic carbon, thereby contributing to the mitigation of greenhouse gases emissions. Agroforestry has various economic, environmental and socioeconomic advantages. For instance, in agroforestry systems, trees enhance their soil fertility by managing erosion, maintaining the soil organic matter and physical properties, increasing the N accretion, nutrient extraction from the deep soil horizons and encouraging closer cycling.

9.11 Zero/Reduced Tillage Agriculture

The agricultural practice associated with minimal soil disturbance and permanent soil cover combined with appropriate crop rotations is known as conservation agriculture. Conservation agriculture provides both adaptation and mitigation benefits and also sustains the agricultural productivities under the adverse effects of climate change and global warming as it lowers the greenhouse gas emissions. Zero tillage with crop residue mulching of the soil reduces the evaporation losses of water by increasing water retention in the soil, thereby requiring less water in conservation agriculture compared to the conventional agriculture. Soil tillage is one of the most critical, costly and fuel-consuming processes. The use of new biotechnology has helped to reduce air pollution in the cropping systems. This technology (which uses bio-materials according to a particular targeted methodology) replaces soil properties and decreases fuel consumption in farm operations, thereby reducing environmental pollution. Carbon dioxide emissions are mainly associated with the decomposition of crop residues which is enhanced by soil disturbance through different cultural practices such as tillage practices, excessive fertilizer applications, use of heavy farm machinery. Accordingly, conservation agriculture can also affect the influxes of carbon dioxide, nitrous oxide and methane into the soil.

9.12 Cropping Systems

Rotation of crops could prevent the formation and activation of pathogens and pests, as soil incorporation of crop residues modifies the plant–microbial associations through improved nutrient use efficiency. Crop rotation also helps fight against the effects of soil erosion. Crop rotations tend to promote soil fertility by switching between deep-rooted and low-rooted crops. Through improving the tilth of soil and microbial activities, it will contribute to a reduction in soil erosion by the establishment of a more stable soil structure.

Intercropping, which can be an effective strategy in the current situation of climate change, can be accompanied by favorable outcomes. There are potential benefits of grain legume intercropping such as stable yields, better resource use, the reduction of insect pests, weeds and diseases, an increase in cereal protein contents and reduced nutrient leaching in relation to mono cropping. Cover crops are considered as a viable technique to mitigate the emissions of greenhouse gases. Cover crops include crimson clover, oats, annual ryegrass, oil-seed radishes, and other legumes which can grow in any cropping system. Cover crops are grown to protect the land, which would otherwise be subjected to erosion and nutrient losses. These crops have the ability to immobilize the nitrogen so that it will remain in the soil for the next main crop after harvesting. Cover crops are an important sequestration strategy for soil carbon. Bacteria, fungi, earthworm and other soil organisms depend on these cover crops for their feed and in turn raise the soil carbon concentrations with time.

9.13 Conclusion

The complexity of microbial communities living below ground and the different pathways around them makes it difficult to detect the different response reactions caused by global warming to soil microorganisms. Whether the microbial processes contribute to climate change or the negative feedback reduces its effects is clearly an ecosystem-level response to future climate and climate change, but microbes have a huge impact. Soil respiration plays an important role in soil emissions due to the large amount of CO_2 and CH_4 emissions produced during respiration, the dependence of carbon reserves in soils on the rate of respiration, and the initial sensitivity of soil respiration to atmospheric temperatures. Further studies on the effects of soil respiration on climate change contribute to our understanding of the overall effects of climate change, including the ability of terrestrial forests to overcome excess CO_2 from the atmosphere. As we seek to reduce greenhouse gas emissions and adapt to the effects of climate change, we turn to micro-life, which is found beneath the surface, perhaps leading us to a macro and global future. It may help to better prepare for the changes. Climate change is likely to have a significant impact on the soil, affecting all the services that soil biodiversity provides. However, all mitigation and attenuation measures taken to mitigate global climate change are expected to have beneficial effects on soil biodiversity conservation, soil performance and related services.

REFERENCES

1. Plaas E, Meyer-Wolfarth F, Banse M, Bengtsson J, Bergmann H, Faber J, et al. (2019) Towards valuation of biodiversity in agricultural soils: a case for earthworms. *Ecological Economics 159*:291–300.
2. Hirschfeld MNC, Cares JE, Esteves AM (2019) Land use, soil properties and climate variables influence the nematode communities in the Caatinga dry forest. *Applied Soil Ecology* 103474.
3. Castro HF, Classen AT, Austin EE, Norby RJ, Schadt CW (2010) Soil microbial community responses to multiple experimental climate change drivers. *Applied and Environmental Microbiology 76*(4):999–1007.
4. Bardgett RD (2005) *The biology of soil: a community and ecosystem approach*. Oxford University Press, UK.
5. Wardle DA, Bardgett RD, Klironomos JN, Setälä H, van der Putten WH, Wall DH (2004) Ecological linkages between aboveground and belowground biota. *Science 304*(5677):1629–1633.
6. van Eekeren N, van Liere D, de Vries F, Rutgers M, de Goede R, Brussaard L (2009) A mixture of grass and clover combines the positive effects of both plant species on selected soil biota. *Appl Soil Ecol 42*(3):254–263.
7. van der Putten WH, Bardgett RD, de Ruiter PC, Hol WHG, Meyer KM, Bezemer TM, et al. (2009) Empirical and theoretical challenges in aboveground–belowground ecology. *Oecologia 161*(1):1–14.
8. Melillo J, Steudler P, Aber J, Newkirk K, Lux H, Bowles F, et al. (2002) Soil warming and carbon-cycle feedbacks to the climate system. *298*(5601):2173–2176.
9. Hunter WR, Ogle N, O'Connor NJ (2019) Warming affects predatory faunal impacts upon microbial carbon cycling. *33*(5):924–935.
10. Schlesinger WH (1999) Carbon sequestration in soils. *Science 284*(5423):2095.
11. Gifford R (1992) Implications of the globally increasing atmospheric CO_2 concentration and temperature for the Australian terrestrial carbon budget: integration using a simple-model. *Aust J Bot 40*(5):527–543.
12. Eamus D (1991) The interaction of rising CO_2 and temperatures with water use efficiency. *Plant Cell Environ 14*(8):843–852.

13. Giagnoni L, Maienza A, Baronti S, Vaccari FP, Genesio L, Taiti C, et al. (2019) Long-term soil biological fertility, volatile organic compounds and chemical properties in a vineyard soil after biochar amendment. *344*:127–136.
14. Maitra S, Palai JB, Manasa P, Kumar DP (2019) Environment, biotechnology. potential of intercropping system in sustaining crop productivity. *12*(1):39–45.
15. Marquardt K, Vico G, Glynn C, Weih M, Eksvärd K, Dalin P, et al. (2016) Farmer perspectives on introducing perennial cereal in Swedish farming systems: a sustainability analysis of plant traits, farm management, and ecological implications. *40*(5):432–450.
16. Weißhuhn P, Reckling M, Stachow U, Wiggering HJS (2017) Supporting agricultural ecosystem services through the integration of perennial polycultures into crop rotations. *9*(12):2267.
17. Cox TS, Glover JD, Van Tassel DL, Cox CM, DeHaan LR (2006) *Prospects for developing perennial grain crops*. American Institute of Biological Sciences.
18. Pugliese JY (2017) *Above-and belowground response to managing Kernza (Thinopyrum intermedium) as a dual-use crop for forage and grain*. The Ohio State University.
19. Favre JR, Castiblanco TM, Combs DK, Wattiaux MA, Picasso VD (2019) Technology. Forage nutritive value and predicted fiber digestibility of Kernza intermediate wheatgrass in monoculture and in mixture with red clover during the first production year. *JAFS 258*:114298.
20. Kader MA, Singha A, Begum MA, Jewel A, Khan FH, Khan, Nazrul IKNI (2019) Mulching as water-saving technique in dryland agriculture. *Bulletin of the National Research Centre 43*(1):1–6.
21. Qin W, Hu C, Oenema OJSr (2015) Soil mulching significantly enhances yields and water and nitrogen use efficiencies of maize and wheat: a meta-analysis. *Scientific Report 5*:16210.

10

Soil Fertility Decline Under Climate Change

Abdel Rahman Mohammad Al-Tawaha
Al Hussein Bin Talal University, Jordan

Hikmet Günal
Gaziosmanpaşa University, Turkey

Josef Křeček
Czech Technical University, Czech

Rares Halbac Cotoara Zamfir
Politehnica University Timisoara, Romania

Patel H K, and Vyas R V
Anand Agricultural University, India

Cristina Halbac Cotoara Zamfir
Politehnica University Timisoara, Romania

Ismail Celik
Cukurova University, Turkey

Amanullah and Shah Khalid
The University of Agriculture Peshawar, Pakistan

Alla Aleksanyan
Institute of Botany aft. A.L. Takhtajyan NAS RA, Armenia

Abdel Razzaq Al-Tawaha
Universiti Putra Malaysia, Malaysia

David L. McNeil
The UWA Institute of Agriculture, Western Australia

Imran
The University of Agriculture Peshawar, Pakistan

Abdur Rauf
University of Swabi, Pakistan

Jeyabalan Sangeetha
Central University of Kerala, India

Shah Fahad
The University of Haripur, Pakistan

Laila Trioui
University Hassan II of Casablanca, Morocco

Ahmed Abu Zaiton
Al A lbayt University, Jordan

Ezz Al-Dein Al-Ramamneh
Al-Balqa Applied University, Jordan

CONTENTS

10.1 Introduction

Global warming is defined as an increase in combined air temperatures over land and sea surfaces averaged over the globe for more than a 30-year period (Allen et al. 2018). The impacts of global warming on several ecosystems at regional and local scales have begun to be felt, and the ongoing effects of continuing climate change in the future may worsen the negative influence on ecosystem services (IPCC 2014). The average amount of warming from the pre-industrial age (the period between 1850 and 1900) level to the decade of 2006–2015 is determined to be 0.87°C. However, the temperature increase in some of the continental regions is greater than the global average, while the level of warming in most ocean regions is lower. Between 20 and 40% of the human population of the world live in regions where the temperature increase was 1.5°C above the pre-industrial age in at least one season of the decade 2006–2015. The temperature change affects the ecosystems by changing both the total amount of precipitation of a region and the distribution of precipitation throughout the year (Brevik 2013). In addition to greenhouse gas emissions our environment is also deteriorating due to the emissions of other pollutants, which have risen substantially since the 1800s. While now declining in most developed areas, they are at unacceptable levels and rising in many developing areas (Khan et al. 2016). UN Environment estimates, for example, that ground-level ozone pollution reduces crop yields by up to 50 million tons p.a. (UN Environment 2019).

In this chapter, we have used the climate change definition of the Intergovernmental Panel on Climate Change (IPCC 2012). This definition includes both natural variabilities and human-induced climate change. Anthropogenic activities include the direct release of pollutants such as inorganic acids, ozone, ammonia, hydrocarbons, and halogens as well as CO_2. As well as direct pollution effects (e.g. forest acidification), greenhouse gas release is considered to be the main driver for increased weather-related natural hazards, such as floods, droughts and windstorms. Extreme weather events have substantial impacts on both human society and the environment, including loss of life, structural damage, losses in agricultural production, natural capital degradation etc.

Climate changes commonly lead to losses and damages which can be classified as either economic or non-economic. This last category involves "products" (e.g. loss of habitat and biodiversity, damage to ecosystem services) which are not traded and capitalized in markets. They thus increase the importance of climate change impacts and must be addressed (UNFCCC 2013; Costanza et al. 2014; Morrissey and Oliver-Smith 2013; Rahman et al. 2016).

The degradation of ecosystems continues worldwide due to the direct effects of anthropogenic activities (such as overexploitation and pollution), these activities combine within sufficient knowledge and understanding of the services existing ecosystems provide to generate adverse effects (WWAP 2015; MEA 2005). The impacts of climate change on biodiversity, ecosystems and associated services are increasing and will continue to increase worldwide. These effects may compound with the effects of other anthropogenic activities to accelerate environmental degradation. Climate change thus adds an additional profound challenge to the existing socio-economical, environmental and cultural issues (Groffman et al. 2014; Mechler et al. 2019; Miranda et al. 2019) which are degrading our ecosystems.

Ecosystems play a key mediating role (van der Geest et al. 2019) in affecting the human environment; therefore, researchers focus mainly on climate change impacts on human society and less on loss of ecosystem services and the cascading impacts on human societies resulting from this (Zommers et al. 2014). However, climate change impacts on ecosystems will increase their vulnerability to negative outcomes of extreme events, exacerbate the degradation of ecosystems and reduce the efficiency and capacity of ecosystems to buffer the impacts of these events (Staudt et al. 2013; Peters et al. 2011; Staudinger et al. 2012; Bangash et al. 2013; Lorencová et al. 2013). The IPCC (2014) states that there is strong evidence for the severity of climate change impacts on natural systems with limited adaptation options. The limitation is enhanced by additional factors like reduced financial commitments, lack of knowledge, understanding and motivation (Ayeb-Karlsson et al. 2016).

10.1.1 Climate Change Effect on Agricultural Ecosystem

Agriculture, a nature-based and climate-dependent economic sector, provides the world population with not only food but also livelihood security (de Schutter 2008; Alexandratos and Bruinsma 2012; IPCC 2014, Carter et al. 2017). Moreover, in many developing countries, agriculture is a way of life representing the most important economic sector (Stabinsky and Ching 2012). This fundamental human activity is currently at risk due to the unequivocal consequences of climate changes, and in the absence of sustainable mitigation options, agriculture will experience a series of predominantly negative effects from the current and future climate changes (IPCC 2007; Tubiello et al. 2008; Lepetz et al. 2009; Bocci and Smanis 2019). Agricultural ecosystems will need to deal with the double pressure resulting from the negative effects of projected climate changes and also from the increased need for food, fuel and fiber. Inadequate agricultural practices, in combination with extreme climatic events, will lead to an acceleration of natural resources degradation (Tubiello et al. 2008). Among the key anticipated impacts of climate change on agricultural ecosystems are: the loss of biodiversity in fragile environments; unpredictable changes in annual rainfall volumes and patterns; the increased frequency of extreme climatic events (droughts and flooding); the increase/decrease of growing season lengths and reliability; significant reductions in crop and livestock production; increased erosion; increased competition from weeds, especially invasive species and increased risks of pests, and so on (Alexandrov and Hoogenboom 2000; Bruinsma 2003; Melillo et al. 2014; Mekonnen 2018; Easterling et al. 2000; Bocci and Smanis 2019; Alexandratos and Bruinsma 2012; Adnan et al. 2016 IPCC 2014; Schlenker and Lobell 2010; Semenov and Halford 2009)

10.1.2 Climate Change Effect on the Forest Ecosystem

Forests have a significant importance in hosting the largest part of terrestrial biodiversity, playing a pivotal role in mitigating climate change effects, enhancing the abilities of people and ecosystems to mitigate effects of climate variability and providing a wide range of ecosystem services to society from cleaning air and water resources to climate regulation and forest products (MEA 2005; FAO 2018;

Morin et al. 2018). Climate changes have direct (e.g. increased atmospheric CO_2 can alter photosynthesis and nutrient and water demand) and indirect impacts on forest development and productivity as well as on the ecosystems which have strong relationships with the forests. The most important factors which lead to changes in structures of forest ecosystems and functions are CO_2 concentration, temperature and precipitation. These factors may cause increases in drought severity as well as the risk of extreme precipitation events. Ultimately, the results may be any of: the shifting of the geographical ranges of tree species and altered competition species (Lidner et al. 1997; Lexer et al. 2002; Parmesan 2006; Rouault et al. 2006; Lavergne et al. 2010; Morin et al. 2008; Suttle et al. 2007); the disappearance of less resilient tree species (Thuiller et al. 2005; Lindner et al. 2008; Lindner et al. 2010); changes in the distribution of tree species (CCSP 2008; USGCRP 2009; Garzón et al. 2011); changes in tree growth and productivity (Bergh et al. 2003; Loustau et al. 2005; Tilman et al. 2014); changes in the frequencies of pest outbreaks (Volney and Fleming 2000); and deflagration and propagation of forest fires (*Mukhopadhyay* 2009).

10.1.2.1 Forest Soils

Soils are one of the key constituents of life on Earth (Platt, 2004) and are considered part of the natural capital for the sustainability of human beings. Forest soils provide various environmental services to human societies, such as recharging water resources, timber resources, landscape stabilization, wildlife etc. (MEA 2005). The formation of forest soils has been influenced by forest vegetation which varies greatly around the world. Forest vegetations, in general, are characterized by massive root zones, significant litter layers and recycling of organic matter and nutrients (Hillel 2005). Soil fertility is being partly maintained by the litter fall. Coniferous forests are considered to be less-suitable nutrient cycling maintainers compared to broadleaf forests (Andersson 2005).

10.1.2.2 Atmospheric Acid Deposition

Soil acidification is defined as a decrease in the acid neutralization capacity of soil solids. Forest soils generally become acidic under humid climates (Hillel 2005). The acidification of soils can be intensified by the atmospheric deposition of industrial compounds such as sulfuric and nitric acids and ammonia. The serious acidification of sensitive forest soils by industrial emissions in Europe has occurred since at least the 1880s. The relative importance of other processes (e.g. the tree uptake of excess basic cations) depends mainly on the rate of the deposition and forestry practices. Tree uptake may be the dominant cause of forest soil acidification in northern Scandinavia and parts of North America. On the other hand, the toxicity of dissolved aluminum to many plant and animal species is the main deleterious effects of the acidification of the soil environment. The threshold pH value, associated with a rapid mobilization of the toxic aluminum in the soil water, is considered to be 5.3 (Bache 1986). In Central Europe, long-term environmental monitoring has provided strong evidence of the base cation depletion accompanied with high Al levels and unfavorable Mg/Al and Ca/Al ratios of the soil solutions. The acid atmospheric deposition culminated there in the 1980s causing the widespread acidification of freshwaters with serious damage to drinking water quality and populations of fish and other aquatic organisms (Křeček and Hořická 2006).

Since the early 1990s, with the increase in international assistance to decline atmospheric emissions there have been signs of a recovery in acidified European headwater regions as reported by Křeček et al. (2019). But the emission of acidifying compounds in Europe needs further substantial reduction. For example, mean annual pH at the headwater of the Jizerka catchment (Northern Czech Republic) increased from 4.0 (1982–1985) to 5.3 (1990–1994), but the repetitive episodic acidification after snowmelt (pH values below the threshold 5.3) still affects recovery of the biota. Moreover, the environmental recovery is decelerated by desorption of previously stored sulphate in the soil.

10.1.2.3 Effects of Forestry Practices

Forestry observations and forest defense strategies are among the key aspects for the management of their beneficial environmental functions. From the silviculture point of view, nutrient cycling is essential

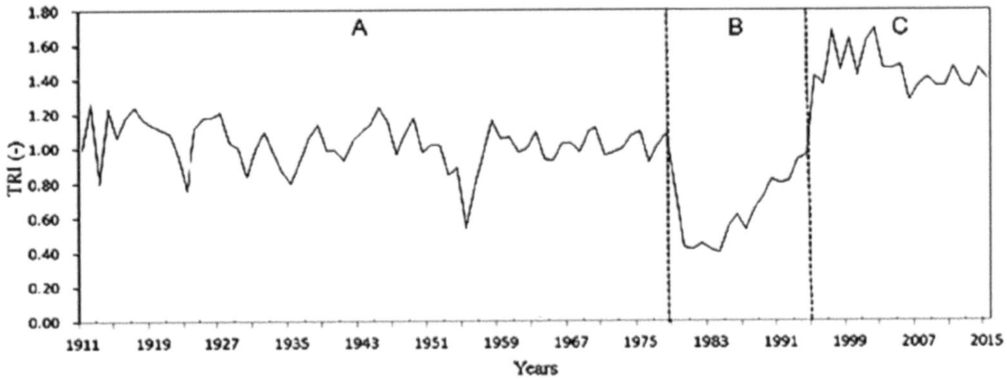

FIGURE 10.1 Mean average standard chronology in spruce stands of the Jizera Mountains (Czechia), TRI – tree-ring index; A – normal; B – reduced by acidification; C – intensive growth periods (Vrtiška et al. 2018).

to optimize tree species composition using appropriate techniques for forest regeneration and thinning. Kubin et al. (2017) documented the long-term impact of soil preparation techniques on the leaching of nitrogen from forest soils in the Boreal climate.

In regions of significant atmospheric pollutant emissions, deciduous or mixed forests in comparison with mature spruce forests could decrease the acid load to forest soils due to their lesser leaf area and surface roughness, mainly in the dormant season (Křeček et al. 2019). Similarly, the alternative herbaceous canopy at harvested sites decreases the acid atmospheric load and acidification of soil and water. These findings correspond with the Ellenberg indicators calculated from vascular plants of the herbaceous understory (Křeček et al. 2010). On the other hand, intensive forestry (with its extended network of skid trails and forest roads) can drastically increase soil erosion and sedimentation (Akbarimehr and Naghdi 2012).

10.1.2.4 Climate Change Impact on Forest Ecosystem

Climate and forests are linked with each other; therefore, global climate changes can exert stress on forests through elevated temperatures, altered precipitations, and more frequent extreme weather events (IPCC 2007). Climate change models indicated that the changes are expected to continue during the 21st century. In particular, the rainfall changes (including storm intensity), along with expected changes in temperature, evapotranspiration demand, solar radiation, and atmospheric CO_2 concentrations are expected to combine to cause an increase in soil erosion rates. Vrtiška et al. (2018) reported impacts of acidification and climate change on standard chronology of Norway spruce (*Piceaabies*) in the upper plain of the Jizera Mountains, Czechia (Figure 10.1).

10.2 The Impact of Climate Change on Soil Functions

Soils ensure the sustainability of ecosystem services through several characteristics and functions, including: enabling productivity, possessing storage and filtering capacity, providing habitat and gene pool variability, containing a carbon pool, acting as a physical and cultural environment for human beings, being both a source of raw materials and also an archive for the geological and archaeological heritage. The functions of soils in the production of food and fiber, the prevention and mitigation of climate change and the provision of other ecosystem services to living organisms are extremely important.

Soil fertility is defined as the capability of soil to provide essential nutrients and water needed for healthy plant growth and reproduction, and the maintenance of biological diversity (Mader et al. 2002). Since the concept of soil fertility is primarily related to the supply of nutrients and water required for crops, the definitions often focus on the chemical and physical aspects of the soil (Bünemann et al. 2018).

Organic matter type and amount has a strong influence on most of the soil physical, chemical and biological properties and consequently the productivity function of soils through improving the stability of aggregates, increasing water-holding capacity and enhancing the biological activity of soils. Increased temperature and decreased moisture tend to decrease productivity function of soils and will accelerate the decomposition of organic matter (Bot and Benites 2005). Nutrient and carbon cycling are dependent on the habitat of soil biota; therefore, changes in temperature may affect the diversity, abundance, and composition of species (Mau et al. 2018). An increase in air temperature, and consequently higher soil temperature, will increase the losses of organic matter by stimulating respiration, altering plant productivity and altering the structure of soil fauna and flora (Mau et al. 2018). These changes will have direct and indirect impacts on land productivity through playing essential roles in nutrient cycling, decomposition and incorporation of crop residue (Barrios 2007; Lehman et al. 2015). Losses of soil organic matter will have negative impacts on productivity functions by decreasing: the stability of aggregates; biological activity; nutrient cycling; and the water-holding capacity of soils. Overall, warming may lead to decreased organic carbon stocks in soils in addition to an increase in atmospheric emissions of CO_2 (Lal 2004; Davidson and Janssens 2006). Decreasing the organic matter content of soils causes a weakening of the stability of soil aggregates, a decrease in the infiltration rate and an increasing susceptibility to compaction and erosion (Kay 2018) (Figure 10.2).

The increase in temperature (and hence increased evapotranspiration demand) along with the decrease in precipitation in some local areas will have a negative impact on soil water budgets (Várallyay 2010), which are essential for many vital soil processes related to productivity. The water budget of the soil regulates water supply to plants, regulates the gas exchange and heat uptake and loss of the soil environment, the availability of plant nutrients and the level of biological activity. Deficiency in available water reduces

FIGURE 10.2 Severe water erosion consequence of low organic matter of soils.

crop productivity as well as crop yield by decreasing the availability and transport of plant nutrients to the root zone (Barber, 1995).

Prolonged periods of heat and drought between rainy periods in arid and semi-arid environments may lead to secondary salinization and alkalization which eventually reduces the productivity of soils (Lal 2009; Várallyay 2010). Climate change is also expected to lead to rising sea levels (by ice melt and heat-induced expansion of sea water) which may cause saline water intrusion to agricultural fields in coastal areas. Salinization due to saline water intrusion will reduce the productivity functions of soils (Renaud et al. 2015). In contrast to reduced precipitation, the increase in precipitation that will arise globally and locally in some areas may lead to increased leaching of nutrients and acidification.

Changes in temperature (up) and amount of precipitation (up or down) will have a remarkable impact on vegetation cover, available water content, infiltration of water, runoff and erosion (Li and Fang 2016). Lu et al. (2013) reported a linear relationship between the increase in precipitation and sediment loss and water discharge. Unfortunately, the reverse may not happen where locally precipitation decreases in response to climate change as adverse effects of reduced rainfall in ecosystems adapted to higher rainfall may include reduced vegetation cover and increased susceptibility to erosion. More intense rainfall events may lead to the redistribution of water within the basin, where there is a decrease in the storage of water. The decrease in storage of water in soil profiles reduces water security and available water for crop production. More frequent or intense rainfall events may cause water-logging in poorly drained agricultural areas (Eekhout et al. 2018) (Figure 10.3).

In contrast to the negative effects of climate change on productivity functions of soils, biological production at higher latitudes is expected to be improved from temperature increases compared to the situation at low latitudes (Hoegh-Guldberg et al. 2018). Yield increases are expected for winter wheat, sunflower and sugar beet with the combined effects of increased CO_2 concentration, improved efficiency in radiation use and extended growing periods of C3 crops leading to increased photosynthesis and growth provided other nutrients are available to support the additional growth. In addition to increases in crop yield, greater amounts of carbon will be added to soil and this will have positive effects on soil functions (Lobell and Gourdji 2012). Climate change is expected to reduce the coverage area of current arable lands and convert some grasslands and forests into arable lands. Land use changes cause significant alterations in soil biological activity, which has strong influences on aggregate stability, biodiversity,

FIGURE 10.3 Water logging after intensive and frequent precipitation in Carsamba Plain, Samsun, Turkey.

plant–soil interactions and the availability of nutrients (Karmakar et al. 2016). Thus, in some instances it is difficult to foresee exact consequences of climate change due to there being a period of instability as physical and biological adaptation occurs in the region to a different (and potentially less stable) set of climate values.

10.3 Impacts of Climate Change on Soil Fertility

Long-term climate change alters the structure of the terrestrial ecosystem and disturbs the equilibrium which has been established for centuries. Increasing temperatures, extending the duration and severity of drought, increased precipitation intensity and increased evaporative demand in many regions are the major outcomes of climate change. Two major parameters of climate (temperature and rainfall) have a decisive role in many soil processes controlling the soil fertility. These will also interact with pollution effects and management of other human inputs. Temperature- and water-dependent soil processes are expected to be significantly modified by further changes in climate (Penuelas et al. 2018). Mineral nutrients are considered a very important factor that limits plant biomass and productivity in many ecosystems (Turk and Tawaha 2001; Turk and Tawaha 2002a, 2002b; Turk and Tawaha 2002a; Turk and Tawaha 2002b; Tawaha et al. 2003; Turk et al. 2003a; Turk et al. 2003b; Turk et al. 2003c; Nikus et al. 2004a; Nikus et al. 2004b; Tawaha and Turk 2004; Turk and Tawaha 2004; Abebe et al. 2005; Abera et al. 2005; Al-Tawaha et al. 2005; Al-Kiyyam et al. 2008; Al-Ajlouni et al. 2009; Al-Tawaha et al. 2010b; Al-Juthery et al. 2018).

The response of soil organic matter content to changes in the C and N cycles will have a significant impact on the productivity function of soils (Brevik 2013). Therefore, the potential soil fertility decline under climate change will be explained by concentrating on climate change effects on the N cycle, C storage in soils and erosion processes.

10.3.1 Climate Change Effect on Nutrient Cycles

Soils are essential components of nutrient cycles, especially the C and N cycles, which are important for the interactions of soils and climate change because both are the major components of the Soil Organic Matter (SOM) (Brady and Weil 2008). Given the existing soil chemistry the soil moisture regime determines the availability status of soil nutrients for plants. Soil temperature along with the moisture regime are the main determinants of root growth and availability of the existing nutrients (Brouder and Volenec 2008; Pareek 2017). Therefore, the decline in soil moisture due to climate change will have a significant influence on the nutrient cycles of soils. The increase in the number of drought and storm events, frequency and magnitude will increase the vulnerability to nutrient losses and will have a negative effect on soil microbial communities (Karmakar et al. 2016). Important nutrient losses (e.g. nitrate leaching) can also be triggered by excessive precipitation (Sun et al. 2007; Tang et al. 2008). The effects of increased rainfall resulting from climate change, in combination with a poor drainage in some areas, will lead to water-logging which can generate hypoxicity and nutrients deficiency (Drew 1988). However, there are also several positive effects of climate change on nutrient cycles. The increase in soil temperature will result in a root surface area increases which will improve nutrient uptake by plants. Higher temperatures will also affect both transpiration rates (leading to larger and faster acquisitions of nutrients) (Bassirirad, 2000) and plant phenology (Nord and Lynch 2009). Higher photosynthesis rates (Bazzaz 1990) for nitrogen-fixing plants (Li et al. 2017) may provide more carbohydrates for fixation and hence increased nitrogen inputs into some environments.

Elevated CO_2 can have significant effects on soil nitrogen cycling and amounts. This can be both through the process of Progressive Nitrogen Limitation (PNL), leading to higher inputs of carbon into the soil and increasing ammonium immobilization (Hungate et al. 1999). Thus, both the nitrogen stimulation effect of fixation and the nitrogen immobilization effect of the PNL can operate in opposition and their relative balance will determine the balance of climate change effects on soil nitrogen.

10.3.2 Climate Change Effect on Organic Matter Content of Soils

The SOM has a strong influence on soil quality. Water-holding capacity, nutrient storage, and soil stability are among the key factors driven by SOM and affect the quality of soils (Karmakar et al. 2016; Pareek 2017). Changes in precipitation and temperature influence soil temperature and moisture which have great impact on SOM. The real impact of climate change on SOM is still under debate, considering that global warming might have both positive and negative effects. Nevertheless, there is an agreement about the necessity of implementing mitigation actions considering the positive effects of climate changes on SOM (increased N mineralization, increased biological nitrogen fixation, increased plant growth) will be likely overrun by negative effects (loss of SOM, reductions in soil organic C, increased released CO_2 (Gorissen et al. 2004; Niklaus and Körner 2004; Norby and Luo 2004; Beedlow et al. 2004).

10.3.3 Climate Change Effect on Erosion

Climate change is expected to influence and to increase soil erosion especially by water-driven processes (Brevik 2013). There are several driving climatic factors for soil erosion, the most important being rainfall intensity as dominant factor and wind (Lee et al. 1996; Nearing et al. 2004; Karmakar et al. 2016). Both of these will be increased by the increase in energy retained in the atmosphere due to temperature and evaporation increases due to greenhouse gas increases. In addition, other factors influencing the process of soil erosion are land use, land management, soil type, topography etc. Therefore, the climate change effect on erosion depends on the combined impacts of climate, relief, vegetation and the features of soil erodibility (Karmakar et al. 2016).

10.4 Biofertilizers to Increase the Resilience of Soil Productivity Function Under Climate Change

Microbial biodiversity is essential for the maintenance of soil fertility. The wealth of biodiversity below ground is vast, with hundreds of millions of creatures living and reproducing in the top soil, which is an ecosystem which plays an essential role in providing food for living beings on Earth. Plants and microbes have evolved together and worked with each other over millions of years. As an example mycorrhizal fungi co-evolved with plants at least 400 million years ago and enabled plant colonization of low-fertility environments. The endophytic, ectophytic and epiphytic colonization by microflora, including bacteria, fungi and actinobacteria, are influenced by abiotic environmental factors around the rhizosphere. Phenotypic agility of changing soil abiotic factors has either a positive or a negative impact on the shift of microbial communities, which influences plant phenotypic and genotypic traits.

10.4.1 Soil Fertility

Soil fertility, or its ability to improve natural and agricultural crops, is needy upon three interacting and jointly dependent components: physical, chemical and biological fertility (Al-Tawaha et al. 2005; Yang et al. 2005; Lee et al. 2005; Sulpanjan et al. 2005; Tawaha et al. 2005; Assaf et al. 2006; Turk et al. 2006; Hameed et al. 2008; Al-Tawaha et al. 2010a; Ananthi et al. 2017; Al-Tawaha et al. 2017; Abu Obaid et al. 2018; Al-Tawaha et al. 2018a, 2018b; Amanullah et al. 2019a, b; Hani et al. 2019).

10.4.2 Soil Microorganisms

Bacteria: These are the most predominant group of microorganisms in the soil and represent one-half of the microbial biomass in soils. They are found in neutral to slightly alkaline soils and the genus Bacillus has the largest representation in soils in terms of species. Soils with low organic matter and sandy texture contain very low populations of bacteria, which indirectly affects soil fertility and health.

Actinomycetes: They are bacteria, but act as a connecting link between bacteria and fungi due to their general branching mycelial growth habit and belong principally to the genera Streptomyces and Nocardia. Among soil actinomycetes 70% are reported as Streptomyces species. They are predominantly found in neutral to slightly alkaline soils.

Fungi: They are most numerous in the surface layers of well-aerated and cultivated soils. Common genera present in soil are Aspergillus, Mucor, Penicillium, Alternaria, Rhizopus. Soil fungi can grow in a wide range of soil pH, but interestingly their population is higher under acidic conditions.

Algae: Abundant in habitats exposed to light and sufficient moisture. Important genera are Anabaena, Nostoc and Aulosira.

Protozoa: Abundant in surface soils and derive their nutrition by devouring soil bacteria. They regulate the biological equilibrium in soil. Their population is high in soils with high organic matter content.

10.4.3 Microbial Fertigation for Soil Fertility (Figure 10.4)

10.4.3.1 Rhizobium

Rhizobium, which belongs to the family Rhizobiaceae, is symbiotic in nature and fixes nitrogen (commonly 50–100 kg ha^{-1}) in legumes. The rhizobium colonizes the roots of specific legumes to form tumor-like growths called root nodules, which act as factories of ammonia production. The differentiated, bacteroid forms within these nodules fix atmospheric nitrogen and the resultant ammonia is used as a source of fixed nitrogen. This symbiosis gives the bacteria with an exclusive niche and, in return, the plants obtain a personalized nitrogen source along with gibberellins and auxin synthesized by bacteria (Andrew et al. 2007). When Rhizobium inoculants are applied with organic manure such as Farm Yard

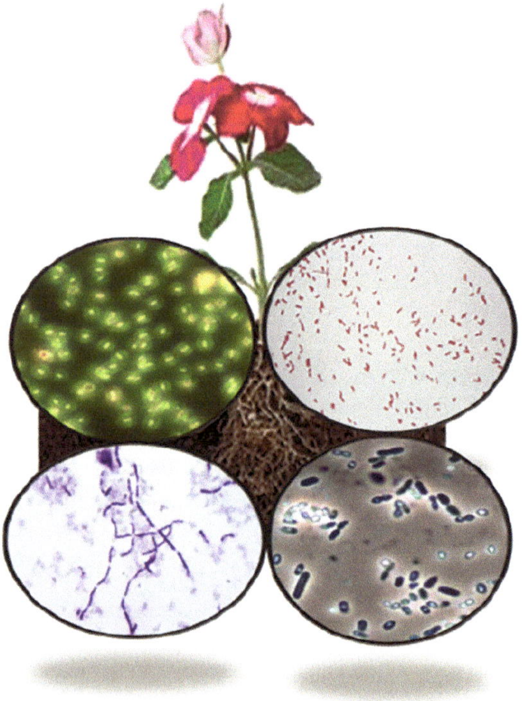

FIGURE 10.4 Plant growth promotion by biofertilizers.

TABLE 10.1

Efficacy of native Rhizobium isolates on yield of summer groundnut

Treatments	Pod yield (kg ha^{-1})	Haulm yield (kg ha^{-1})	Oil content (%)
50 % RDN + GNR I	3163	5274	46.87
50 % RDN + GNR II	2990	5205	46.98
50 % RDN + GNR STD	3101	5288	47.20
75 % RDN + GNR I	2917	5372	47.07
75 % RDN + GNR II	3101	5361	47.15
75 % RDN + GNR STD	3195	5368	46.74
RD FYM + 50 % RDN + GNR I	3327	5375	47.23
RD FYM + 50 % RDN + GNR II	3351	5354	47.16
RD FYM + 50 % RDN + GNR STD	3313	5389	47.08
RD FYM + 75 % RDN + GNR I	3177	5358	47.02
RD FYM + 75 % RDN + GNR II	3319	5347	47.05
RD FYM + 75 % RDN + GNR STD	3389	5399	47.31
100 % RD	3233	5483	47.02
Absolute Control	1888	4354	40.53
S.Em. ±	356.97	373.61	0.23
CD at 0.05%	NS	NS	0.658
CV %	19.91	12.25	0.842

GNR I: *Rhizobium huautlense* GNR I & GNR II: *Rhizobium giardinii*GNR II.
RD FYM: Recommended dose, farmyard manure, RDN: Recommended Dose Organic Nitrogen.

Manure (FYM) it can replace up to 50% of chemical fertilizer without compromising groundnut quality and yield (Table 10.1) in a sustainable manner (Patel et al. 2016). Equally native rhizobia in the soil can provide increased fixation and increased nitrogen cycling under conditions of enhanced photosynthesis arising from elevated CO_2. This would particularly be the case in local areas where increased rainfall and/or increased temperatures stimulate nitrogen fixation in naturally nodulated legumes or non-legume nitrogen fixers (Cernusak et al. 2011; Rogers et al. 2009).

10.4.3.2 Azotobacter

Azotobacter is a free-living nitrogen-fixing, obligate aerobic and heterotrophic bacteria, which can grow under low oxygen concentration. Desai et al. (2015) reported that combined the application of inorganic fertilizer with FYM, bio-fertilizer (Azotobacter + Phosphate solubilizing microorganisms (PSM)) and sulfur in the form of gypsum gave significantly higher wheat yields than recommended chemical fertilizers.

10.4.3.3 Azospirillum

Azospirillum belongs to the family Spirilaceae, which are heterotrophic and associative in nature. *A. lipoferum* is reported to be associated with the roots of C4 plants like maize, while *A. brasilense* is associated with C3 plants like rice and wheat. When Azospirillum colonize the roots they not only remain on the root surface, but a sizable proportion of them penetrate into the root tissues and live in harmony within the plants. Patel et al. (1991) found that there was a significant impact of culture treatments of Azospirillum on nitrogen content of pearl millet. Nitrogen content of plants treated with Azospirillum treatments was higher than that of plants without the treatment. Thus, Azospirillum inoculation may be one way of overcoming PNL arising from increased CO_2 fixation under climate change scenarios. They may benefit also from the increased availability of carbohydrate in the soil and roots to support their nitrogen fixation in the same way that legumes do.

10.4.3.4 Phosphate Solubilizing Microorganisms (PSM)

Several strains of Pseudomonas, Bacillus, Rhizobium, Enterobacter from bacterial genera and Aspergillus, and Penicillium from fungal genera, were reported worldwide as potent phosphate solubilizers. Phosphate-solubilizing microorganisms inoculation (or natural occurrence in the soil) may be one way of overcoming P deficiency arising from P dilution due to increased CO_2 fixation under climate change scenarios (O'Neill et al. 1987). PSM may benefit from the increased availability of carbohydrate in the soil and roots to support their nitrogen fixation in the same way that legumes do.

10.4.3.5 Mycorrhizae

Mycorrhiza is the mutualistic symbiosis (non-pathogenic association) between soil-borne fungi with the roots system of plants. In such kinds of relation the immobility of nutrients such as phosphorus, zinc, copper and other nutrients such as cadmium are also made available by the mycorrhizae network (Liu et al. 2002). Thus, mycorrhiza inoculation (or natural occurrence in the soil) may be one way of overcoming deficiency arising from plant nutrient dilution due to increased CO_2 fixation under climate change scenarios. However, it is equally possible that nutrient limitation may reduce mycorrhizal growth (Staddon et al. 1999).

10.5 Conclusion

Climate changes induced by elevated greenhouse gases, particularly the increase in temperature, will lead to locally warmer and drier or warmer and wetter conditions. These can trigger disturbances in ecosystems and impair the services provided by the ecosystems. Therefore, temperature-related variables and water availability are the most prominent climatic drivers in the degradation of agricultural and forestry ecosystems. However, degradation can also arise from other forms of atmospheric pollution. For example, direct negative effects result for plants due to increases in ground-level ozone. Soil acidification from fossil fuel burning depositing inorganic acids may harm the uptake of some nutrients (e.g. Mo), but make others more available (e.g. Mn). There is a potential for beneficial effects on soils due to elevated photosynthesis at enhanced CO_2 levels leading to greater carbohydrate availability for roots and soils. However, this seems to be largely defeated in many ecosystems due to Progressive Nitrogen Limitation (PNL) as well as the dilution of other elements such as P. Legume and non-legume nitrogen fixation may overcome PNL. There is a growing body of evidence for this possibility. Other element limitations may also be overcome by additional carbohydrates, for example stimulating PSM or mycorrhiza. However, there is little evidence to support this at present.

Overall, the increase in occurrence and severity of drought, pathogens, insect outbreaks, nutrient deficiency, acidification and fire may surpass the resilience thresholds of ecosystems, resulting in severely altered ecosystems. Increasing the average temperatures and changes in precipitation patterns will markedly decrease soil organic matter, which will negatively affect nutrient cycles, the formation and stability of aggregates, water- and nutrient-holding capacities and cation exchange capacity of soils.

The influence of climate extremes on ecosystems can thus be summarized under direct, indirect and interaction effects of which the direct impact is the most prominent agent causing the abiotic disturbances. The interactions of drought and wind may cause the impacts of other disturbance agents, such as insects and fire, to increase; thus, the interactions may increase the deleterious effect of a single agent. Existing research investigating the effects of climate change on ecosystems reveals that ongoing climate changes are likely to affect several ecosystems at the global scale, though the expected alterations in agricultural and forestry ecosystems are going to be greater in a warming world compared to the other ecosystems. On the other hand, the potential of microorganisms to form natural nutrient cycles in soil is amazing and potentially enormously beneficial. They may thus present economic opportunities relative to costly chemical fertilizers as a means to overcome some of the limitations that occur naturally or due to climate change. Microbial inoculants have been established as eco-friendly and environmentally safe, low-cost inputs the world over for the improvement of soil fertility to achieve the best productivity and to reduce agro-ecosystem pollution. Thus, microbes play an essential and vital role in the maintenance of physio-chemical and biological properties of soil for human, animal and plant life on our globe.

REFERENCES

Abebe G, Hattar B, Al-Tawaha A (2005) Nutrient availability as affected by manure application to cowpea (Vigna unguiculata L. Walp.) on calacarious soils. *J Agri Soc Sci 1*(1):1–6.

Abera T, Feyisa D, Yusuf H et al. (2005) Grain yield of maize as affected by biogass slurry and NP fertilizer rate at Bako, Western Oromiya Ethopia. *Biosci Res 2*:31–38.

Abu Obaid AM, Melny AV, Onichko VI et al (2018) Evaluation of six sunflower cultivar for forage productivity under salinity condition. *Adv Environ Biol 12*(7):13–15.

Adnan M, Shah Z, Ullah H, Khan B, Arshad M, Mian IA, Khan WU (2016) Yield response of wheat to nitrogen and potassium fertilization. *Pure Appl Biol 5*(4):1.

Akbarimehr M, Naghdi R (2012) Reducing erosion from forest roads and skid trails by management practices. *J Forest Sci 58*:165–169.

Al-Ajlouni MM, Al-Ghzawi AA, Al-Tawaha AM (2009) Crop rotation and fertilization effect on barley yield grown in arid conditions. *J Food Agri Enviro 88*(3):869–872.

Alexandratos N, Bruinsma J (2012) World agriculture towards 2030/2050: the 2012 revision. ESA Working paper No. 12-03. Rome: FAO.

Alexandrov VA, Hoogenboom G (2000) Vulnerability and adaptation assessments of agricultural crops under climate change in the southeastern USA. *Theor. Appl. Climatol. 67*:45–63.

Al-Kiyyam MA, Turk M, Al-Mahmoud M, Al-Tawaha AR (2008) Effect of plant density and nitrogen rate on herbage yields of marjoram under Mediterranean conditions. *J American-Eurasian J Agri Environ Sci 3*:153–158.

Allen MR, Dube OP, Solecki et al. (2018) *Framing and context. In: Global warming of 1.5°C. An IPCC Special Report on the impacts of global warming of 1.5°C above pre-industrial levels and related global greenhouse gas emission pathways, Assessment*, Island Press, Washington, DC, 155.

Al-Tawaha AM, Turk MA, Lee KD et al. (2005) Impact of fertilizer and herbicide application on performance of ten barley genotypes grown in northeastern part of Jordan. *Int J Agri Biol 7*(2):162–166.

Al-Tawaha AM, Turk M, Hameed K et al. (2010a) Effects of application of Olive Mill by-products on lentil yield and their symbiosis with mycorrhizal fungi under Arid conditions. *Crop Res 40*(3):66–75.

Al-Tawaha AM, Yadav SS, Turk M et al. (2010b) Crop Production and Management Technologies for Drought Prone Environments. *Chapter in Book. Climate Change and Drought Management in Cool Season Grain Legume Crops. (Springer).*

Al-Tawaha AR, Turk MA, Abu-Zaitoon YM et al. (2017). Plants adaptation to drought environment. *Bulg J Agri Sci 23*(3):381–388.

Al-Tawaha AR, Al-Karaki GN, Al-Tawaha AR et al. (2018a) Effect of water flow rate on quantity and quality of lettuce (*Lactuca sativa* L.) in nutrient film technique (NFT) under hydroponics conditions. *Bulg J Agri Sci 245*:791–798.

Al-Tawaha AR, Al-Tawaha AR, Alu'datt M et al. (2018b) Effects of soil type and rainwater harvesting treatments in the growth, productivity and morphological trains of barley plants cultivated in semi-arid environment. *Aust J Crop Sci 12*(06):975–979.

Amanullah KN, Khan MI et al. (2019a) Wheat biomass and harvest index increases with integrated use of Phosphorus, Zinc and beneficial microbes under semiarid climates. *J Microbiol Biotechnol Food Sci 9*(2):242–247.

Amanullah, Khalid S, Imran et al. (2019b) Organic matter management in cereals based system: symbiosis for improving crop productivity and soil health. In: Lal R, Francaviglia R (eds) *Sustainable agriculture reviews 29. vol 29*. Springer, Cham. https://doi.org/10.1007/978-3-030-26265-5_3.

Ananthi T, Amanullah M, Al-Tawaha AM (2017) A review on Maize- Legume ıntercropping for enhancing the productivity and soil fertility for sustainable agriculture ın India. *Adv Environ Biol 11*(5):49–63.

Andersson FA (2005) *Coniferous forests*. Springer BW, Amsterdam, 637 p.

Andrew JW, Jonathan D, Andrew R et al. (2007) Living without Fur: the subtlety and complexity of iron-responsive gene regulation in the symbiotic bacterium Rhizobium and other α-proteobacteria. *Biometals 20*:501–511.
Assaf TA, Hameed KM, Turk MA, Tawaha AM (2006) Effect of soil amendment with Olive Mill by-products under soil solarization on growth and productivity of Faba Bean and their symbiosis with mycorrhizal fungi. *W J Agri Sci 2*(1):21–28.
Ayeb-Karlsson S, van der Geest K, Ahmed I et al. (2016) A people-centred perspective on climate change, environmental stress, and livelihood resilience in Bangladesh. *Sustain Sci 11*(4):679–694.
Bache BW (1986) Aluminium mobilization in soils and waters. *J Geologic Soc 143*(4):699–706.
Bangash RF, Passuello A, Sanchez-Canales M et al. (2013) Ecosystem services in Mediterranean river basin: climate change impact on water provisioning and erosion control. *Science Total Environ 458*:246–255.
Barber SA (1995) *Soil nutrient bioavailability: a mechanistic approach.* John Wiley & Sons.
Barrios E (2007) Soil biota, ecosystem services and land productivity. *Ecologic Econ 64*(2):269–285.
Bassirirad H (2000) Kinetics of nutrient uptake by roots: responses to global change. *New Phytol 147*:155–169.
Bazzaz FA (1990) The response of natural ecosystems to the rising global CO_2 levels. *Ann Rev Ecol System 21*(1):167–196.
Beedlow PA, Tingey DT, Phillips DM et al. (2004) Rising atmospheric CO and carbon sequestration in forests. *Front Ecol Environ 2*:315–322.
Bergh J, Freeman M, Sigurdsson B et al. (2003) Modelling the short-term effects of climate change on the productivity of selected tree species in Nordic countries. *Forest Ecol Manage 183*:327–340.
Bocci M, Smanis T (2019) Assessment of the impacts of climate change on the agriculture sector in the Southern Mediterranean. Union for the Mediterranean. (accessed at https://ufmsecretariat.org/wp-content/uploads/2019/04/Climate-Change-impact-on-Agriculture.pdf on July 26th, 2019).
Bot A, Benites J (2005) *The importance of soil organic matter: key to drought-resistant soil and sustained food production (No. 80).* Food & Agriculture Org.
Brady NC, Weil RR (2008) *The nature and properties of soils*, 14th ed.; Pearson Prentice Hall, Upper Saddle River, NJ.
Brevik EC (2013) The potential impact of climate change on soil properties and processes and corresponding influence on food security. *Agriculture 3*:398–417.
Brouder SM, Volenec JJ 2008 Impact of climate change on crop nutrient and water use Efficiencies. *Physiologia Plantarum 133*(4):705–724.
Bruinsma J (2003) *World agriculture: towards 2015/2030.* An FAO Perspective. Earthscan Publications Ltd., London.
Bünemann EK, Bongiorno G, Bai Z et al. (2018) Soil quality–A critical review. *Soil Biol Biochem 120*:105–125.
Carter C, Cui X, Ghanem D (2017) Identifying the economic impacts of climate change on agriculture. *Ann Rev Res Econ 10*:361–380.
CCSP (2008) The Effects of Climate Change on Agriculture, Land Resources, Water Resources, and Biodiversity in the United States. Chapter 3: Land Resources: Forest and Arid Lands. A Report by the U.S. Climate Change Science Program and the Subcommittee on Global Change Research. Environmental Protection Agency, Washington, DC, USA.
Cernusak LA, Winter K, Martínez C, Correa E et al. (2011) Responses of legume versus nonlegume tropical tree seedlings to elevated CO_2 concentration. *Plant Physiol 157*(1):372–385.
Costanza R, de Groot R, Sutton P et al. (2014) Changes in the global value of ecosystem services. *Glob Environ Change 26*:152–158.
Davidson EA, Janssens IA (2006) Temperature sensitivity of soil carbon decomposition and feedbacks to climate change. *Nature 440*:7081–7165.
De Schutter O (2008) Building resilience: a human rights framework for world food and nutrition security. Report of the special rapporteur on the right to food, Olivier de Schutter, to the UN General Assembly (A/HRC/9/23).
Desai HA, Dodia NI, Desai Ck et al. (2015) Integrated Nutrient Management in Wheat (*Triticum aestivum* L.) *Trends Biosci 8*(2):472–475.
Drew M (1988) Effects of flooding and oxygen deficiency on the plant mineral nutrition. In: Tinker E, Tinker PB, Lauchli A (eds) *Advances in plant nutrition*, Greenwood Publishing Group, Westport Connecticut, pp 115–159.

Easterling D, Meehl G, Parmesan C et al. (2000) Climate extremes: observations, modeling and impacts. *Science 289*(5487):2068–2086.
Eekhout JP, Hunink JE, Terink et al. (2018) Why increased extreme precipitation under climate change negatively affects water security. *Hydrol Earth Syst Sci 22*(11):5935–5946.
FAO (2018) Climate change for forest policy-makers – An approach for integrating climate change into national forest policy in support of sustainable forest management – Version 2.0. FAO Forestry Paper no.181. Rome, 68 pp. Licence: CC BY-NC-SA 3.0 IGO.
Garzón BM, Alía R, Robson MT et al. (2011) Intra-specific variability and plasticity influence potential tree species distributions under climate change. *Glob Ecol Biogeograph 20*:766–778.
Gorissen A, Tietema A, Joosten N et al. (2004) Climate change affects carbon allocation to the soil in shrublands. *Ecosystems 7*:650–661.
Groffman PM, Kareiva P, Carter S et al. (2014) Ch. 8: Ecosystems, Biodiversity, and Ecosystem Services. *Climate Change Impacts in the United States: The Third National Climate Assessment*, JM Melillo, Terese Richmond, GWm Yohe, U.S. Global Change Research Program, 195–219.
Hameed K, Turk M, Assaf T, Tawaha AM (2008) Effect of soil amendment with olive mill by-products under soil solarization on the growth and productivity of chickpea and their symbiosis with mycorrhizal fungi under arid conditions. *Int J Plant Prod 2*(4):341–352.
Hani NB, Al-Ramamneh EAD, Haddad M et al. (2019) The Impact Of Cattle Manure On The Content Of Major Minerals And Nitrogen Uptake From 15n Isotope-Labeled Ammonium Sulphate Fertilizer In Maize (*Zea Mays* L.) Plants. *Pakistan J Bot 51*(1):185–189.
Hillel D (2005) *Encyclopedia of soils in the environment*. Elsevier, Amsterdam, 2200 p.
Hoegh-Guldberg O, Jacob D et al. (2018) Impacts of 1.5°C global warming on natural and human systems
Hungate BA, Dijkstra P, Johnson DW et al. (1999) Elevated CO_2 increases nitrogen fixation and decreases soil nitrogen mineralization in Florida scrub oak. *Glob Change Biol 5*(7):781–789.
IPCC (2007) Climate Change 2007: Synthesis report. Contribution of Working Groups I, II and III to the Fourth Assessment Report of the Intergovernmental Panel on Climate Change [Core Writing Team, Pachauri RK and Reisinger A (eds)] IPCC. Geneva.
IPCC (2012) Managing the risks of extreme events and disasters to advance climate change adaptation. In: Field CB, Barros V, Stocker TF, Qin D, Dokken DJ, Ebi KL, Mastrandrea MD, Mach KJ, Plattner GK, Allen SK, Tignor M, Midgley PM (eds) *A special report of working groups I and II of the intergovernmental panel on climate change*. Cambridge University Press, Cambridge, UK, and New York, NY.
IPCC (2014) Summary for policymakers, In: Climate Change 2014: Impacts, Adaptation, and Vulnerability. Part A: Global and Sectoral Aspects. Contribution of Working Group II to the Fifth Assessment Report of the Intergovernmental Panel on Climate Change., Cambridge University Press, Cambridge, UK and New York, NY, 1–32, 2014.
Juthery HW, Habeeb KH, Altaee FJ K et al. (2018) Effect of foliar application of different sources of nano-fertilizers on growth and yield of wheat. *Biosci Res 15*(4):3976–3985.
Karmakar R, Das I, Dutta D et al. (2016) Potential effects of climate change on soil properties: a review. *Sci Int 4*(2):51–73.
Kay BD (2018) Soil structure and organic carbon: a review. In *Soil processes and the carbon cycle*. CRC Press, pp. 169–197.
Khan A, Shah N, Gul A, Ismail A, Aziz F, Farooq M, Rizwan, M. (2016). Comparative Study of Toxicological Impinge of Glyphosate and Atrazine (Herbicide) on Stress Biomarkers; Blood Biochemical and Hematological Parameters of the Freshwater Common Carp (Cyprinus carpio). *Pol J Environ Stud 25*(5):1993–1999.
Křeček J, Hořická Z (2006) Forests, air pollution and water quality: influencing health in the headwaters of Central Europe's" Black Triangle". Unasylva (FAO).
Křeček J, Novakova J, Hořická Z (2010) Ellenberg's indicator in water resources control: the Jizera Mountains, Czech Republic. *Ecol Eng 36*(9):1112–1117.
Křeček J, Palán L, Stuchlík E (2019) Impacts of land use policy on the recovery of mountain catchments from acidification. *Land Use Pol 80*:439–448.
Kubin E, Křeček J, Palán L (2017) Effects of forest practices on water resources recharge in the Boreal climate. *Environ Process 4*(3):509–522.
Lal R (2004) Soil carbon sequestration to mitigate climate change. *Geoderma 123*(1–2):1–22.
Lal R (2009) Soil degradation as a reason for inadequate human nutrition. *Food Secur 1*(1):45–57.

Lavergne S, Mouquet N, Thuille et al. (2010). Biodiversity and climate change: integrating evolutionary and ecological responses of species and communities. *Annual Review Ecology, Evol System 4*:321–350.
Lee JJ, Phillips DL, Dodson RF (1996) Sensitivity of the US corn belt to climate change and elevated CO: II. Soil erosion and organic carbon. *Agri Syst 52*:503–521.
Lee KD, Tawaha AM, Yang S (2005) Effect of phosphorus application on yield, mineral contents and active components of *Chrysanthemum coronarium* L. *Biosci Res 2*(3):118–124.
Lehman RM, Acosta-Martinez V, et al. (2015) Soil biology for resilient, healthy soil. *J Soil Water Conserv 70*(1):12–18.
Lepetz V, Massot M, Schmeller DS et al. (2009) Biodiversity monitoring: some proposals to adequately study species' responses to climate change. *Biodivers Conserv 18*:3185–3203.
Lexer MJ, Honninger K, Scheifinger H et al. (2002) The sensitivity of Austrian forests to scenarios of climatic change: a large-scale risk assessment based on a modified gap model and forest inventory data. *Forest Ecol Manage 162*:53–72
Li Z, Fang H (2016) Impacts of climate change on water erosion: a review. *Earth Sci Rev 163*:94–117.
Li Y, Yu Z, Liu X et al. (2017) Elevated CO_2 increases nitrogen fixation at the reproductive phase contributing to various yield responses of soybean cultivars. *Front Plant Sci 8*:1546.
Lindner M, Bugmann H, Lasch P et al. (1997) Regional impacts of climatic change on forests in the state of Brandenburg, Germany. *Agriculture and Forest Meteorology 84*:123–135.
Lindner M, Garcia-Gonzalo J, Kolstrom M et al. (2008) Impacts of climate change on European forests and options for adaptation. AGRI-2007-G4-06. Report to the European Commission Directorate-General for Agriculture and Rural Development.
Lindner M, Maroschek M, et al. (2010) Climate change impacts, adaptive capacity, and vulnerability of European forest ecosystems. *Forest Ecology and Management 259*:698–709.
Liu A, Hamel C, Elmi A et al. (2002) Concentrations of K, Ca and Mg in maize colonised by arbuscular mycorrhizal fungi under field conditions. *Canadian Journal of Soil Science 82*(3):271–278.
Lobell DB, Gourdji SM (2012) The influence of climate change on global crop productivity *Plant Physiology 160*(4):1686–1697.
Lorencová E, Frélichová J, Nelson E (2013) Past and future impacts of land use and climate change on agricultural ecosystem services in the Czech Republic. *Land Use Policy 33*:183–194.
Loustau D, Bosc A, Colin A et al. (2005) Modeling climate change effects on the potential production of French plains forests at the sub-regional level. *Tree Physiology 25*:813–823.
Lu XX, Ran LS, Liu S (2013) Sediment loads response to climate change: a preliminary study of eight large Chinese rivers. *International Journal of Sediment Research 28*(1):1–14.
Mäder P, Fliessbach A, (2002) Soil fertility and biodiversity in organic farming. *Science 296*:1694–1697.
Mau RL, Dijkstra P, Schwartz E et al. (2018) Warming induced changes in soil carbon and nitrogen influence priming responses in four ecosystems. *Applied Soil Ecology 124*:110–116.
MEA (Millennium Ecosystem Assessment) (2005) *Ecosystems and human well-being: current state and trends.* Island Press, Washington DC.
Mechler R, Bouwer LM, (2019) *Loss and damage from climate change. Concepts, methods and policy options.* Springer. https://doi.org/10.1007/978-3-319-72026-5.
Mekonnen Z (2018) Observed and Projected Reciprocate Effects of Agriculture and Climate Change: implications on Ecosystems and Human Livelihoods, *Climate Change and Global Warming, Ata Amini*, IntechOpen. DOI:10.5772/intechopen.79118.
Melillo JM, Terese TCR, Gary WY (2014) *Highlights of climate change impacts in the United States: The third national climate assessment.* U.S. Global Change Research Program, Washington, DC, pp 148.
Miranda LS, Imperatriz-Fonseca VL, Giannini TC (2019) Climate change impact on ecosystem functions provided by birds in southeastern Amazonia. *PLoS One 14*(4):e0215229. https://doi.org/10.1371/journal.pone.0215229.
Morin X, Viner D, Chuine I (2008) Tree species range shifts at a continental scale: new predictive insights from a process-based model. *Journal of Ecology 96*:784–794.
Morin X, Fahse L, Jactel H et al. (2018) Long-term response of forest productivity to climate change is mostly driven by change in tree species composition, *Scientific Reports 8*:5627.
Morrissey J, Oliver-Smith A (2013) *Perspectives on Non-Economic Loss and Damage: understanding values at risk from climate change.* Loss and Damage Series, United Nations University Institute for Environment and Human Security, Bonn.

Mukhopadhyay D (2009) *Impact of climate change on* forest *ecosystem and* forest *fire in India. Earth and Environmental Science 6*(3):20–27.
Nearing MA, Pruski FF, O'Neal MR (2004) Expected climate change impacts on soil erosion rates: a review. *Journal of Soil and Water Conservation 59*:43–50.
Niklaus PA, Körner C (2004) Synthesis of a six-year study of calcareous grassland responses to in situ CO_2 enrichment. *Ecological Monographs 74*:491–511.
Nikus O, Al-Tawaha AM, Turk MA (2004a) Effect of manure supplemented with phosphate fertilizer on the fodder yield and quality of two sorghum cultivars (*Sorghum bicolor* L.). *Bioscience Research 1*:1–7.
Nikus O, Turk MA, Al-Tawaha AM (2004b) Yield response of sorghum (*Sorghum bicolor* L.) to manure supplemented with phosphate fertilizer under semi-arid Mediterranean conditions. *International Journal of Agriculture and Biology 6*:889–893.
Norby RJ, Luo Y (2004) Evaluating ecosystem responses to rising atmospheric CO2 and global warming in a multi-factor world. *New Physiologist 162*:281–293.
Nord EA, Lynch JP (2009) Plant phenology: a critical controller of soil resource acquisition. *Journal of Experimental Botany 60*:1927–1937.
O'Neill EG, Luxmoore RJ, Norby R J (1987) Elevated atmospheric CO_2 effects on seedling growth, nutrient uptake, and rhizosphere bacterial populations of *Liriodendron tulipifera* L. *Plant and Soil 104*(1):3–11.
Pareek N (2017) Climate change impact on soils: adaptation and mitigation. *MOJ Ecology & Environmental Sciences 2*(3):136–139.
Parmesan C (2006) Ecological and evolutionary responses to recent climate change. *Annual Review of Ecology, Evolution and Systematics 37*:637–669.
Patel IG, Kalyansundaram NK, Patel BT (1991) Iron content and uptake by pearl millet as influenced by Azospirillum inoculation. *Journal of the Indian Society of Soil Science 39*:394–395.
Patel HK, Patel SR, Vyas RV (2016) *Diversity and efficacy study of native Rhizobium sp. on summer groundnut from sukhi river command area of middle Gujarat.* Paper presented at the *3rd International Conference on Food, Water, Energy Nexus in Arena of Climate Change Anand Agricultural University*, Anand (India), 14-16 June, 2016.
Peñuelas J, Sardans J, Filella I. et al. (2018) Assessment of the impacts of climate change on Mediterranean terrestrial ecosystems based on data from field experiments and long-term monitored field gradients in Catalonia. *Environmental and Experimental Botany 152*:49–59.
Peters DPC, Lugo AE, Chapin FS et al. (2011) Cross-system comparisons elucidate disturbance complexities and generalities. *Ecosphere 2*:1–26.
Platt RH (2004) *Land use and society*. Island Press, Washington, 479 p.
Rahman I, Ali S, Rahman I, Adnan M, Ullah H, Basir A, Malik FA, Shah AS, Ibrahim M, Arshad M (2016). Effect of pre-storage seed priming on biochemical changes in okra seed. *Pure Appl Biol 5*(1):165–171.
Renaud FG, Le TTH, Lindener C et al. (2015). Resilience and shifts in agro-ecosystems facing increasing sea-level rise and salinity intrusion in Ben Tre Province, Mekong Delta. *Climatic Change 133*(1):69–84.
Rogers A, Ainsworth EA, Leakey AD (2009) Will elevated carbon dioxide concentration amplify the benefits of nitrogen fixation in legumes. *Plant Physiology 151*(3):1009–1016.
Rouault G, Candau JN, Lieutier F et al. (2006). Effects of drought and heat on forest insect populations in relation to the 2003 drought in Western Europe. *Annals of Forest Science 63*:613–624.
Schlenker W, Lobell DB (2010) Robust negative impacts of climate change on African agriculture. *Environmental Research Letters, 5*. doi:10.1088/1748-9326/5/1/014010.
Semenov MA, Halford NG (2009) Identifying target traits and molecular mechanisms for wheat breeding under a changing climate. *Journal of Experimental Botany 60*(10):2791–2804.
Stabinsky D, Ching (2012) Ecological Agriculture, Climate Resilience and a Roadmap to Get There. TWN Environment & Development Series No. 14, Third World Network, Penang.
Staddon PL, Fitter AH, Graves JD (1999) Effect of elevated atmospheric CO2 on mycorrhizal colonization, external mycorrhizal hyphal production and phosphorus inflow in Plantago lanceolata and Trifolium repens in association with the arbuscular mycorrhizal fungus Glomus mosseae. *Global Change Biology 5*(3):347–358.
Staudinger M, Grimm NB, Staudt A et al. (2012) Impacts of climate change on biodiversity, ecosystems, and ecosystem services: technical input to the 2013 National Climate Assessment (pp i–A). United States Global Change Research Program.

Staudt A, Leidner AK, Howard J et al. (2013) The added complications of climate change: understanding and managing biodiversity and ecosystems. *Frontiers in Ecology and the Environment 11*:494–501.

Sulpanjan A, Yang MS, Tawaha ARM, Lee KD (2005) Effect of Magnesium application on yield, mineral contents and active components of *Chrysanthemum coronarium* L. under Hydroponics Conditions. *Bioscience Research 2*(2):73–79.

Sun Y, Solomon S, Dai AG et al. (2007) How often will it rain? *Journal of Climate 20*:4801–4818.

Suttle KB, Thomsen MA, Power ME (2007) Species interactions reverse grassland responses to changing climate. *Science 80*(315):640–642.

Tang JL, Zhang B, Gao C. et al. (2008). Hydrological pathway and source area of nutrient losses identified by a multi-scale monitoring in an agricultural catchment. *Catena 72*:374–385.

Tawaha AM, Turk MA (2004) Field pea seeding management for semi-arid Mediterranean conditions. *Journal of Agronomy and Crop Science 190*:86–92.

Tawaha AM, Singh VP, Turk MA et al. (2003) A review on growth, yield components and yield of barley as influenced by genotypes, herbicides and fertilizer application. *Research on Crop 4*(1):1–9.

Tawaha ARM, Turk MA, Lee KD (2005) Adaptation of chickpea to cultural practices in a Mediterranean type environment. *Research Journal of Agriculture and Biological Science 1*(2):152–157.

Thuiller W, Lavorel S, Araújo M et al. (2005) Climate change threats to plant diversity in Europe. *Proceedings of the National Academy of Sciences of the United States 102*:8245–8250.

Tilman D, Isbell F, Cowles JM (2014) Biodiversity and ecosystem functioning. *Annual Review of Ecology, Evolution, and Systematics 45*:471–493.

Tubiello F, Schmidhuber J, Howden M et al. (2008). *Climate change response strategies for agriculture: challenges and opportunities for the 21st century*. The World Bank, Washington, DC.

Turk MA, Tawaha AM (2001) Common vetch (*Vicia sativa* L.) productivity as influenced by rate and method of phosphate fertilization in a Mediterranean environment. *Agricultura Mediterranea 131*:108–111.

Turk MA, Tawaha AM (2002a) Onion (*Allium cepa* L.) as influenced by rate and method of phosphorus placement. *Crop Research 23*(1):105–107.

Turk MA, Tawaha AM (2002b) Impact of seeding rate, seeding date and method of phosphorous application in faba bean (*Vicia faba* L. minor) in the absence of moisture stress. *Biotechnology, Agronomy, Society and Environment 6*(3):171–178

Turk M, Tawaha AR (2004) Effect of variable sowing ratios and sowing rates of bitter vetch on the herbage yield of oat-bitter vetch mixed cropping. In Proceedings 7th International Oat Conference/Pirjo Peltonen-Sainio and Mari Topi-Hulmi (eds) MTT.

Turk MA, Tawaha AM, Shatnawi M (2003a) Lentil (*Lens culinaris* Medik) Response to plant density, sowing date, phosphorus fertilization and Ethephon application in the absence of moisture stress. *Journal of Agronomy and Crop Science 189*(1):1–6.

Turk MA, Tawaha AM, Samara N (2003b) Effects of seeding rate and date, and phosphorus application on growth and yield of narbon vetch (*Vicia narbonensis*). *Agronomie 23*:1–4.

Turk MA, Hameed KM, Aqeel AM, Tawaha AM (2003c) Nutritional status of durum wheat grown in soil supplemented with olive mill by-products. *Agrochimica 47*(5–6):209–219.

Turk MA, Assaf TA, Hameed KM, Tawaha AM (2006) Significance of mycorrhizae. *World Journal of Agricultural Sciences 2* (1):16–20.

UN Environment (2019) Air pollution: know your enemy https://www.unenvironment.org/news-and-stories/story/air-pollution-know-your-enemy, accessed September 2019.

UNFCCC (2013) Decision 2/CP.19, Warsaw International Mechanism for loss and damage associated with climate change impacts, UN Doc FCCC/CP/2013/10/Add.1.

USGCRP (2009). Global climate change impacts in the United States. In Karl TR, Melillo JM, Peterson TC (eds) *Climate change impacts by sectors: ecosystems*. United States Global Change Research Program. Cambridge University Press, New York, NY.

Van der Geest K, de Sherbinin A, Kienberger S et al. (2019) The impacts of climate change on ecosystem services and resulting losses and damages to people and society. Loss and damage from climate change. Concepts, methods and policy options. Springer.

Várallyay G (2010) The impact of climate change on soils and on their water management. *Agronomy Research 8*(2):385–396.

Volney WJA, Fleming RA (2000) Climate change and impacts of boreal forest insects. *Agriculture, Ecosystems and Environment 82*:283–294.

Vrtiška J, Křeček J, Tognetti R (2018). Indication of environmental changes in mountain catchments by dendroclimatology. *Soil and Water Research 13*(4):208–217.
WWAP (United Nations World Water Assessment Programme). (2015) *The United Nations world water development report 2015: water for a sustainable world.* UNESCO, Paris.
Yang MS, Tawaha AM, Lee YD (2005) Effects of ammonium concentration on the yield, mineral content and active terpene components of Chrysanthemum coronarium L. in a hydroponic system. *Research Journal of Agriculture and Biological Science 1*(2):170–175.
Zommers Z, Wrathall D, van der Geest K (2014) Loss and damage to ecosystem services UNUEHS Working Paper Series, No. 12. Bonn, United Nations Institute for Environment and Human Security.

11

Plant Diversity of the Cholistan Desert in Pakistan: Anthropogenic Factors and Conservation

Hafiz Muhammad Wariss
COMSATS University Islamabad, Pakistan

Muhammad Asad Salim, Saeed Ahmad, Khurshid Alam, Muhammad Abbas Qazi, Shazia Anjum, and Muhammad Akram
Chinese Academy of Sciences, China

CONTENTS

11.1 Introduction

Deserts usually regarded as bare and rather uniform areas that exhibit relatively low diversity in comparison to other biomes (Middleton et al. 2011). They are distributed along the northern and southern hemispheres (Ezcurra 2006). Deserts and dry arid land regions are astonishingly rich, represent about 41% of the Earth's land surface, inhabited by 2 billion people and embracing about 25% of continental vertebrate species and comprising highly adapted and specialized species that are endemic (Brito and

Pleguezuelos 2019). They show high rates of endemism owing to the adaptive processes of organisms to the extreme environmental conditions, as well as locally endangered hotspots of biodiversity (Murphy et al. 2012; Brito et al. 2014; Vale and Brito 2015; Rossini et al. 2018). The extreme climatic conditions experienced by these areas create sharp landscape and ecological gradients that may be connected with the range limits of many species (Brito et al. 2014). These characteristics make deserts and arid regions excellent natural laboratories to study the effects of extreme environments on the distribution of biodiversity patterns (Ward 2016). Biodiversity is locally endangered through increased human exploitation activities and climate change (Santarém et al. 2019). Fortunately, these threats are now raising international awareness for desert biodiversity conservation (Ward 2016; Santarém et al. 2019; Brito and Pleguezuelos 2019).

11.2 Cholistan Desert

The Cholistan desert, which covers an area of 26,000 km^2, lies within South of Bahawalpur in the Punjab extending through the Nara and Thar deserts of Sindh between 27°42' and 29°45' North latitude and 69°52' and 75°24' East longitude (Figure 11.1; Akhter and Arshad 2006; Wariss et al. 2013). The word 'Cholistan' is derived from a Turkish word, 'Chol', meaning a desert, while some historians believe that this name has been distorted from an Iraqi (Kurdish) word, 'Chilistan', meaning waterless wasteland (Ahmad 1999; Ahmad et al. 2005; Auj and Auj 1991). Traditionally, Cholistan, known as 'Rohi' in Saraiki, a local dialect, is steeped in ancient history which resonates with the folklore, poetry, handicrafts, dances and myths which form the narrative of the people of the Rohi. In a dialect that still spoken in some parts of Tibet, 'roh' means a hill, from which the name Rohilla has been attributed. In fact, Rohi has been derived from the Pushto word 'roh', meaning sandy desert (Ahmad et al. 2005).

Historically, the Cholistan desert is a cradle of the Hakra River Civilization which flowed through the area during 1200 BC regularly and become irregular about 600 BC. Its civilization can be compared with the Mesopotamian, Egyptian and Harappan civilizations. This great civilization vanished due to a

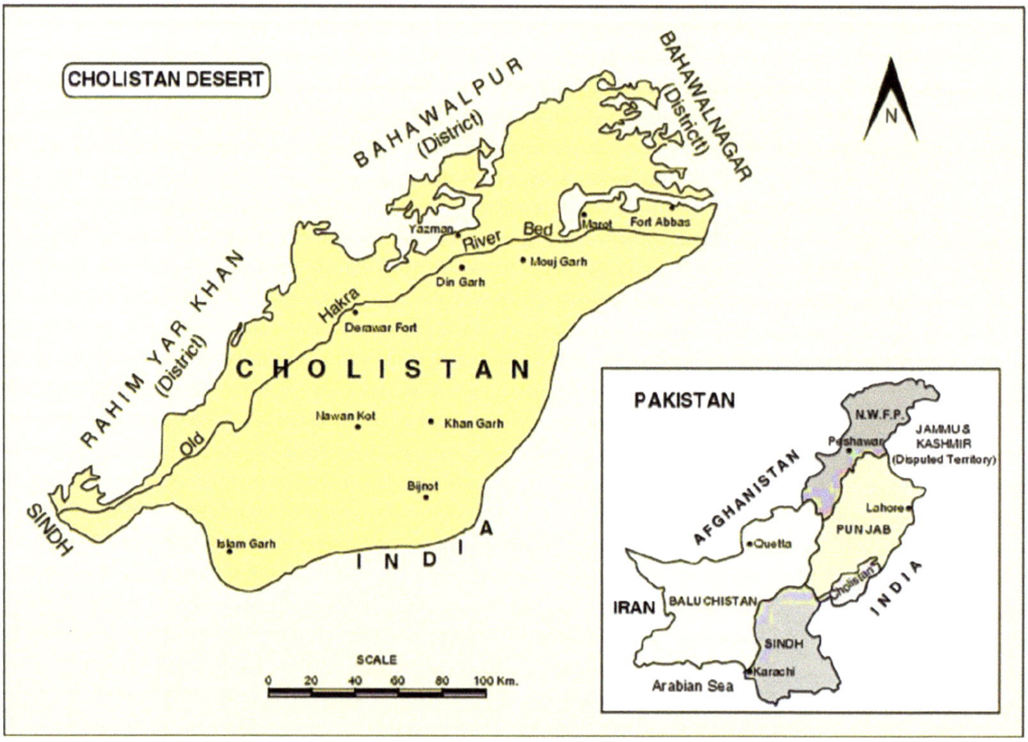

FIGURE 11.1 Geographical location of the Cholistan Desert.

succession of hostile invasions (Ahmad 1999). The origin of the desert is not old. During the assault of Alexander the Great (300 BC), the region north of the Rann of Kutch and east of the Indus in the south of the Punjab was fertile, lush-green and well-populated. The old river flows through the area, now visible only in the form of their dried--up channels (Wariss et al. 2013). The Cholistan desert received heavy monsoon downpours along with the Indus valley civilization including Mohenjo Daro and Harappa of world's oldest civilization about 5000 years ago. The Cholistan desert was created during the Pleistocene and recent periods by the thick mantle deposition of sands (Mughal 1997). A gradual change in monsoon winds, in combination with other causes, increases the aridity and has ultimately converted the area into a desert (Leopold 1963). In the Cholistan desert, the remnants of many forts can be found, displaying the cultural and historical traits of the area. Bijnot Fort, which was built around 757 CE, is considered as the oldest amongst a series of forts from the medieval period of the Rohi. These forts are reportedly build on older foundations or remains, yet the latest construction dates back to the 1770s. The construction of forts and different other structures was carried out during the Nawabs regime, which began in 1805 and continued until 1954. Derawar Fort, comprising of necessary structures, such as a palace, a mosque, markets and housing continued to be used by the Nawabs of Bahawalpur until the 1970s and was constructed on the right bank of Hakra River. Its strategic location and infrastructure ensured that Derawar Fort remained in use until the very end of the Nawabs' regime and therefore in a much better condition in comparison to the other forts (Akhter and Arshad 2006; Hameed et al. 2011; Mughal 1997; Akbar et al. 1996).

11.2.1 Climate

The climate of the Cholistan desert is sub-tropical, arid and semi-arid, scorchingly harsh, with monsoon rainfall influenced by periodic long droughts. The relative humidity in the area is very low with a high rate of evaporation (Arshad et al. 2008). The mean annual rainfall varies between 100 mm in the west and 250 mm in the east, with high rainfall during July to September in the monsoon season and January to March during winter. The annual temperature of the area is high in summer and low in winter, with no frost. The mean summer temperature is 34–38°C, and the winter temperature is 15–20°C, with the highest temperatures reaching over 51.6°C (Arshad et al. 2006). June is the hottest month, with temperature normally exceeding from 45°C and sometimes even topping 50°C (Arshad et al. 2007).

11.2.2 Topography

Topographically, the area can be divided into two geomorphic regions based on parent material, soil and vegetation. The old Hakra River is the dividing line between the two eco-regions of the desert. The northern region, which constitutes the desert margins adjoining with canal irrigated areas, consists of a series of saline alluvial flats alternating with low sand ridges or dunes, covering about 7,770 Km^2, and is known as Lesser Cholistan. The wind-swept sandy desert comprising of a number of old Hakra River terraces with various forms of sand ridges and inter-ridges valleys, covering about 18,130 km^2 in the southern region known as Greater Cholistan. It extends about 480 km in length and 32 to 192 km in width (Akbar 2000; Akbar et al. 1996; Chaudhry 1992.

11.2.3 Soil

The soil of the Cholistan desert is very poor in having organic matters. The Lesser Cholistan consists of saline alluvial, compact hard plains (locally called 'Dahars'), alternating with low sandy ridges. Sand dunes are stabilized, semi-stabilized or less frequently shifting, while the interdunal areas are covered with sand. The Greater Cholistan consists of wind-shifting sandy dunes and ridges, with large interdunal flat areas. The sand dunes of longitudinal and transverse type varying in size, with less than 100 m in Lesser Cholistan and about 100 m in Greater Cholistan. The soils are of four types: sand dunes, sandy soils, loamy soils and saline sodic clayey soils. The sand dune consists of undulating and steep slopes shifted by strong winds, very excessively drained and coarse textured. Sandy soils are nearly level to smoothly sloping, deep to very deep, excessively drained, calcareous, coarse textured. Loamy

TABLE 11.1
Types of soils and wind erosion in the Cholistan desert

Soil Types	Extent (Hec.)	Percentage	Wind erosion	Extent (Hec.)	Percentage
Saline sodic clayey soils (Dhars)	441,900	17.0	Non or slight	441,900	17.0
Loamy soils	58,700	2.0	Moderate	58,700	2.0
Sand dunes	1,133,900	44.0	Severe	2,079,400	81.0
Sandy soils	945,500	37.0			
Total	2,580,000	100.0	Total	2,580,000	100.0

soils are leveled to nearly level with hummocks of fine sand on the surface, moderately deep, somewhat excessively drained to well-drained, calcareous, moderately coarse textured to medium textured (Wariss et al. 2013). Clayey soils are mostly leveled, moderately deep, poorly drained, calcareous, saline-sodic (Table 11.1) and moderately fine textured to fine textured. The pH ranges between 8.6 and 10.0 saline and saline-sodic respectively (Baig et al. 1980).

11.2.4 Water Resources

In the Cholistan desert, there are two sources of water: rainfall and sub-soil water. Rainwater is collected into "Tobas" man-made ponds or natural depression (Figure 11.2. Tobas are made in clayey flats locally called "Dahars" which act as a good source of rainwater storage area to avoid the loss of runoff and water percolation. There are no permanent natural bodies of surface water. Low rainfall, high rates of evaporation, and high rates of water percolation into the sand, leads the area into aridity. The second source of water is underground water, which is to be found at a depth of 30 to 90 m. It is brackish and unfit for drinking and agriculture because of containing total dissolved salts about 9000–27,000 ppm mg/L (Baig et al. 1980). In Cholistan, two aquifers of sweet water are present surrounded by saline water. The first aquifer extends from Fort Abbas towards Mouj Garh, ranging 80 km long and 10–15 km wide. It has an estimation volume of 10,000 million liters which lies at a depth of 40–100 m below the surface. The second aquifer occupies an area of 50 km^2 towards northwest of Derawar Fort and is at a depth of 25 m below the surface (Wariss et al. 2013).

11.2.5 Cultural Heritage

The people of Cholistan desert comprises about 110,000 nomadic pastoralists, who are battling for survival in a harsh and hostile environment. The majority of the people live on periphery of the desert; by contrast, the interior of the desert is thinly populated (Akhter and Arshad 2006; Arafat et al. 2016). The local communities live in small families/tribes in traditional settlements. 'Gopa' and 'Sal' are the two traditional house forms. Gopa is a circular mud house with a thatched roof and the Sal is a rectangular thatched room. Gopas and Sal are found mixed in the peripheral desert villages, whereas in Greater Cholistan only Gopas are found. They are constructed from mud, twigs and local plant species, particularly *Crotalaria burhia, Leptadenia pyrotechnica, Aerva javanica, Tamarix aphylla* and *Prosopis cineraria*, etc.

11.2.6 Pastoralism

The economy of the desert is mainly based around a pastoral and nomadic lifestyle, showing little change over centuries. They have smaller to large herds of camel, cows, sheep and goats. The monsoon onset and rainfall distribution mainly dictates the pattern of movement of nomadic herders. During the months of March or April, nomadic herders with their animals move towards nearby irrigated areas due to the shortage of water and forage in the desert. In the onset of monsoon during July or August, they return

(a)

(b)

FIGURE 11.2 Water resources of Cholistan desert: (a) Man-made Tobas (Ponds for rainwater harvesting) (b) Natural depression (Tobas) used for water collection and this water used for animals and human beings drinking and house-hold usage.

to the desert with their herds. Natural vegetation and Tobas water are the source of fodder and drinking, respectively. In October and November, due to depleted feed and water resources, the nomads move to semi-permanent settlements, where primitive unlined wells and kunds (usually lined) are available. At the ending of these resources, the nomads move to peripheral areas of the desert (Akbar et al. 1996; Akbar 2000; Dasti and Agnew 1994; Ahmad 1999; Ahmad 2007; Ali et al. 2016).

11.2.7 Customs and Crafts

The people of the Cholistan desert have a rich and unique culture. The lifestyle, music, dance, storytelling traditions, wonderful crafts of Ralli (sheep wool carpet) and Falasi (camel wool carpet) desert community culture expressions. The dress and wearing of desert women are very striking and express their culture as embroidered calico ghaggras, cholis (short blouses) and chaddars (veils), Katmala (necklace), kangan (bangles) and pazeb (anklets). The women also make most exquisite gindi/rallis (bed/floor spreads), khalto (multicoloured embroidered purse) during their spare time. Khawaja Ghulam Fareed, the Sufi poet, spent some 18 years in the Cholistan desert, and his poetry is imbued with the spirit of the desert. The Urs of Khawaja Ghulam Fareed and Mela of Channan Peer are the annual festivals of the desert (Mughal 1997).

11.3 Plants Diversity

Deserts are an important component of terrestrial ecosystems. Nevertheless, compared to forests and grassland ecosystems, relatively little is known about potential ecological influences on desert plants diversity (Bertiller et al. 2009; Bisigato et al. 2009; Miriti et al. 2007; Wang et al. 2013; Wu and Yang 2013). Current biodiversity loss in deserts is alarming as compared to forests and grassland ecosystem. The exploration of changes in species distributions and diversity are crucial for protecting biodiversity in deserts (Báez and Collins 2008; Berry et al. 2006; Butterfield et al. 2010; Munson et al. 2012; Wariss et al. 2014b). As global climate warming continues, desert plant communities may become less stable as interspecific interactions lead to declines in biodiversity (Báez and Collins 2008; Wassenaar et al. 2007). Interactions among species composition, community structure and their controlling factors within ecosystems are the product of ecological processes operating over a wide range of spatial and temporal scales. The distribution and diversity of plants species within desert communities have most often been related to climatic, geographic and edaphic heterogeneity factors (Enright et al. 2005; Guisan and Thuiller 2005).

The Cholistan desert encompasses sets of interlinked habitats in a relatively large area. These habitats range from sand dunes, sandy soils, loamy soils and saline sodic clayey soils, saline and non-saline depressions, inland ridges and manmade rain-fed depression (Toba) (Akhter and Arshad 2006). These habitats support diverse floras and faunas, including 156 species of flowering plants, 10 amphibians and reptiles, 19 birds, 44 herpetofauna and 14 mammals, some of which are endemic and threatened (Arshad et al. 2008; Sial et al. 2012; Wariss et al. 2013; Wariss et al. 2016; Baig et al. 2008; Dasti and Agnew 1994; Arshad and Rao 1994; Arshad and Akbar 2002; Sial and Arshad 2003; Sial and Chaudhry 2012). Floristically, Cholistan desert belongs to 146 plant species of 98 genera and 35 families distributed into 31 families of dicotyledons and 4 families of monocotyledons (for more details, (Wariss et al. 2013).

The vegetation of Cholistan desert comprises of xerophytic species adapted to a wide range of severe temperatures, moisture and edaphic conditions. The distribution pattern of vegetation depends on the topography and the chemical composition of the soil in the area. The vegetation is dominated by perennial shrubs and scattered small trees. The Greater Cholistan is dominated by shrubs which are very sparse and there is a very low of species diversity. The vegetation cover is at a low level in the Greater Cholistan on sand dunes and the unstable sand dunes lack any vegetation. In the Lesser Cholistan, vegetation is dominated by several species of shrubs, perennial grasses, scattered trees with annuals and ephemerals.

Generally, true xerophytes, including *Haloxylon salicornicum, Haloxylon stocksii, Suaeda fruticosa*, etc., which are leafless or possess reduced and modified leaf- like structures, form the dominant species in the hardy saline plains and Dahars. The Dahars with sandy cover are dominated by *Aeluropus lagopoides, Sporobolus ioclados, Prosopis cineraria, Capparis decidua, Ochthochloa compressa*, and

Crotalaria burhia etc. The sand dunes and interdunal sandy areas dominated with *Calligonum polygonoides, Dipterygium glaucum, Panicum turgidum, Lasiurus scindicus*, and with ephemerals as *Mollugo cerviana, Limeum indicum, Gisekia pharnaceoides, Aristida funiculata, Stipagrostis plumosa* etc. The moist habitat-dominated plant species are *Cyperus rotundus* (Rao et al. 1989; Arshad and Akbar 2002; Arshad et al. 2008; Wariss et al. 2015). Following these earlier studies, and our exploration and documentation of the flora of the Cholistan desert (Wariss et al. 2013), the following major habitat with indicator species are recognized:

11.3.1 Recognized Habitat of the Area and Vegetation

The Cholistan desert area can be divided into four habitats on the basis of topography, edaphic conditions and soil pattern. The distribution of plant species/communities established in the study area is as follows: Sand dunes, Sandy plains (Dahar), Compact hard plains with gravels, Saline/Sodic areas (Saline Dahar)

11.3.2 Sand Dunes Vegetation

Sand dunes play a very important role in the distribution of the desert vegetation. They vary in height range from a few meters to a hundred meters. The vegetation cover on sand dunes is very poor. The vegetation composition of the plant community on the sand dunes is as follows: *Aerva javanica, Aristida adscensionis, Aristida funiculata, Calligonum polygonoides, Cenchrus ciliaris, Cyperus conglomeratus, Dipterygium glaucum, Gisekia pharnaceoides, Haloxylon salicornicum, Lasiurus scindicus, Leptadenia pyrotechnica, Limeum indicum, Mollugo cerviana, Panicum turgidum* and *Stipagrostis plumosa*, etc.

11.3.3 Sandy Plains (Dahars)

The sandy plains, which are found alongside the sand dunes, have almost the same plant species as sand dunes vegetation, albeit with a few changes. The major plant species to be found there are the following: *Aerva javanica, Aristida adscensionis, Aristida funiculata, Calligonum polygonoides, Capparis decidua, Cenchrus ciliaris, Cenchrus prieurii, Citrullus colocynthis, Crotalaria burhia, Dactyloctenium aegyptium, Dipterygium glaucum, Gisekia pharnaceoides, Haloxylon salicornicum, Heliotropium crispum, Indigofera argentea, Indigofera sessiliflora, Leptadenia pyrotechnica, Limeum indicum, Mukia maderaspatana, Prosopis cineraria, Stipagrostis plumosa* and *Tribulus longipetalus*, etc.

11.3.4 Compact Hard Plains with Gravels

The soil of the compact hard plain is very hard, gravel and with or without sand cover. The vegetation of the compact hard plains or 'Dahars' consists of following major species: *Aerva javanica, Aristida hystricula, Calotropis procera, Capparis decidua, Cenchrus ciliaris, Cleome brachycarpa, Cleome scaposa, Dactyloctenium aegyptium, Haloxylon stocksii, Heliotropium strigosum* subsp. *strigosum, Leptadenia pyrotechnica, Mollugo nudicaulis, Ochthochloa compressa, Panicum antidotale, Polygonum plebejum, Prosopis cineraria, Saccharum bengalense, Salsola imbricata, Seetzenia lanata, Tribulus longipetalus, Tribulus terrestris* and *Zaleya pentendra*, etc.

11.3.5 Saline/Sodic Areas (Saline daharas)

The saline or saline sodic habitat of the Cholistan desert supports the halophytic species. This habitat found in interdunal flat areas of the desert and contains the following plants: *Aeluropus lagopoides, Calotropis procera, Capparis decidua, Cleome brachycarpa, Cleome scaposa, Cressa cretica, Cymbopogon jwarancusa, Eragrostis japonica, Glinus lotoides, Haloxylon salicornicum, Haloxylon stocksii, Mollugo nudicaulis, Ochthochloa compressa, Prosopis cineraria, Pulicaria crispa, Saccharum bengalense, Salsola imbricata, Sporobolus ioclados, Suaeda fruticosa, Tamarix aphylla, Tamarix dioica, Trianthema triquetra, Zaleya pentendra* and *Zygophyllum simplex*, etc.

In the Cholistan desert aquatic vegetation is very rare due to the absence of permanent water bodies. There is only rainfall is the main source of water, which is collected in natural depressions or manmade Tobas, which have been specifically created for the collection of water. The most common aquatic plant species were *Cyperus rotundus, Typha domingensis, Tamarix dioica* and *Saccharum bengalense* along the water bodies.

11.4 Ethnobotanical Resources of Cholistan Desert

Human societies rely mainly on the biodiversity of the surrounding ecosystem, especially plant resources, in order to sustain their livelihoods (Carpenter et al. 2006; Díaz et al. 2006; Kaimowitz and Sheil 2007; Saxena et al. 2003). Traditional ecological knowledge (TEK) of plant resources and their utilization has been globally recognized (Gemedo-Dalle et al. 2005; Leduc et al. 2006; Hussain et al. 2018; Harun et al. 2017; Malik et al. 2015; Bano et al. 2014). Ethnobotany is the link between plant species and the communities exploring these resources and adjusting their livelihoods according to the uses for these plants (Malik et al. 2015; Abbasi et al. 2013; Ahmad et al. 2014). Ethnobotanical studies focus on the connections between botanical diversity and socio-cultural systems (Salim et al. 2019; Wariss et al. 2014a; Akhtar et al. 2013). The global research community has recently shifted their emphasis towards the use of different analytical techniques for understanding the TEK related to botanical resources of different indigenous communities and their localities (Amjad et al. 2015) (Table 11.2).

Documentation of this knowledge is vital for exploration of new drugs, establishment of value chains contributing to regional, national and international markets and sustainable use and conservation of these natural resources (Gemedo-Dalle et al. 2005; Salim et al. 2019) (Balick and Cox 1997; Hameed et al. 2011; Qureshi et al. 2010; Farooq et al. 2008).

Medicinal plants from the Cholistan desert are a source of income, and collecting these plants is among the major economic activities in the local community (Table 11.1). These medicinal valued plants are used to treat many diseases in both local communities and those in urban areas in the surrounding of desert. Studies on indigenous medicinal uses indicate the reliance of traditional medication due to ease and economic availability of medicinal resources. We found that folk medicine is still the most important procedure for medication and therapy (Hameed et al. 2011; Malik et al. 2015; Ahmad et al. 2014).

Mostly woody perennial species, including *Acacia nilotica, Acacia jacquemontii, Prosopis cineraria, Prosopis juliflora, Tamarix aphylla and Zizyphus nummularia* harvested by local communities for use as fuel wood. Some species are preferred over others and commonly harvested, which places these preferences under pressure. The livestock depends entirely on natural vegetation supplemented with fodder, including major fodder trees, shrubs, herbs and grasses, such as *Acacia jacquemontii, Acacia nilotica, Prosopis cineraria, Prosopis juliflora, Zizyphus nummularia, Zizyphus spina-christi, Tamarix aphylla, Salvadora oleoides, Aeluropus lagopoides, Stipagrostis plumosa, Cenchrus ciliaris, Ochthochloa compressa, Lasiurus scindicus, Panicum antidotale, Panicum turgidum, Sporobolus iocladus, Cyperus conglomeratus, Eragrostis barrelieri, Tribulus longipetalus* and *Tribulus terrestris*, etc.

Moreover, local communities exploit many plant species for other economic uses, e.g. food, fodder, house construction and various other uses. Seeds and fruits of *Prosopis cineraria* are edible and extensively used in a number of local dishes (Arshad et al., 2006). A herbal aqueous extract of *Cymbopogon jwarancusa* is commonly used for relaxing and reducing thirst during summer (Arshad et al. 2002; Malik et al. 2015). Along with these *Acacia jacquemontii, Acacia nilotica, Calligonum polygonoides, Capparis decidua, Capparis spinosa, Caralluma edulis, Cencherus ciliaris, Cenchrus biflorus, Chenopodium album, Cucumis melo* var. *agrestis, Panicum antidotale, Panicum turgidum, Pennisetum divisum, Portulaca oleracea, Prosopis cineraria, Salvadora oleoides* having nutritional value and used in different ways, such as pickles, bread making, vegetables and fruits (Malik et al. 2015; Wariss et al. 2014a; Ahmad et al. 2014).

Many plants species have been used for domestic purposes and construction of "Gopa", "Sal" (Local traditional houses), baskets, for example, *Calotropis procera* fruit floss is used for stuffing in pillows and cushions. *Abutilon muticum, Acacia nilotica, Aerva javanica, Calligonum polygonoides, Calotropis procera, Capparis decidua, Crotalaria burhia, Haloxylon stocksii, Haloxylon salicornicum,*

TABLE 11.2

Ethnobotanical uses of plants of Cholistan desert

Botanical name	Common name	Family	Medicinal uses	Other uses
Abutilon muticum (Del.ex DC.) Sweet	Kanghi-buti	Malvaceae	Anti-inflammatory, indigestion and kidney stones	Firewood and insect repellent
Acacia jacquemontii Benth.	Banwli	Mimosaceae	Pyrexia and acute viral diseases	Firewood, forage and house construction
Acacia nilotica (L.) Delile	Babul or Kikar	Mimosaceae	Hypertension, aphrodisiac, diabetes, bleeding disorder, pyrexia and tonic.	Timber, firewood and fodder
Aerva javanica var. javanica	Bui	Amaranthaceae	Renal lithiasis, respiratory infectious diseases, poor digestion, tapeworms and constipation	Firewood, clay huts and fencing, cattleyards
Alhagi maurorum Medic	Jawansa	Papilionaceae	Intermittent fever, antiseptic, respiratory infection, diaphoretic, dryness of alimentary canal and purgative	Forage and making tatties in summer (outer curtain) dyeing purpose
Arnebia hispidissima (Lehm.) A. DC.	Surkhi booti	Boraginaceae	Dementia, muscular spasm and menstrual disorder	
Blepharis scindica T. Anders.	Gandi-boti	Acanthaceae	Reliving pain and inflammation.	Little grazing sheep's and goats
Boerhavia procumbens Banks ex Roxb	Biskhipra	Nyctaginaceae	Kidney failure, dysmenorrhea, hepatitis and diabetes.	Fodder
Calligonum polygonoides L.	Phog	Polygonaceae	Conjunctivitis, throat infection. indigestion, sore throat and pain.	Food, forage, firewood and fencing of cattleyards
Calotropis procera subsp. hamiltonii (Wight) Ali	Ak	Asclepiadaceae	Dysmenorrhea, endometriosis, prolapse of uterus, asthma, gastric pain, poor digestion, snake bite, hemorrhoids, leprosy, venereal disease and joint pain.	Firewood
Capparis decidua (Forsskal.) Edgew	Karir	Capparidaceae	Pyrexia, hemorrhoids, menstrual disorders, hypertension, glycosuria, obesity, dyspepsia and bone fractures aphrodisiac, diaphoretic and purgative.	Food, fire wood, fencing cattle yards and pickle making
Capparis spinosa L.	Kubber	Capparidaceae	Arthralgia, myalgia, asthma, digestive problems, and hepatic disorders, tonic for premature ejaculation.	Food and fodder
Caralluma edulis (L.) Benth ex Hook. f.	Seetoo	Asclepiadaceae		Used as vegetable
Cassia italica subsp. italica	Ghoray-wall or Sana	Caesalpinaceae	Rheumatism, pain, pyrexia and dyspepsia	Fodder
Cenchrus biflorus Roxb.	Mohabat Boti	Poaceae	Anti-parasitic	Food and fodder
Cenchrus ciliaris L.	Daman	Poaceae	Anti-parasitic, milk promoter and anti-inflammatory.	Fodder
Cenchrus setigerus Vahl.	Chuti Daman	Poaceae	Allergic rhinitis, hay fever and purgative.	Fodder

(*Continued*)

TABLE 11.2 (*Continued*)

Ethnobotanical uses of plants of Cholistan desert

Botanical name	Common name	Family	Medicinal uses	Other uses
Chrozophora sabulosa Kar. & Kir.	Nilakari	Euphorbiaceae	Allergic rhinitis, skin diseases, throat infection	
Citrullus colocynthis (L.) Schrad	Tummaor Korr- tumma	Cucurbitaceae	Hypertensive, menstrual disorder, neurasthenia, hyperglycemia, and gastric pain.	Roots as teeth cleaning
Cleome brachycarpa Vahl. ex. DC.	Noli or kastoori	Capparidaceae	Anti-parasitic, diabetes and hepatic disorder.	
Cleome scaposa DC	Khastoori boti	Capparidaceae	Scabies, increase sugar level, itchiness, high blood sugar, used as diuresis, biliary dysfunction antipyretic and pain.	
Convolvulus prostratus Forssk	Hiran-booti	Convolvulaceae	Anorexia, flatulence, diuresis, diabetes, pyrexia excessive thirst, heartburning, as diuretic, anti- and severe constipation.	Fodder
Corchorus depressus (L.) Stocks	Bhaon- phali	Tiliaceae	Diuresis, diabetes, premature ejaculation and aphrodisiac.	Sheep and goats grazing
Cressa cretica L.	Ooini	Convolvulaceae	Dyspnoea, cough, skin diseases and pyrexia.	
Crotalaria burhia Buch.-Ham. ex Benth	Chag	Papilionaceae	Antipyretic, arthralgia, leukoderma, gastric disorder and myalgia.	Fencing yards of sheep and goats, Gopa and Sal (traditional house making)
Cucumis melo var. agrestis Naudin	Chibbarr	Cucurbitaceae	Gastric problems and constipation.	Food and fodder
Cuscuta reflexa Roxb.	Akashbail	Cuscutaceae	Hair loss, hair growth, alopecia, Jaundice, lumbago, orchitis, headache, pyrexia, gastrointestinal disorder and joint, muscular pain.	
Cymbopogon jwarancusa (Jones) Schult.	Katrin or Khavi	Poaceae	Indigestion, pyrexia, diuresis, joint inflammation.	Fodder
Cyperus rotundus L.	Moothaor	Cyperaceae	Poor digestion, heart burning, diuresis, hepatitis, skin diseases, hyperglycemia and anorexia.	Fodder
Dipterygium glaucum Decne.	Phel	Capparidaceae	Diseases of skin, ulcer, eczema and antipyretic.	Fodder
Echinops echinatus Roxb.	Unt-katara	Asteraceae	Aphrodisiac, hepatic disorder, gastrointestinal derangement and skin diseases.	
Euphorbia granulata Forssk.	Dudheli	Euphorbiaceae	Psoriasis, eczema, hypertensive, hyperglycemia, gastrointestinal diseases.	Fodder
Euphorbia prostrata Ait.	Hazar-dani	Euphorbiaceae	Skin allergy, urticaria, piles, aphrodisiac.	Fodder
Fagonia cretica L.	Dhmasa	Zygophyllaceae	Hepatic diseases, anemia, antipyretic, and as blood purifier and enhancing immunity.	
Farsetia hamiltonii Royle	Lathia or Farid-booti	Brassicaceae	Gastric disorder, arthralgia, myalgia hyperglycemia.	Fodder

(*Continued*)

TABLE 11.2 (*Continued*)

Ethnobotanical uses of plants of Cholistan desert

Botanical name	Common name	Family	Medicinal uses	Other uses
Gisekia pharnaceoides L.	Buloka-sag	Aizoaceae	Hyperbilirubinemia, indigestion, pyrexia and pain.	Fodder
Glinus lotoides L.	Gandi-booti	Molluginaceae	Intestinal irritation, indigestive, hepatic disorder, ulcer and wound.	
Haloxylon recurvum Bunge. ex. Boiss.	Khar or Sajji	Chenopodiaceae	Gastritis and renal calculi. Ash of the plant is used for heartburn	Making washing soda and firewood
Haloxylon salicornicum (Moq.) Bunge	Lana	Chenopodiaceae	Gingival bleeding, dyspepsia and insect bites and Hepatoprotective.	Firewood, making washing soda and areal parts as fodder
Heliotropium crispum Desf.	Kali-lani	Boraginaceae	General weakness, gastritis and lethargy.	
Heliotropium strigosum subsp. strigosum	Gorakh-Pan	Boraginaceae	Aching eyes, wounds healing, sore throat, tenderness and ulceration nipples of breasts, jaundice, used as blood purifier and to cure cough and cold.	
Indigofera argentea Burm. f.	Neel	Papilionaceae	Intestinal worms and patchy skin.	Fodder
Launaea residifolia Less	Dudhkal	Asteraceae	Fever, bowel syndrome and gastric problems.	Fodder
Leptadenia pyrotechnica (Forsskal.) Decne	Khip	Asclepiadaceae	Abdominal pain, constipation, dysmenorrhea, obesity and hypertension.	Firewood and house preparation
Mollugo cerviana (L.) Seringe	Padi or Sarr	Molluginaceae	Pyrexia, dysuria and venereal diseases. Pruritus and as blood purifier.	
Mollugo nudicaulis Lamk.	Gandi-buti	Molluginaceae	Respiratory tract infection, whooping cough; leaves are applied as poultice on carbuncles.	
Mukia maderaspatana (L.) M.J. Roem	Gawala-kakri	Cucurbitaceae	Pyrexia, muscular fatigue, Jaundice, and lower backache.	
Neurada procumbens L.	Chhapri	Neuradaceae	Erectile dysfunction, premature ejaculation and general weakness and nervine tonic	Food
Oligochaeta ramosa (Roxb.) Magenitz	Birham dandi	Asteraceae	Hepatoprotective, anti-inflammatory, anti-pyretic, arthritis and memory enhancer,	
Oxystelma esculentum (Linn. f.) R. Brown	Dudhani	Asclepiadaceae	Dysuria, venereal diseases, pimples and itching.	
Panicum antidotale Retz	Murrot or Bansi ghaa	Poaceae	Throat infection, pyrexia, small pox and respiratory tract infection	Fodder
Peganum harmala L.	Harmal	Zygophyllaceae	Behavioral disorders, dysmenorrhea, convulsion, insanity and itchy skin. Abdominal cramps and smoke has insect-repellent properties.	Smoke is an insect repellent
Pergularia daemia (Jacq.) N. E. Brown.	Karial	Asclepiadaceae	Intestinal worms (or intestinal infestation), antipyretic, flatulence, chest congestion, stomachache and gynecological disorders	
Polygonum plebejum R. Br	Chiri-Hatta	Polygonaceae	Chest infections, indigestion, vomiting and diarrhea	

(*Continued*)

TABLE 11.2 (*Continued*)
Ethnobotanical uses of plants of Cholistan desert

Botanical name	Common name	Family	Medicinal uses	Other uses
Prosopis cineraria (L.) Druce	Jandi or Jand or kunda	Mimosaceae	Wounds, healing, contraceptive, protein deficiency anemia, lack of protein, dysmenorrhea and arthritis, muscle pain.	Timber, food and fodder.
Pulicaria crispa (Cass.) Benth. & Hook. f.	Bui	Compositae	Pyrexia, headache, severe cold, cough and jaundice.	
Salsola baryosma (Roem. ex. Scult.) Dany.	Lani	Chenopodiaceae	Itching, dyspepsia, and sores.	Washing soda and firewood
Salvadora oleoides Decne.	Pilu	Salvadoraceae	Lack of nutrients, loss of appetite, skin ulceration, hypoglycemia, gingival bleeding and stomach ache.	Food, fodder, basket making and teeth cleanings.
Solanum surattense Burm. F.	Kanderi	Solanaceae	Arthritis, pyrexia, detoxifies, dyspnea, severe headache, leprosy, as diuretic, hair tonic and cure abdomen pain, bowel gases, chronic cough and pain.	
Sporobolus ioclados (Nees ex Trin.) Nees		Poaceae	Antipyretic, headache, nausea and vomiting.	Fodder
Suaeda fruticosa Forssk. ex J. F. Gmelin	Kali lani	Chenopodiaceae	Constipation, dysmenorrhea, conjunctivitis, gastritis and wound healing.	Washing soda preparation and fodder.
Tamarix aphylla (L.) Karst	Jhao and Ukan or Frash	Tamaracaceae	Hepatic diseases, indigestion, stomach ache, leucorrhoea, sexual dysfunction and dermatitis	Timber, fodder
Tribulus longipetalus subsp. longipetalus	Tirkandi or Bakharra	Zygophyllaceae	Renal calculi, erectile dysfunction, anemia and debility.	
Tribulus longipetalus subsp. Macropterus	Tirkindi or Bakharra	Zygophyllaceae	Erectile dysfunction, azoospermia, psoriasis, chest pain, piles, epitaxia and pain.	
Withania coagulens (Stocks) Dunal	Paneer	Solanaceae	Jaundice, loss appetite and skin problems.	Insect repellent
Withania somnifera (L.) Dunal	Asgandh	Solanaceae	Carbuncles, nerve stimulant, joint pain and as sexual tonic.	
Ziziphus nummularia (Burm. f.) Wight & Arn	Beri	Rhamnaceae	Skin diseases, cold, cough stomachache, diarrhea, hair roughness and hypoglycemia, help wound healing.	Timber, fodder and fruit.
Zygophyllum simplex L.	Lunak	Zygophyllaceae	Patchy skin, wounds, acne and bleeding.	

Leptadenia pyrotechnica, Prosopis cineraria, Prosopis juliflora, Salvadora oleoides, Tamarix aphylla, Zizyphus spina-christi, used in different ways in common house hold uses and as well as industrial uses as *Haloxylon salicornicum* have contained chemical content beneficial for soap industry (Ahmad et al. 2014; Malik et al. 2015; Alam et al. 2016; Wariss et al. 2015; Hameed et al. 2011).

11.5 Anthropogenic Factors and Conservation

The Anthropocene, the current era, is driven by human influence and it has steered in a growing number of direct and indirect challenges that can have substantial impact on the health and prosperity of people and the planet (Ellis 2015). The signatories of the Convention on Biological Diversity (CBD) (1992) have committed to: (a) respect, preserve, and maintain traditional knowledge relevant to conservation and sustainable use of biological diversity; (b) promote wide application of traditional knowledge; and (c) encourage equitable sharing of benefits arising from the use of traditional knowledge (Cámara-Leret et al. 2014). Modernization has enabled communities to drastically change their lifestyles, which has impacted on the way they approach their livelihoods, including the altercations or altogether total abandonment and loss of TEK (Benz et al. 2000; Brosi et al. 2007). Hence the CBD signatories are required to fill in these gaps by taking stock of the available TEK and evaluating the current situation of TEK, ethnic groups and minorities while devising mechanisms to keep track of this knowledge and how it can be sustained and conserved (Cámara-Leret et al. 2014).

The communities in Cholistan rely on a nomadic lifestyle which enables them to move and exploit any available resource in the desert at a particular time. As these communities have grown over time, their impact on the available scant resources is increasingly disturbing the ecosystem, especially the botanical and hydrological resources (Bidak et al. 2015; Abu Zeid 1991). Although there are a few developmental projects, their main focus remains on agricultural diversification. We, on the other hand, are trying to streamline the TEK associated with the ecosystem and the impacts and repercussions of this knowledge is lost completely.

The Cholistan desert is a unique habitat due to its biodiversity and endemism of a number of species (Akhter and Arshad 2006). Habitat degradation due to intensive agricultural practices is a serious threat to the diversity of ethnobotanically important plant species. Agricultural communities may promote the cultivation of desirable species while destroying or ignoring others which they find undesirable. In contrast, some of the local communities may use these "undesirable" species extensively in their daily lives. Furthermore, overgrazing by a large number of ruminants (camels, cattle, goats and sheep) has resulted in habitat degradation. The loss of plant diversity has stimulated an urgent desire to conserve the natural habitat and promote existing knowledge and documentation of medicinally important plant species.

11.5.1 Overharvesting

An increase in the human population in the area has elevated the demand for fuel woods. Collectors usually target larger woody perennials and species that develop woody branches and roots (e.g., *Calligonum polygonoides, Leptadenia pyrotechnica, Prosopis cineraria*, etc.). Individual plants are obviously under threat; their removal also increases soil erosion and overall habitat degradation.

11.5.2 Overgrazing

Overgrazing has led to extreme changes in the community structure and ecosystem functioning of many grasslands (Nautiyal et al. 2004; Diaz et al. 2007; Wan et al. 2015). On a global scale, researchers have estimated that 60% of grasslands suffer from grazing or overgrazing, and that these conditions are more serious in some areas with high populations of grazing animals (Osem et al. 2002). A historically sparse nomadic population had small grazing herds with limited impacts on the scarce resources in the desert region (Heneidy 2012; Heneidy 2002, 2003). Furthermore, overgrazing by a large number of ruminants (camels, cattle, goats and sheep) has resulted in habitat degradation. Loss of plant diversity has stimulated an urgent desire to conserve the natural habitat and promote existing knowledge and documentation of medicinally important plant species. Moreover, local communities exploit medicinally important plant species for other economic uses, e.g. food, fodder, house construction and various other uses.

Current increases in the human population coupled with excessively large herd sizes and stocking rates in some areas has resulted in the depletion of rangeland grasses. Annual species that are heavily browsed are disappearing and are under threat. Motorized trucking now allows local inhabitants to transport their herds as well as milk and their belongings from one grazing site to another, thereby rapidly depleting vegetation in previously inaccessible areas. The adoption of modern techniques, the availability of transportation, and supplementary feed and water allow the locals to maintain larger herds and to take them to previously unreachable areas, where they graze marginal habitats at rates far exceeding ecosystem resilience capabilities (Heneidy 2000; Heneidy 2012).

11.5.3 New Intensive Agriculture Practices

One of the most serious threats to native plants in the region is the complete and irreversible destruction of habitats by modern agricultural activities. Rain-fed agriculture was practiced traditionally by the local inhabitants of the smaller Cholistan areas where they grew winter crops such as wheat, rapeseed (mustard). As the local inhabitants began to settle down, they started to practice modern intensive agriculture over larger areas. The establishment of new irrigation canals in the peripheries of the smaller Cholistan area has made these intensive agricultural practices possible. It is estimated that over the last two decades the terrain used for this type of intensive agriculture have tripled in area at the expense of rangelands and shrublands (El-Kawy et al. 2011). The fragile desert ecosystem is unable to support these types of intensive, unsustainable agricultural practices, which remove all the natural vegetation that previously provided refuge and shelter to wildlife, thereby causing the depletion of soil nutrients, soil erosion, and reduce rangeland resources (Bidak et al. 2015). In consequence, grazing pressure is increasing dramatically in all remaining patches of natural habitats and in marginal areas not suited for cultivation, leading to the further degradation of naturally vegetated areas.

11.5.4 Decline of Traditional Knowledge

Another major noteworthy threat to mention is the decline in the number of local inhabitants holding traditional knowledge on the uses of the native plants in medicine and in other traditional uses. The new generation of local inhabitants is no longer interested in gaining this knowledge and prefers to work in developmental projects in the cities. An action should be directed toward the documentation of this traditional knowledge in order to preserve it for future generations as part of a social heritage and as a component of biodiversity preservation.

11.6 Conclusions and Recommendations

The natural resources of Cholistan desert are one of the major tourist attractions, in order to offering huge potential for scientific research. There is a substantial pool of plant species being utilized by the local inhabitants as food resources, medicine, fuel wood and fodder. Land management, while keeping in view the natural ecosystem supporting ecological and economic values and services, underpins the conservation of desert ecosystem. Such interventions are required on a strategically measured scale in order to ensure the local adaptation and mainstreaming of TEK and associated cultural values of the indigenous communities residing in the region. Medicinal plants are an important source of livelihoods and income generation for the locals. Focusing on their conservation and efficient utilization will ensure long-term benefits for the locals in addition to providing them with opportunities to continue practicing their TEK. As these plants are one of the key contributors to annual income, there is a need for an extensive management program for the local communities which focuses on the following points:

- Training the local communities, especially collectors of medicinal plants and herbal medicine practitioners, in sustainable and efficient harvesting techniques.
- Creating linkages and matchmaking between the local collectors of medicinal plants and other market players, including wholesalers and retailers.

This can be done through extensive value chain analysis of the medicinal plants and their markets from the region. By doing so, the key players within the chain and weak linkages amongst the value chain players will be identified. Such interventions are ideally done by the mutual collaboration of governmental and non-governmental organizations and relevant stakeholders of the value chain. In Pakistan, there are few biodiversity-trained professionals. Hence there is a communication gap between national and international experts. The capacity building of concerned departments is urgently needed to facilitate in bridging these gaps. There is a need for a National Biodiversity Council (NBC) with objectively defined national and international mandates.

Ecotourism is also becoming increasingly significant at the global level. The impacts of ecotourism on environmental and cultural protection have been well documented. Promoting ecotourism may also prove vital for encouraging local communities as well as public authorities to focus on the conservation and management of the natural resources in addition to cultural heritage. Since Cholistan desert has been the least explored in terms of ecotourism, there is a huge potential for its promotion and creating opportunities for income diversification while conserving the environmental resources and the rich TEK and associated cultural heritage of the local indigenous communities.

REFERENCES

Abbasi AM, Khan MA, Shah MH, Shah MM, Pervez A, Ahmad M (2013) Ethnobotanical appraisal and cultural values of medicinally important wild edible vegetables of Lesser Himalayas-Pakistan. *J Ethnobiol Ethnomed* 9. doi:10.1186/1746-4269-9-66.

Abu Zeid A (1991) *Desert Societies in Egypt. North Sinai-National Center for Social & Criminal Research (NCSCR)*, Cairo, Egypt.

Ahmad F (1999) Rainwater utilization and ecological restoration in Cholistan. *People and Rangelands Building the Future, Vols 1 and 2.*

Ahmad F (2007) Geoinformatics application to investigate agricultural potential in Cholistan desert. *J Food Agri Environ 5*(2):310–314.

Ahmad F, Ali Z, Farooq S (2005) Historical and archaeological perspectives of soil degradation in Cholistan. *Sociedade & Natureza 1*(1).

Ahmad S, Alam K, Wariss H, Anjum S, Mukhtar M (2014) Ethnobotanical studies of plant resources of Cholistan desert, Pakistan. *Int J Sci Res 3*:1782–1788.

Akbar G (2000) Developing sustainable strategies for Cholistan desert: opportunities and perspectives. *Sci Vision 5*:77–85.

Akbar G, Khan TN, Arshad M (1996) *Cholistan desert*, Pakistan, Rangelands, pp 124–128.

Akhtar N, Rashid A, Murad W, Bergmeier E (2013) Diversity and use of ethno-medicinal plants in the region of Swat, North Pakistan. *J Ethnobiol Ethnomed 9*. doi:10.1186/1746-4269-9-25.

Akhter R, Arshad M (2006) Arid rangelands in the Cholistan desert (Pakistan). *Science et changements planétaires/Sécheresse 17*(1):210–217.

Alam MK, Ahmed S, Anjum S, Akram M, Shah SMA, Wariss HM, Usmanghani K (2016) Evaluation of antipyretic activity of some medicinal plants from Cholistan desert Pakistan. *Pak J Pharmaceut Sci 29*(2):529–533.

Ali S, Alam M, Basir A, Adnan M, Malik M FA, Shah AS, Ibrahim M. (2016). Effect of seed priming on germination performance and yield of okra (*Abelmoschus esculentus* L.). *Pak J Agri Res 29*(3):253–262.

Amjad MS, Arshad M, Qureshi R (2015) Ethnobotanical inventory and folk uses of indigenous plants from Pir Nasoora National Park, Azad Jammu and Kashmir. *Asian Pac J Trop Biomed 5*(3):234–241. https://doi.org/10.1016/S2221-1691(15)30011-3.

Arafat Y, Shafi M, Khan MA, Adnan M, Basir A, Arshad M, Shah JA (2016). Yield response of wheat cultivars to zinc application rates and methods. *Pure Appl Biol 5*(4):1.

Arshad M, Akbar G (2002) Benchmark of plant communities of Cholistan desert [Pakistan]. *Pak J Biol Sci.*

Arshad M, Akbar G, Rashid S (2002) *Wealth of medicinal plants of Cholistan desert*, Hamdard Medicus, Pakistan.

Arshad M, Anwar ul H, Ashraf MY, Noureen S, Moazzam M (2008) Edaphic factors and distribution of vegetation in the Cholistan desert, Pakistan. *Pak J Botany 40*(5):1923–1931.

Arshad M, Ashraf M, Arif N (2006) Morphological variability of *Prosopis cineraria* (L.) Druce, from the Cholistan desert, Pakistan. *Genet Resour Crop Evol 53*(8):1589–1596. doi:10.1007/s10722-005-8563-5.

Arshad M, Ashraf MY, Ahamad M, Zaman F (2007) Morpho-genetic variability potential of *Cenchrus ciliaris* L., from Cholistan Desert, Pakistan. *Pak J Bot 39*(5):1481–1488.

Arshad M, Rao AUR (1994) Flora of Cholistan Desert (Systematic list of trees, shrubs and herbs). *J Econ Taxon Bot 18*:615–625.

Auj NZA, Auj N (1991) *Cholistan: land and people*. Caravan Book Centre.

Báez S, Collins SL (2008) Shrub invasion decreases diversity and alters community stability in northern Chihuahuan Desert plant communities. *PloS One 3*(6):e2332.
Baig KJ, Masroor R, Arshad M (2008) Biodiversity and ecology of the herpetofauna of Cholistan desert, Pakistan. *Russian J Herpetol 15*(3):193–205.
Baig MS, Akram M, Hassan MA (1980) Possibilities for range development in Cholistan desert as reflected by its physiography and soils. *Pakistan J Forest 30*(2):61–65.
Balick MJ, Cox PA (1997) Ethnobotanical research and traditional health care in developing countries. *Med Plants Forest Conserv Health Care 92*:12–23.
Bano A, Ahmad M, Ben Hadda T, Saboor A, Sultana S, Zafar M, Khan MPZ, Arshad M, Ashraf MA (2014) Quantitative ethnomedicinal study of plants used in the Skardu Valley at high altitude of Karakoram-Himalayan range, Pakistan. *J Ethnobiol Ethnomed 10*. doi:10.1186/1746-4269-10-43.
Benz BF, Cevallos J, Santana F, Rosales J, Graf S (2000) Losing knowledge about plant use in the Sierra de Manantlan biosphere reserve, Mexico. *Econ Bot 54*(2):183–191.
Berry KH, Murphy R, Mack JS, Quillman W (2006) Introduction to the special issue on the changing Mojave Desert. *J Arid Environ 67*:5–10.
Bertiller MB, Marone L, Baldi R, Ares JO (2009) Biological interactions at different spatial scales in the Monte desert of Argentina. *J Arid Environ 73*(2):212–221.
Bidak LM, Kamal SA, Halmy MWA, Heneidy SZ (2015) Goods and services provided by native plants in desert ecosystems: examples from the northwestern coastal desert of Egypt. *Glob Ecol Conserv 3*:433–447.
Bisigato AJ, Villagra PE, Ares JO, Rossi BE (2009) Vegetation heterogeneity in Monte Desert ecosystems: a multi-scale approach linking patterns and processes. *J Arid Environ 73*(2):182–191.
Brito JC, Godinho R, Martínez-Freiría F, Pleguezuelos JM, Rebelo H, Santos X, Vale CG, Velo-Antón G, Boratyński Z, Carvalho SB (2014) Unravelling biodiversity, evolution and threats to conservation in the Sahara-Sahel. *Biological Reviews 89*(1):215–231.
Brito JC, Pleguezuelos JM (2019) Desert Biodiversity—World's Hot Spots/Globally Outstanding Biodiverse Deserts. In: *Reference Module in Earth Systems and Environmental Sciences*. Elsevier. doi: https://doi.org/10.1016/B978-0-12-409548-9.11794-4.
Brosi BJ, Balick MJ, Wolkow R, Lee R, Kostka M, Raynor W, Gallen R, Raynor A, Raynor P, Lee Ling D (2007) Cultural erosion and biodiversity: Canoe-making knowledge in Pohnpei, Micronesia. *Conserv Biol 21*(3):875–879.
Butterfield BJ, Betancourt JL, Turner RM, Briggs JM (2010) Facilitation drives 65 years of vegetation change in the Sonoran Desert. *Ecology 91*(4):1132–1139.
Cámara-Leret R, Paniagua-Zambrana N, Balslev H, Macía MJ (2014) Ethnobotanical knowledge is vastly under-documented in Northwestern South America. *PloS One 9*(1):e85794. doi:10.1371/journal.pone.0085794.
Carpenter S, Bennett E, Peterson G (2006) Scenarios for ecosystem services: an overview. *Ecology and Society 11*(1).
Chaudhry S (1992) The Cholistan desert, A Tokten Consultancy Report. Bahawalpur, Cholistan Institute of Desert Studies, Islamia University, p 34.
Dasti A, Agnew ADQ (1994) The vegetation of Cholistan and Thal deserts, Pakistan. *J Arid Environ 27*(3):193–208. doi:10.1006/jare.1994.1058.
Díaz S, Fargione J, Chapin III FS, Tilman D (2006) Biodiversity loss threatens human well-being. *PLoS Biol 4*(8):e277.
Diaz S, Lavorel S, McIntyre S, Falczuk V, Casanoves F, Milchunas DG, Skarpe C, Rusch G, Sternberg M, Noy-Meir I (2007) Plant trait responses to grazing–a global synthesis. *Glob Change Biol 13*(2):313–341.
El-Kawy OA, Rød J, Ismail H, Suliman A (2011) Land use and land cover change detection in the western Nile delta of Egypt using remote sensing data. *Appl Geograph 31*(2):483–494.
Ellis EC (2015) Ecology in an anthropogenic biosphere. *Ecol Monograph 85*(3):287–331. doi:10.1890/14-2274.1.
Enright N, Miller B, Akhter R (2005) Desert vegetation and vegetation-environment relationships in Kirthar National Park, Sindh, Pakistan. *J Arid Environ 61*(3):397–418.
Ezcurra E (2006) Global deserts outlook. UNEP/Earthprint.

Farooq Z, Iqbal Z, Mushtaq S, Muhammad G, Iqbal MZ, Arshad M (2008) Ethnoveterinary practices for the treatment of parasitic diseases in livestock in Cholistan desert (Pakistan). *J Ethnopharmacol 118*(2):213–219. doi: https://doi.org/10.1016/j.jep.2008.03.015.

Gemedo-Dalle T, Maass BL, Isselstein J (2005) Plant biodiversity and ethnobotany of Borana pastoralists in southern Oromia, Ethiopia. *Econ Bot 59*(1):43–65.

Guisan A, Thuiller W (2005) Predicting species distribution: offering more than simple habitat models. *Ecol Lett 8*(9):993–1009.

Hameed M, Ashraf M, Al-Quriany F, Nawaz T, Ahmad MSA, Younis A, Naz N (2011) Medicinal flora of the Cholistan desert: a review. *Pakistan J Bot 43*:39–50.

Harun N, Chaudhry AS, Shaheen S, Ullah K, Khan F (2017) Ethnobotanical studies of fodder grass resources for ruminant animals, based on the traditional knowledge of indigenous communities in Central Punjab Pakistan. *J Ethnobiol Ethnomed* 13. doi:10.1186/s13002-017-0184-5.

Heneidy S (2000) Palatability, Chemical composition and nutritive value of some common range plants from Bisha, Asir region, southwestern Saudi Arabia. *Desert Institute Bulletin, Egypt 50*(2):345–360.

Heneidy S (2002) Role of indicator range species as browsing forage and effective nutritive source, in Matruh Area, a Mediterranean Coastal Region, NW-Egypt. *Online J Biol Sci 2*(2):136–142.

Heneidy S (2003) Accessible forage biomass of browse species in Matruh area, a Mediterranean coastal region, Egypt. *Pak J Biol Sci 6*:589–596.

Heneidy SZ (2012) Rangelands in arid ecosystem. In: *Diversity of Ecosystems*. IntechOpen.

Hussain W, Ullah M, Dastagir G, Badshah L (2018) Quantitative ethnobotanical appraisal of medicinal plants used by inhabitants of lower Kurram, Kurram agency, Pakistan. *Avicenna J Phytomed 8* (4):313–329.

Kaimowitz D, Sheil D (2007) Conserving what and for whom? Why conservation should help meet basic human needs in the tropics. *Biotropica 39*(5):567–574.

Leduc C, Coonishish J, Haddad P, Cuerrier A (2006) Plants used by the Cree Nation of Eeyou Istchee (Quebec, Canada) for the treatment of diabetes: a novel approach in quantitative ethnobotany. *J Ethnopharmacol 105*(1–2):55–63.

Leopold A (1963) *The desert New York*. Time-Life International, New York.

Malik S, Ahmad S, Sadiq A, Alam K, Wariss HM, Ahmad I, Hayat MQ, Anjum S, Mukhtar M (2015) A comparative ethno-botanical study of Cholistan (an arid area) and Pothwar (a semi-arid area) of Pakistan for traditional medicines. *J Ethnobiol Ethnomedicine 11*. doi:10.1186/s13002-015-0018-2.

Middleton N, Stringer L, Goudie A, Thomas D (2011) The forgotten billion: MDG achievement in the drylands. In: *United Nations Convention to Combat Desertification*, Bonn.

Miriti MN, Rodríguez-Buriticá S, Wright SJ, Howe HF (2007) Episodic death across species of desert shrubs. *Ecology 88*(1):32–36.

Mughal MR (1997) *Ancient Cholistan: archaeology and architecture*. Ferozsons.

Munson SM, Webb RH, Belnap J, Andrew Hubbard J, Swann DE, Rutman S (2012) Forecasting climate change impacts to plant community composition in the Sonoran Desert region. *Glob Change Biol 18*(3):1083–1095.

Murphy NP, Breed MF, Guzik MT, Cooper SJ, Austin AD (2012) Trapped in desert springs: phylogeography of Australian desert spring snails. *J Biogeograph 39*(9):1573–1582.

Nautiyal M, Nautiyal B, Prakash V (2004) Effect of grazing and climatic changes on alpine vegetation of Tungnath, Garhwal Himalaya, India. *Environmentalist 24*(2):125–134.

Osem Y, Perevolotsky A, Kigel J (2002) Grazing effect on diversity of annual plant communities in a semi-arid rangeland: interactions with small-scale spatial and temporal variation in primary productivity. *J Ecol 90*(6):936–946.

Qureshi R, Bhatti GR, Memon RA (2010) Ethnomedicinal uses of herbs from nothern part of Nara desert, Pakistan. *Pakistan J Bot 42*(2):839–851.

Rao A-u-R, Arshad M, Shafiq M (1989) *Perennial grass germplasm of Cholistan desert and its phytosociology*. Cholistan Institute of Desert Studies, Ismalia University.

Rossini R, Fensham R, Stewart-Koster B, Gotch T, Kennard M (2018) Biogeographical patterns of endemic diversity and its conservation in Australia's artesian desert springs. *Divers Distribut 24*(9):1199–1216.

Salim MA, Ranjitkar S, Hart R, Khan T, Ali S, Kiran C, Parveen A, Batool Z, Bano S, Xu J (2019) Regional trade of medicinal plants has facilitated the retention of traditional knowledge: case study in Gilgit-Baltistan Pakistan. *J Ethnobiol Ethnomedicine 15*(1):6. doi:10.1186/s13002-018-0281-0.

Santarém F, Saarinen J, Brito JC (2019) Desert conservation and management: ecotourism. In: *Reference module in Earth systems and environmental sciences*. Elsevier. doi: https://doi.org/10.1016/B978-0-12-409548-9.11827-5.
Saxena K, Rao K, Sen K, Maikhuri R, Semwal R (2003) Integrated natural resource management: approaches and lessons. Integrated natural resource management: linking productivity, the environment and development: 211.
Sial N, Arshad M (2003) Biodiversity in the surface dwelling Fauna from Cholistan Desert, Pakistan. *Pakistan J Biol Sci 6*(13):1195–1198.
Sial N, Chaudhry MS (2012) A preliminary report on the diversity of spiders (Arachnida: Araneae) in the Cholistan Desert, Pakistan. *Pakistan Journal of Zoology 44*(5):1439–1442.
Sial N, Ruby T, Malik S, Mushtaq S (2012) A checklist of the spiders of Cholistan and neighbouring areas. *Pakistan J Agri Sci 49*(3):335–338.
Vale CG, Brito JC (2015) Desert-adapted species are vulnerable to climate change: insights from the warmest region on Earth. *Glob Ecol Conserv 4*:369–379.
Wan H, Bai Y, Hooper DU, Schönbach P, Gierus M, Schiborra A, Taube F (2015) Selective grazing and seasonal precipitation play key roles in shaping plant community structure of semi-arid grasslands. *Landscape Ecol 30*(9):1767–1782.
Wang Y, Yang X, Shi Z (2013) The formation of the patterns of desert shrub communities on the Western Ordos Plateau, China: the roles of seed dispersal and sand burial. *PloS One 8*(7):e69970.
Ward D (2016) *The biology of deserts*. Oxford: Oxford University Press.
Wariss HM, Ahmad S, Anjum S, Alam K (2014a) Ethnobotanical studies of dicotyledonous plants of Lal Suhanra national park, Bahawalpur, Pakistan. *Int J Sci Res 3*(6):2452–2460.
Wariss HM, Mukhtar M, Anjum S, Bhatti GR, Pirzada SA, Alam K (2013) Floristic composition of the plants of the Cholistan Desert, Pakistan. *Am J Plant Sci 4*(12):58.
Wariss HM, Pirzada SA, Alam K, Anjum S, Qureshi R (2014b) Flora of Lal Suhanra National Park, Bahawalpur, Punjab, Pakistan. *Pak J Bot 46*(4):1331–1341.
Wariss HM, Wang H, Ahmad S, Alam K (2015) Taxonomic studies of trees of the cholistan desert, Pakistan. *JBES 6*(1):1–8.
Wariss HM, Wang H, Yi TS, Anjum S, Ahmad S, Alam K (2016) Taxonomic perspective of grasses; a potential resource of cholistan desert, Pakistan. *J Biodivers Environ Sci 9*:26–42.
Wassenaar TD, Ferreira SM, Van Aarde RJ (2007) Flagging aberrant sites and assemblages in restoration projects. *Restor Ecol 15*(1):68–76.
Wu B, Yang H (2013) Spatial patterns and natural recruitment of native shrubs in a semi-arid sandy land. *PloS One 8*(3):e58331.

12

Bio Fertilizer as a Tool for Soil Fertility Management in a Changing Climate

Imran, Amanullah, Shah Khalid, and Muhammad Arif
The University of Agriculture, Pakistan

Shah Fahad
The University of Haripur, Pakistan

Abdel Rahman Mohammad Al-Tawaha
Al Hussein Bin Talal University, Jordan

CONTENTS

12.1 Introduction

Beneficial microbes are essential components of all soils which play a vital role in the retention, break-down and incorporation of plant remains, nutrient cycling and their influence on soil structure and porosity. Climate change has both direct and indirect effects on soil health and on the activities of soil microbes that feedback greenhouse gases into the atmosphere and contribute to global warming. Direct effects include the influence on soil microbes and greenhouse gas production of temperature, changing precipitation and extreme climatic events, whereas indirect effects result from climate-driven changes in plant productivity and diversity which alter soil physicochemical conditions, the supply of carbon to soil and the structure and activity of microbial communities involved in decomposition processes and carbon release from soil. Indirect climate–microbe feedback refers to the effects on soil microbial communities and their activity, and hence the potential for microbial feedback to climate change through its influence on plant growth and vegetation composition. Climate change may have stronger or weaker, permanent or periodical, favourable or unfavourable, harmful (sometimes catastrophic), primary (direct) or secondary (indirect) impact on soil processes. Among these processes the soil moisture regime and beneficial microbes play a central role. This determines the water supply of plants, influences the air and heat regimes, biological activity and the plant nutrient status of soil. In most cases, it determines the agro-ecological potential, the biomass production of various natural and agro-ecosystems and the hazards posed by soil and water pollution.

In a changing climate, excessive precipitation causes a significant source of soil nutrient loss like nitrate leaching. Agricultural areas with poorly drained soils or that experience frequent or intense rainfall events can have waterlogged soils that become hypoxic. The change in soil redox status under conditions of low oxygen can lead to elemental toxicities of Mn, Fe, Al and B that reduce crop yields and produce phytotoxic organic solutes that impair root growth and function. Hypoxia can also result in nutrient deficiency since the active transport of ions into root cells is driven by ATP synthesized through the oxygen-dependent mitochondrial electron transport chain. Significant nitrogen losses can also occur under hypoxic conditions through denitrification as nitrate is used as an alternative electron acceptor by microorganisms in the absence of oxygen. Soil warming can increase nutrient uptake from 100–300% by enlarging the root surface area and increasing rates of nutrient diffusion and water influx. Since warmer temperatures increase rates of transpiration, plants tend to acquire water-soluble nutrients (nitrate, sulfate, Ca, Mg primarily move towards roots through transpiration-driven mass flow) more readily as temperature increases. Temperature increases in the rhizosphere can also stimulate nutrient acquisition by increasing nutrient uptake via faster ion diffusion rates and increased root metabolism. However, any positive effects of warmer temperatures on nutrient capture are dependent on adequate soil moisture. If under dry conditions higher temperatures result in extreme vapor pressure deficits that trigger stomatal closure (reducing the water diffusion pathway in leaves), then nutrient acquisition driven by mass flow will decrease.

Greenhouse gases emissions from crop and livestock agriculture have risen from 4.7 billion tonnes CO_2 equivalent in 2001 to more than 5.3 billion today, an increase of more than 14%. Organic agriculture, including bio fertilization, can help to tackle climate change by reducing greenhouse gas emissions. There is a direct correlation between nitrous oxide emissions and the amount of nitrogen fertilizer applied to agricultural land. Nitrous oxide emissions from managed soils account for almost 40% of agricultural emissions in the European Union. This is particularly important because the impact of 1 kilo of nitrous oxide on warming the atmosphere is about 300 times greater than the impact of 1 kilo of carbon dioxide. Organic farming does not allow the use of synthetic nitrogen fertilizers, focusing instead on establishing closed nutrient cycles, minimizing losses via runoff, volatilization, and emissions, nitrogen levels on organic farms tend to be lower per hectare than on conventional farms, which can contribute to a sustainable climate-friendly production system that delivers enough food. Conventional agriculture uses vast quantities of synthetic fertilizers and pesticides. It takes significant amounts of energy to manufacture these chemicals. Organic agriculture and bio fertilizer application minimizes energy consumption by between 30 and 70% per unit of land by eliminating the energy required to manufacture synthetic fertilizers, and by using internal farm inputs. Organic agriculture can also help combat global warming by storing carbon in the soil. Many management practices used by organic agriculture (e.g. beneficial microbes application, minimum tillage, returning crop residues to the soil, the use of cover crops and rotations, and the greater integration of nitrogen fixing legumes) increase the return of carbon to the soil. This raises productivity and favours carbon storage.

Crop nutrient management is an important aspect provided mainly by chemical fertilizers. However, it is widely accepted that the application of a balanced fertilizer with effective use of organic sources and beneficial microbes is key to achieving higher crop production (Imran et al. 2016a). Repeated applications of chemical fertilizers and heavy metal leads to soil degradation and causing environmental damage. The application of organic and bio fertilizers can reduce these problems and thereby improve soil fertility. Thus, it was considered appropriate to merge the three different sources of nutrients, organic, bio and chemicals to get more efficient and economical result in the long run (Diep et al. 2016). Literature on the use of organic substances along with bio and chemical fertilizers on yield of agronomic crops are very limited. The challenges for agriculture production faced by the harsh and hostile environment with climate change at the twenty-first century for food and fiber could only be overcome through the application of bio fertilizer (Imran 2017). Beneficial microbes favored by the presence of high levels of plant roots, which they colonize readily. Some strains are highly rhizosphere-competent, meaning that they able to colonize and grow on roots as they develop. The most strongly rhizosphere-competent strains can be added to soil or seeds by any method. Once they come into contact with roots, they colonize the root surface or cortex, depending on the strain (Elkoca et al. 2008). Thus, if added as a seed treatment, the best strains will colonize root surfaces even when the roots a meter or more below the soil surface and they can persist at useful numbers up to 18 months after application (FAO 2014). However, most strains lack this ability. Use of mineral fertilizers to increase grain production is faster and safer; however, cost and other restrictions discourage farmers to use mineral fertilizers (Dubey et al. 1997). So farmers need to take full advantage of the potential of alternative sources of plant nutrients. Thus, we should pay more attention on the efficient use of chemical fertilizers and use these low-cost sources of nutrients such as organic matter and biological fertilizer to reduce production costs along with soil productivity and health support (Drevon and Hartwig 1997).

In addition, *Trichoderma* spp. attack, parasitize and otherwise gain nutrition from other fungi. Since *Trichoderma* spp. grow and proliferate best when there are abundant healthy roots, they have evolved numerous mechanisms both to attack other fungi and to enhance plant and root growth. Several new general methods for both biocontrol and for causing the enhancement of plant growth have recently been demonstrated; it is now clear that there must be hundreds of separate genes and gene products involved in these processes (Imran et al. 2015b, 2017; Afzal et al. 2017).

12.2 Climate Change and Soil Health

The Soil Science Society of America (SSSA) defines soil quality as: 'The capacity of a specific kind of soil to function, within natural or managed ecosystem boundaries, to sustain plant and animal productivity, maintain or enhance water quality and support human health and habitation.' Soil quality could, in part, be viewed as a static (qualitative) measure of the capability of soil, whereas 'Soil health' infers a dynamic state, where human impact causes a shift in quality. Bio fertilizers and organic matter is vital because it supports many soil processes that are associated with fertility and physical stability of soil across the various ecosystem services under a changing climate. In particular, organic matter provides an energy source for microbes structurally stabilizes soil particles, stores and supplied plant essential nutrients such as nitrogen, phosphorus and sulphur and provides cation/anion exchange for the retention of ions and nutrients in a harsh environment. Carbon within the terrestrial biosphere can also behave as either a source or sink for atmospheric CO_2 depending on land management, thus potentially mitigating or accelerating the greenhouse effect. The cycling of soil organic carbon is also strongly influenced by moisture and temperature, two factors which are predicted to change under global warming.

12.3 The World Population Depends on Soil Biota and Beneficial Microbes

Microbes quite literally support all of the life on Earth. They are the very bottom of the food chain, no matter where you look, and regulate a number of processes critical to all of us. They are also the vast majority of life forms on the planet, outnumbering every other living thing by staggering amounts. In the vast majority of cases, however, we have yet to understand how they will be impacted by climate change—and how in turn they will affect how climate change unfolds. Soil fertility is vital to help feed growing global populations in the face of a changing climate and pressures on land. Soil is a non-renewable, finite natural asset. Yet the world's expanding populations are destroying land capital at alarming rates. This reduces opportunities for food production, spaces for recreation and the wilderness needed by nature. We all benefit from conserving soil, while agriculture actually depends on it. In turn, fertility relies on the populations of microbes of which our understanding is very limited. Soils are systems of pores and aggregated mineral and organic particles in differing sizes. Within the pores microbes are continuously eating, respiring, reproducing, competing, cooperating, and responding to their environment. They concentrate mainly around roots, which exude sugars, thereby providing them with energy. Soil temperature, moisture content, acidity/alkalinity and nutrient availability regulate microbial activity, which changes according to the seasons and the way in which land is used. Adding nitrogen decreases populations of fungi compared with bacteria, while liming tends to favor fungi. Fungi are encouraged by good practices such as reducing the intensity of cultivation, increasing rotations and using cover cropping. Intensive cropping excludes all but one plant from an area of land and reduces its microbial diversity. Subsequent husbandry should aim at restoring fertility because the numbers of beneficial and benign microbes far outweigh and suppress those causing diseases. Additionally, beneficial microbes enhance crop water-use efficiency and nutrient uptake, particularly phosphorus. Maintaining soil fertility reduces the need for synthetic fertilizers and pesticides and saves growers money. Degrading dead plants and animals into useful organic matter is a key microbial function. Decomposition underpins sustainable land use in all managed, agricultural and horticultural, semi-natural and natural soils, part of the natural carbon and nitrogen cycles that are driven by free solar energy captured by plant photosynthesis.

12.4 Bio Fertilizers Applications

The commercial use of beneficial bacteria and fungi is gaining ground as one element of integrated pest management strategies. Belchim's Contans contains *Coniothyrium minitans*, which helps control soil-borne pathogens such as *Sclerotinia sclerotiorum* and *S. minor*. According to specialist Fargro technical and energy director Paul Sopp: 'Growers are experimenting with biopesticides such as Serenade (Bacillus subtilis QST713) because this widens their options for pest and disease control. This is especially the case with diseases such as downy mildew on brassicas and onions.' Several formulations of beneficial microbes antagonistic to pathogens are available including *Agrobacterium radiobacter* active against *A. tumefaciens* (crown gall, NoGall), *Pseudomonas fluorescens* active against Sclerotinia, Rhizoctonia and Pythium (Biomonas), *Streptomyces griseoviridis* active against Fusarium, Phytophthora, Pythium and Rhizoctonia (Mycostop), *Trichoderma atroviride* active against Sclerotium cepivorum (Plantmate) and *T. harzianum* active against Pythium, Rhizoctonia and Fusarium (Rootshield, PlantShield).

12.5 Plant, Microbe's Interaction

Microbes and plant roots communicate by exchanging chemical signals. When necessary, they activate their own defences and those of host plants—for example, in mycorrhizal combinations. Exploiting these natural communications offers prospects for the development of new generations of environmentally benign fertilisers, crop stimulants and protectants. For example, Isopyrazam (IZM), Syngenta's new cereal fungicide, is a move in this direction.

Good soil care, and consequently the conservation of beneficial microbes which can maintain soil fertility, is of prime importance. We must feed a world population of 10 billion by 2080 from diminishing land assets and in increasingly adverse climates. Microbes and soil fertility are among our best friends in grappling with these problems. Microbes are also known as microorganisms. They are very small living organisms which cannot be seen with the naked eye. They are everywhere in this world; in the air, water, and soil, and even in animal and human bodies. Their numbers in this world are much higher than any other organism and they play important roles in our daily lives. Among the most well-known microbes groups are bacteria, viruses, fungi and algae. Some of the microbes are beneficial to us and some of them are our enemy. In term of agriculture, microbes play a very important role in plant growth development. Plants even help in taking care of them. For example, 50% of carbon created from the photosynthesis process is released through the plants' root for the microbes to be consumed as food. Back to our main topic: the question now is how do microbes contribute to the soil fertility?

12.6 Emerging Trends of Bio Fertilization

The current generation of commercially available microbes are actively antagonistic to specific disease-causing pathogens. An alternative and more effective approach is to use combinations with differing modes of action. For instance, this might include combining non-pathogenic forms such as strains of *F. oxysporum* that compete with pathogens for energy sources, along with bacterial antagonists that block access to iron. Alternatively, there is scope for integrating microbial treatments with beneficial husbandry. This approach could be valuable where biofumigant crops are used. A natural suppression of soil-borne pests and pathogens follows the incorporation of green manures or harvest residues from brassicas with high-glucosinolate contents such as Caliente mustard into soil. Adding compost containing growth-promoting rhizobacteria strengthens the control of diseases such as Fusarium wilt and Rhizoctonia damping-off and helps to further build soil fertility.

12.6.1 Microorganisms Convert Soil Carbon into Stable Forms

The stable forms of soil carbon such as humus and glomalin are manufactured by microorganisms. They convert the carbon compounds that are readily oxidised into CO_2 into stable polymers that can last thousands of years in the soil. The process of making composts uses microbes to build humus and other stable carbons. The microorganisms that create compost continue working in the soil after compost applications, converting the carbon gifted by plant roots into stable forms. The regular applications of compost or compost teas will inoculate the soil with beneficial organisms that build humus and other long-lasting carbon polymers. Synthetic nitrogen fertilizers are one of the major causes of the decline of soil carbon. This is because it stimulates a range of bacteria that feed on nitrogen and carbon to form amino acids for their growth and reproduction. These bacteria have a carbon to nitrogen ratio of around 30 to 1. In other words, every ton of nitrogen applied results in the bacteria consuming 30 tonnes of carbon. The rapid addition of these nitrogen fertilizers causes the nitrogen-feeding bacteria to rapidly multiply, consuming the soil carbon to build their cells.

12.6.2 Soil Microbes and Carbon, Nitrogen Cycles

Carbon from dead plants and animals is recycled naturally back into the atmosphere as carbon dioxide by microbes using chemical energy. Nitrogen is taken by free-living soil bacteria from organic matter and added to the soil. Associations of bacteria living with legume roots add further nitrogen supplies into soils. Fixation of atmospheric nitrogen by legume-rhizobia deposits between 33 and 46 million tonnes of nitrogen annually into the world's soils. At current fertilizer prices, this is worth 50–70 billion US dollars yearly. Using processes understood only in outline, microbes make soil inorganic phosphorus soluble and available to plant roots. Additionally, the biochemical diversity and adaptability of microbes can be used in repairing polluted land. These processes are further stimulated by adding substances such as active charcoal or biochar.

12.6.3 Bio Fertilizer Act as a Suppressing Agent for Pests and Pathogens

Fertile soils rich in microbes suppress plant pathogens. Actively beneficial bacteria include *Bacillus* spp, *Enterobacter* spp, *Flavobacterium balustinum* and *Pseudomonas* spp and fungi such as *Penicillium* spp, *Gliocladium virens* and several *Trichoderma* spp. Microbes such as *Trichoderma* spp. offer several beneficial effects. They compete with pathogens for nutrients and soil space, parasitize pathogens and induce plant disease resistance. The soil-borne diseases that are reduced and crops that benefit this natural biocontrol include Pythium in turf grass, snow moulds in turf grass and Fusarium, Pythium and Rhizoctonia in tomato and cucumber. The suppression of *R. solani*-attacking cucumber resulted from inoculating seedling compost with *Trichoderma asperellum*, while a reduction of bacterial leaf spot of radish, lettuce and tomato followed the inoculation of compost with *T. hamatum*. Some free-living soil fungi diminish soil nematode populations very effectively. Their fungal hyphae form nooses that capture the nematodes and they then exude enzymes that dissolve them.

12.6.4 Beneficial Microbes Application Enhance Nitrogen Capturing and Fixation

There are some microbes that are helping to prepare the nitrogen element within the soil for plants to use. These types of microbes contribute and play important roles in ensuring the healthy growth of plants. An example of a very well-known and common microbe that is doing this type of job are the nitrogen fixation bacteria. This type of bacteria helps to fix the nitrogen in the atmosphere into a form that can be absorbed by the plants. This is because plants cannot absorb the nitrogen in its original form and can only absorb the nitrogen in the form of ammonia or nitrate after the fixation process. Two examples of bacteria species that perform this type of role are Azotobacter and Clostridium. Nitrogen fixation bacteria play a very important role in the nitrogen circle in the atmosphere as 90% of nitrogen fixation is being contributed by these microbes.

12.6.5 Beneficial Microbes' Application Improve Soil Structure

There are also some microbes that help to improve the soil structure. As we all know, soil structure plays a very important role in securing healthy and vigorous plant growth. The good news is that microbes can also help you to achieve this. An example of a soil microbe that can help in improving the soil structure is algae. Algae is a microbe that can be categorized as microflora. As with any other floras, algae can produce their own food through the photosynthesis process. In terms of helping to improve the soil structure, algae will usually produce some type of slimy substance that can fix the soil particle together that is water-stable and soluble. Another type of microbe that can help in improving soil structure same as algae is the cyanobacteria. The soil particles that were fixed together with the help of the above said microbes help to increase soil moisture content and also reduce the effect of erosion from the running water.

12.6.6 Beneficial Microbes' Application Digest Nutrients in the Soil

Another function of soil microbes is to digest the nutrients in the organic matters within the soil before releasing them back in the form that can be taken by the plants. It is very well known that microbes plays major roles in the decaying process. The nutrients that are available in the organic matter will be unavailable to the plants until that organic matter is fully decayed and releases all the nutrients that they contain back into the soil. The decaying process is highly depending on the availability of the microbes in the soils. The more beneficial microbes you have in your soil, the faster will be the process of decay.

12.6.7 Beneficial Microbes' Application Preventing Diseases and Pests Attack

This might be a shocking fact for some of you as we are usually informed that microbes are usually the source of many diseases, not only to humans but also to plants. However, many studies have shown that there are some microbes that can actually help in preventing some type of diseases and pests from ruining your plants. In the wild, there are bacteria that feed on other bacteria. Drawing on this knowledge, we can leverage it by introducing beneficial bacteria that feed on other bacteria into the soils that are highly populated by harmful bacteria so that they can feed on them. Studies also revealed that some microbes could actually excrete some sort of antibiotic substance that helps plants to fight pathogens that cause certain diseases. Mycorrhizal fungi are one of the microbes that function in this way. There are also studies that have been done that show plants and microbes are actually working closely with each other for mutual benefits.

From the studies, the researchers have found that plants can purposely excrete substances that are attracting beneficial microbes to be near them and protect them from the attack of other microbes or pests. It is some sort of a symbiotic relationship where plants will excrete the substance to be as the food for the beneficial microbes and the microbes will return the favor by protecting the plants from their enemies.

12.6.8 Beneficial Microbes' Application Create Organic Matter for Soil

As mentioned earlier, microbes plays an important role in the decaying process, meaning they actually help to prepare organic matter for the soils. As we all know, organic matters help to increase the soil fertility by increasing the water abortion capacity of the soil and improving air circulation within the soils. It is apparent that soils which contain more beneficial microbes in them will be more fertile than the soils that contain less beneficial microbes.

12.7 Bio Fertilizers, Nutrients Availability and Crop Responses

Microbes are also important in preparing the availability of other soil nutrients in the soil such as phosphorus and potassium in their daily activities. Some of them can also help in the mobilization of nutrients in the soil. Other microbes can also excrete some type of hormone that can stimulate the plant's growth.

12.7.1 Bio Fertilizers and Cereal Crops

Bio Power vaccination was assessed for the growth and yield of wheat and maize by Bekere et al. (2012) in field conditions. Bio Power by half and full rate of N fertilizer increased grain yields of wheat and maize. It has been found that beneficial microbes or commercial fertilizers greatly increase maize yield and yield-relating attributes. Bellore and Mall (1975) reported that the inoculation of *Rhizobia* alone, or in combination with chemical fertilizers, produced a much larger number of nodules, number of pods, pod weight and yields as compared with those were not vaccinated. Combining treatments with beneficial microbes gave the highest yield. Berg and Lynd (1985) studied the effectiveness of bio fertilizer in improving the availability of phosphorus. The results indicated that P concentration in maize leaf, stalk, grain and total P uptake was much higher than among non-inoculated plants. A comparison of control with treated plots found that wheat production has increased 21 per cent and 35 per cent, respectively, due to the increased availability of phosphorus. Maize also responded positively to inoculation in terms of an increasing residual biomass of plants, the number of cobs and grain yield (Bhattacharya et al. 2010; Fazlullah et al. 2018).

Gao et al. (2007) and Khan et al. (2017) determined the effect of AMF inoculation on growth performance and Zn uptake by rice genotypes. A pot experiment was conducted with six aerobic rice genotypes inoculated with *Glomus mosseae* or *G. etunicatum*, or without AMF on a low Zn soil. Plant growth, Zn uptake and mycorrhizal responsiveness were determined. AMF-inoculated plants produced increased biomass and took up more Zn than nonmycorrhizal controls. Mycorrhizal inoculation, however, significantly increased Zn uptake only in genotypes that had a low Zn uptake in the nonmycorrhizal condition (Cassman et al. 1981a; Brady 2002). They concluded that genotypes that are less efficient in Zn uptake are more responsive to AMF inoculation.

12.7.2 Bio Fertilizers and Pulses

Mungbean (green gram) inoculation with VAM gave the highest green forage, dry matter and crude protein. It has been reported that highest number of plants m^{-2}, plant height (cm), number of leaves $plant^{-1}$, pods length (cm), numbers of seeds pod^{-1}, number of pods $plants^{-1}$, thousand seed weight, seed yield and harvest index was noted with the inoculation of bio fertilizer (PSB) (Imran et al. 2016a). Alam et al. (2007) studied the impact of beneficial microbe inoculation with three different cultivars of mung bean and reported that growth and yield components were the highest in cv. NM-92 with the inoculation of *Trichoderma*. Recently, the author (Imran) conducted two years (2016 and 2017) consecutive experiment on soybean, maize and subsequent wheat to evaluate soybean, maize and subsequent wheat response to bio fertilizers (PSB, Trichoderma) and their interactions with organic sources and phosphorous levels (Figures 12.1–12.3).

The author has not yet published this work and will be submitted to the Higher Education Commission of Pakistan. He reported that soybean phenological and growth characterises was holistically improved through the inoculation of *Trichoderma*. Dry matter portioning was much higher with PSB inoculation. He concluded that *Trichoderma* are favored by the presence of high levels of plant roots, which they

FIGURE 12.1 Tangible view of mungbean (green gram) experimental plots at ARI with (+) and without (–) bio fertilization.

FIGURE 12.2 Tangible view of soybean experimental plots at ARI with (+) bio fertilization.

colonize readily. Some strains are highly rhizosphere-competent, i.e., able to colonize and grow on roots as they develop. The most strongly rhizosphere-competent strains can be added to soil or seeds by any method (Ansari and Sukhraj 2010). Once they come into contact with roots, they colonize the root surface or cortex, depending on the strain. Thus, if added as a seed treatment, the best strains will colonize root surfaces even when the roots are a meter or more below the soil surface and they can persist at useful numbers up to 18 months after application. However, most strains lack this ability. By fixing atmospheric nitrogen and solubilizing phosphates, pulses contribute to reducing the need for synthetic fertilizers and, in doing so, greatly contribute to reducing the risk of soil and water pollution, supporting soil biodiversity, and combating and building resilience to climate change (Aziz et al. 2016) (Figure 12.4).

FIGURE 12.3 Tangible view of maize experimental plots at ARI with (+) and without (–) bio fertilization.

FIGURE 12.4 Schematic diagram of soil phosphorus mobilization and immobilization by bacteria.

Biological nitrogen fixation is particularly important for global agricultural productivity and might be considered one of the most important biological processes on the planet. It provides *circa* 100 million metric tonnes of N which leads to an annual saving of around US$10 billion in N fertilizer (Arancon et al. 2005) . Lentils alone could fix nitrogen in the range of 35–100 kg ha^{-1}. Furthermore, the reduced need for (or use of) synthetic fertilizers indirectly reduces the amount of greenhouse gases released into the atmosphere. Pulses also promote soil carbon sequestration and, ultimately, reduce soil erosion when included in intercropping farming systems and/or used as cover crops (Anchal et al. 1997). Furthermore, due to their high nutritional value, pulses are also valuable allies in fighting hunger worldwide.

12.7.3 PSB on Crop Production

Phosphate rock minerals are often too insoluble to provide sufficient P for crop uptake. The use of PSMs can increase crop yields by up to 70% (Carsky et al. 2001). Combined inoculation with arbuscular mycorrhiza and PSB give a better uptake of both native P from the soil and P coming from the phosphatic rock (Carvalho et al. 2011). Higher crop yields result from the solubilization of fixed soil P and applied phosphates by PSB (Cassman et al. 1981b). Microorganisms with phosphate-solubilizing potential increase the availability of soluble phosphate and enhance plant growth through improving biological nitrogen fixation (Chaturvedi 2006). *Pseudomonas* spp. enhanced the number of nodules, dry weight of nodules, yield components, grain yield, nutrient availability and uptake in soybean crop (Chauhan et al. 1992; Chela et al. 1993). Phosphate-solubilizing bacteria enhanced the seedling length of *Cicer arietinum* while the co-inoculation of PSM and PGPR reduced P application by 50% without affecting corn yield. Inoculation with PSB increased sugarcane yield by 12.6% (Chen et al. 2006). Sole application of bacteria increased the biological yield, while the application of the same bacteria along with mycorrhizae achieved the maximum grain weight (Chen et al. 2008). Single and dual inoculation along with P fertilizer was 30–40% better than P fertilizer alone for improving the grain yield of wheat, and dual inoculation without P fertilizer improved grain yield up to 20% against sole P fertilization. Mycorrhiza, along with *Pseudomonas putida*, increased leaf chlorophyll content in barley. Rhizospheric microorganisms can interact positively in promoting plant growth, as well as improving N and P uptake. The seed yield of green gram was enhanced by 24% following triple inoculation with *Bradyrhizobium* + *Glomus fasciculatum* + *Bacillus subtilis* (Chiezey and Odunze 2012). Growth and phosphorus content in two alpine Carex species increased by inoculation with *Pseudomonas fortinii*. The integration of a half-dose of NP fertilizer with biofertilizer gives the same crop yield as with the full rate of fertilizer; and through the reduced use of fertilizers the production cost is minimized and the net return maximized (Crews 1993). The use of compost on farm crops may be the subject to some limitations. For example, compost could be effectively used during the establishment of nursery in the field to increase the stock and maintain the health of the plant. Crop nutrient management through an integrated approach is an important aspect provided mainly by bio, organic and chemical fertilizers. However, it is widely accepted that the application of a balanced fertilizer with the effective use of organic sources and beneficial microbes is key to achieving higher crop production and net return (Imran et al. 2016b). Agricultural scientists have been exploring an alternative source to increase farm products with low input cost and maintain soil health without having an effect on soil and plant environment. Plants obtain nutrients from two natural sources: (1) organic matter and minerals and (2) chemical fertilizers. Farmers only rely on chemical fertilizer which not only deteriorates quality of soil but also increases the cost of production. Soil amendments and bio fertilization is an integrated approach to reduce dependency on chemical fertilizers with the advantage of having a low cost of production (Imran 2017). Crop response to biological fertilizers depends largely on crop species/strains of microbe and application method along with climatic conditions. It has been reported that using beneficial microbes reduce requirement of nitrogen 50–70% and an increase yield up to of 20%. Bacterial inoculation not only provides nitrogen, phosphorus and growth hormones, but also makes the plant healthy and less susceptible to pathogens. The use of beneficial microbes, along with the optimal dose of fertilizers, can save about half the recommended dose of chemical fertilizers (Imran et al. 2016a; Imran et al. 2015a; Ahmad et al. 2017).

12.7.4 Mycorrhizas and Crop Production

Mycorrhizas are mutually beneficial associations of fungi with roots. There are classified into two broad forms: endomycorrhiza (particularly arbuscular types), where fungal hyphae inhabit plant cells; and ectomycorrhiza, where a hyphae sheath forms round the root surface. Arbuscular mycorrhizas are crucially important for sustainable cropping. Mycorrhizas increase root access to nutrients, particularly phosphorus, improve water uptake and defend them against invading pathogens.

Commercial formulations are available as soil improvers that enhance amenity and fruit tree and shrub establishment. These are useful for controlling specific replant disorders caused by *Pythium* spp and other pathogens in top and stone fruit and roses. Mycorrhizal preparations are available from Dragonfli or David Austin Roses.

12.8 Conclusion

In conclusion, we can say that microbes play a major role in soil fertility. In fact, there are many other ways that they can contribute in helping to improve the soil fertility in addition to what have been listed here. Future climate change poses serious interlinked challenges with reference to scale and scope, which were never anticipated in the last century. More or less the most important change in soils expected as a result of these changes would be a gradual improvement in fertility and physical conditions of soils in humid and sub-humid climates, change from one major soil-forming process to another in certain fragile tropical soils and changes in soil property due to the poleward retreat of the permafrost boundary. Again changes due to climate change are expected to be relatively well buffered by the mineral composition, the organic matter content or the structural stability of many soils. As a matter of fact, the impact of climate change on soil systems should be monitored in different agro-ecological regions on a regular basis. Climate change and land degradation are closely linked issues and conservation farming has shown promise in minimizing land degradation. Hence, the potential of conservation agriculture in minimizing the impact of climate change needs thorough investigation. There is a need for the harmonization of a data base on land degradation, bearing in mind the productivity and economic losses vis-à-vis climate change effects.

REFERENCES

Afzal F, Adnan M, Rahman IU, Noor M, Khan A, Iqbal S, Nawaz, S (2017) Growth response of olive cultivars to air layering. *Pure Appl Biol 6*(4):1403–1409.

Ahmad I, Jadoon SA, Said A, Adnan M, Mohammad F, Munsif F (2017) Response of sunflower varieties to NPK fertilization. *Pure Appl Biol 6*(1):272.

Alam MN, Jahan MS, Ali MK, Ashraf MA, Islam MK (2007) Effect of vermicompost and chemical fertilizers on growth, yield and yield components of potato in barind soils of Bangladesh. *J App Sci Res 3*:1879–1888.

Anchal D, Kharwara PC, Rana SS, Dass A (1997) Response of gram varieties to sowing dates and phosphorus levels under on farm condition. *Himachal J Agric Res 23*:112–115.

Ansari AA, Sukhraj KK (2010) Effect of vermiwash and vermicompost on soil parameters and productivity of okra in Guyana. *Afr J Agric Res 5*:1794–1798.

Arancon NQ, Edwards CA, Bierman P, Metzger JD, Lucht C (2005) Effects of vermin composts produced from cattle manure, food waste and paper waste on the growth and yield of peppers in the field. *Pedobio 49*:297–306.

Aziz ALA, Ahiabor BDK, Opoku A, Abaidoo RC (2016) Contributions of rhizobium inoculants and phosphorus fertilizer to biological nitrogen fixation, growth and grain yield of three soybean varieties on a fluvic luvisol. *Am J Exp Agric 10*(2):1–11.

Bekere W, Endalkachew W, Tesfu K (2012) Growth and nodulation response of soybean (*Glycine max* L.) to Bradyrhizobium inoculation and phosphorus levels under controlled condition in South Western Ethiopia. *Afric J Agric Rese 7*(30):4266–4270.

Bellore SK, Mall LR (1975) Chlorophyll content as an ecological index of dry matter production. *Indian J Bot Soc 54*(1–2):75–77.

Berg RK, Lynd JQ (1985) Soil fertility effects on growth, yield, nodulation and nitrogenase activity of Australian winter pea. *J Plant Nutr 8*:131–145.

Bhattacharyya RP, Chandra S, Kundu S, Supradip S, Mina BL, Srivastva AK, Gupta H (2010) Fertilization effects on yield sustainability and soil properties under irrigated wheat-soybean rotation of an Indian Himalayan Upper Valley. *Nutr Cyc Agroeco 86*:255–268.

Brady N (2002) Phosphorus and potassium. In: *The nature and properties of soils*. Prentice-Hall of India Pvt. Limited, New Delhi, India, p 352.

Carsky RJ, Singh BB, Oyewole R (2001) Contribution of early–season cowpea to late season maize in the savanna zone of West Africa. *Biol Agric Hort 18*:303–315.

Carvalho ER, Rezende PM, Andrade MJB, Passos AMA, Oliveira JA (2011) Fertilizante mineral The nature and properties of soils. *Revista Ciência Agro 42*:930–939.

Cassman KG, Whitney AS, Fox RL (1981a) Phosphorus requirements of soybean and cowpea as affected by mode of nitrogen nutrition. *Agric J 73*:17–22.

Cassman KG, Munns DN, Beck DP (1981b) Phosphorus nutrition of *Rhizobium japonicum*: strain differences in phosphate storage and utilization. *Soil Sci Amern J 45*:517–520.

Chaturvedi I (2006) Effects of phosphorus levels alone or in combination with phosphate-solubilizing bacteria and farmyard manure on growth, yield and nutrient up-take of wheat (*Triticum aestivum*). *J Agric Soc Sci 2*(2):96–100.

Chauhan YS, Johanson C, (1992) Effect of phosphorus deficiency on phenology and yield components of short duration pigeonpea. *Trop Agri 69*:235–238.

Chela GS, Tiwana MS, Thind IS, Puri KP, Kaur K (1993) Effect of bacterial cultures and nitrogen fertility on the yield and quality of maize fodder (*Zea mays* L.). *Ann Biol 9*:83–86.

Chen YP, Rekha PD, Arunshen AB, Lai WA, Young CC (2006) PSB from subtropical soil and their tri-calcium phosphate solubilizing abilities. *Appl Soil Ecol 34*:33–41.

Chen Z, Shengwu M, Lily L (2008) Studies on phosphorus solubilizing activity of a strain of hosphobacteria isolated from chestnut type soil in China. *Bio Tech 99*:6702–6707.

Chiezey UF, Odunze AC (2012) Soybean response to application of poultry manure and phosphorus fertilizer in the Sub-humid Savanna of Nigeria. *J Eco Natu Envi 1*(2):025–031.

Crews TE (1993) Phosphorus regulation of nitrogen-fixation in a traditional Mexican agroecosystem. *Biogeochem 21*:141–166.

Diep CN, Duong BT, Phan VHL (2016) Effects of rhizobia and phosphate-solubilizing bacteria on soybean (*Glycine max* L.) cultivated on ferralsols of daklak province, Vietnam. *World J Phar Pharmaceu Sci 5*(4):318–333.

Drevon JJ, Hartwig UA (1997) Phosphorus deficiency increases the argon-induced decline of nodule nitrogenase activity in soybean and alfalfa. *Planta 201*:463–469.

Dubey SK, Balasundaram VR, Pant LM, Jaysheela N, Kawale BR, Mishra B (1997) Effect of phosphate dissolving bacteria applied with rock phosphate on nodulation and yield of rainfed soybean under different agro-climatic conditions. *J Indian Soc Soil Sci 45*:503–505.

Elkoca E, Kantar F, Sahin F (2008) Influence of nitrogen fixing and phosphorus solubilizing bacteria on the nodulation, plant growth, and yield of chickpea. *J Plant Nutr 31*:157–171.

FAOS (Food and Agriculture Organization of the United Nations) (2014) http://faostat.fao.org/site/567/DesktopDefault.aspx?PageID=567#ancor. Accessed 12 May 2015.

Fazlullah, Adnan M, Fahad S, Iqbal S, Arshad M, Muhammad D, Wahid F, Noor MN (2018) Integrated application of phosphorus (P) and phosphate solubilizing bacteria (PSB) improve maize yield. *Pure Appl Biol 7*(2), 494–499.

Gao X, Kuyper TW, Zou C, Zhang F, Hoffland E (2007) Mycorrhizal responsiveness of aerobic rice genotypes is negatively correlated with their zinc uptake when nonmycorrhizal. *Plant Soil 290*:283–291.

Imran (2017) Climate Change Is a Real Fact confronting to Agricultural Productivity. *Int J Environ Sci Nat Res 3*(3):555613. doi:10.19080/IJESNR.2017.03.555613.

Imran, Asad AK, Fayaz A (2015a) Phenology of Various Rice Genotypes as Affected by Different Transplanting Dates under Cold Climatic Region of Khyber Pakhtunkhwa-Pakistan. *J Environ E Sci*. ISSN 2224-3216 (Paper) ISSN 2225-0948.

Imran, Asad AK, Inamullah, Luqman (2015b) Weeding stages and their effect on yield and yield components of rice in upper Swat, Pakistan. *Pak J Weed Sci Res 21*(4):555–563.
Imran, Asad AK, Inamullah, Ahmad F (2016a) Yield and yield attributes of Mungbean (*Vigna radiata* L.) cultivars as affected by phosphorous levels under different tillage systems. *Cogent Food Agric 2*:1151982.
Imran, Asad AK, Khan IU, Shahida N (2016b) Weeds density and late sown maize productivity influenced by compost application and seed rates under temperate environment. *Pak J Weed Sci Res 22*(1):169–181.
Imran, Nasar J, Ashfaq A, Asad AK (2017) Grain Yield, Yield Attributes Of Wheat And Soil Physio-Chemical Characteristics Influenced By Biochar, Compost And Inorganic Fertilizer Application. *Agri Res Tech* doi:10.19080/ARTOAJ.2017.10.555795.
Khan A, Gul F, Jamal Y, Ali, Adnan M, Khan MA, Bacha Z (2017) Weeds management in maize hybrids with night plowing. *Pak J Weed Sci Res 22*(2):179–191.

Index

N

O

P

R

S

T

V

W